PLANT CONSERVATION: AN ECOSYSTEM APPROACH

Plant Conservation: An Ecosystem Approach

Editor:

Jaspreet Banga, Ph.D.

Delve Publishing

www.delvepublishing.com

Plant Conservation: An Ecosystem Approach

Editor: **Jaspreet Banga, Ph.D.**

708 3rd Avenue, 6th Floor
New York, NY 10017
United States of America
www.delvepublishing.com

ISBN: 978-1-68095-762-4

Library of Congress Control Number: 2016951518

About the Editor

Jaspreet Banga, Ph.D.

Jaspreet Banga has completed PhD from Central Drug Research Institute, India and Master of Technology from Indian Institute of Technology. Jaspreet's interests range from Drug development, Purification, Fermentation technology, Plant Biotechnology and Oncology.

Contents

List of Contributors

Carlos Alberto Cruz-Cruz
Faculty of Chemistry Sciences, University of Veracruz, Prolongación Oriente 6, No. 1009, Orizaba, Veracruz 94340, México

María Teresa González-Arnao
Faculty of Chemistry Sciences, University of Veracruz, Prolongación Oriente 6, No. 1009, Orizaba, Veracruz 94340, México

Florent Engelmann
UMR DIADE, Joint Research Unit "Diversity, Adaptation and Development of Plants", IRD (Research Institute for Development), 911 Avenue Agropolis, BP 64501, 34032 Montpellier cedex 5, France

Luís Silva
InBIO, Rede de Investigação em Biodiversidade, Laboratório Associado, CIBO, Centro de Investigação em Biodiversidade e Recursos Genéticos, PoloAçores, Departamento de Biologia, Universidade dos Açores, 9501-801 Ponta Delgada, Açores, Portugal

Elisabete Furtado Dias
InBIO, Rede de Investigação em Biodiversidade, Laboratório Associado, CIBO, Centro de Investigação em Biodiversidade e Recursos Genéticos, PoloAçores, Departamento de Biologia, Universidade dos Açores, 9501-801 Ponta Delgada, Açores, Portugal

Julie Sardos
Bioversity-France, Parc Scientifique Agropolis II, 34397 Montpellier Cedex 5, France

Eduardo Brito de Azevedo
Research Center for Climate, Meteorology and Global Change (CMMG - CITAA), Departamento de Ciências Agrárias, Universidade dos Açores, Angra do Heroísmo, Açores, Portugal

Hanno Schäfer
Plant Biodiversity Research, Technische Universität München, D-85354 Freising, Germany

Mónica Moura
InBIO, Rede de Investigação em Biodiversidade, Laboratório Associado, CIBO, Centro de Investigação em Biodiversidade e Recursos Genéticos, PoloAçores, Departamento de Biologia, Universidade dos Açores, 9501-801 Ponta Delgada, Açores, Portugal

Jennifer J Swenson
NatureServe, 4600 North Fairfax Drive, Floor 7, Arlington, VA 22203, USA
Nicholas School of the Environment, Duke University, Box 90328, Durham, NC 27708, USA

Bruce E Young
NatureServe, 4600 North Fairfax Drive, Floor 7, Arlington, VA 22203, USA

Stephan Beck
Herbario Nacional de Bolivia, Universidad Mayor de San Andrés, La Paz, Bolivia

Pat Comer
NatureServe, 4600 North Fairfax Drive, Floor 7, Arlington, VA 22203, USA

Jesús H Córdova
Museo de Historia Natural, Universidad Nacional Mayor de San Marcos, Apartado 140434, Lima-14, Perú

Jessica Dyson
NatureServe, 4600 North Fairfax Drive, Floor 7, Arlington, VA 22203, USA
The Nature Conservancy, 99 Bedford St., 5th Floor, Boston MA 02111 USA

Dirk Embert
Fundación Amigos de la Naturaleza, km 7,5 Doble Vía la Guardia, Santa Cruz de la Sierra, Bolivia, Casilla 2241

Filomeno Encarnación
Instituto de Investigaciones de la Amazonía Peruana, Iquitos, Perú

Wanderley Ferreira
Rumbol, S.R.L. Av. Dorbigni 1608, Cochabamba, Bolivia

Irma Franke
Museo de Historia Natural, Universidad Nacional Mayor de San Marcos, Apartado 140434, Lima-14, Perú

Dennis Grossman
NatureServe, 4600 North Fairfax Drive, Floor 7, Arlington, VA 22203, USA
The Nature Conservancy, 4245 Fairfax Drive, Arlington, VA 22203 USA

Pilar Hernandez
NatureServe, 4600 North Fairfax Drive, Floor 7, Arlington, VA 22203, USA
Ontario Ministry of Natural Resources, 50 Bloomington Road W, Aurora, ON L4G 3G8

Sebastian K Herzog
Asociación Armonía, BirdLife Internacional, Avenida Lomas de Arena 400, Casilla 3566, Santa Cruz de la Sierra, Bolivia

Carmen Josse
NatureServe, 4600 North Fairfax Drive, Floor 7, Arlington, VA 22203, USA

Gonzalo Navarro
Rumbol, S.R.L. Av. Dorbigni 1608, Cochabamba, Bolivia

Víctor Pacheco
Museo de Historia Natural, Universidad Nacional Mayor de San Marcos, Apartado 140434, Lima-14, Perú

Bruce A Stein
NatureServe, 4600 North Fairfax Drive, Floor 7, Arlington, VA 22203, USA
National Wildlife Federation, 901 E Street, NW Suite 400, Washington DC, 20004 USA

Martín Timaná
NatureServe, 4600 North Fairfax Drive, Floor 7, Arlington, VA 22203, USA
Departamento de Ciencias, Pontificia Universidad Católica del Perú, Av. Universitaria 1801, Lima 32, Peru

Antonio Tovar
Centro de Datos para la Conservación, Departamento de Manejo Forestal, Facultad de Ciencias Forestales, Universidad Nacional Agraria La Molina, Apartado 456, Lima 100, Perú

Carolina Tovar
Centro de Datos para la Conservación, Departamento de Manejo Forestal, Facultad de Ciencias Forestales, Universidad Nacional Agraria La Molina, Apartado 456, Lima 100, Perú

Julieta Vargas
Museo Nacional de Historial Natural, Colección Boliviana de Fauna, Casilla 8706, La Paz, Bolivia

Carlos M Zambrana-Torrelio
Herbario Nacional de Bolivia, Universidad Mayor de San Andrés, La Paz, Bolivia
EcoHealth Alliance - 460 W 34th Street, 17th Floor, New York, NY 10001, USA

Zuhal Dilaver
Ankara University, Faculty of Agriculture, Department of Landscape Architecture, Turkey

Shailesh Shukla
Natural Resources Institute, University of Manitoba, Winnipeg, MB, R3T2N2, Canada

James Gardner
Natural Resources Institute, University of Manitoba, Winnipeg, MB, R3T2N2, Canada

Peterson W Wambugu
Queensland Alliance for Agriculture and Food Innovation, The University of Queensland, Brisbane, St Lucia, Qld, Australia
Kenya Agricultural Research Institute, Nairobi, Kenya

Agnelo Furtado
Queensland Alliance for Agriculture and Food Innovation, The University of Queensland, Brisbane, St Lucia, Qld, Australia

Daniel LE Waters
Southern Cross Plant Science, Southern Cross University, Lismore, NSW, Australia

Desterio O Nyamongo
Kenya Agricultural Research Institute, Nairobi, Kenya

Robert J Henry

Queensland Alliance for Agriculture and Food Innovation, The University of Queensland, Brisbane, St Lucia, Qld, Australia

Nikos Krigas
Laboratory of Systematic Botany & Phytogeography, Department of Botany, School of Biology, Aristotle University of Thessaloniki, Greece
Department of Ecology, School of Biology, Aristotle University of Thessaloniki, Greece

Kimon Papadimitriou
Office for Sustainability, Aristotle University of Thessaloniki, Greece

Antonios D. Mazaris
Department of Ecology, School of Biology, Aristotle University of Thessaloniki, Greece

Mohammed Kasso
Department of Zoological Sciences, Addis Ababa University, P.O. Box 1176, Addis Ababa, Ethiopia

Mundanthra Balakrishnan
Department of Zoological Sciences, Addis Ababa University, P.O. Box 1176, Addis Ababa, Ethiopia

Colin K. Khoury
International Center for Tropical Agriculture, Cali, Colombia
Centre for Crop Systems Analysis, Wageningen University, Wageningen, Netherlands

Bettina Heider
International Potato Center, CGIAR Research Program on Roots, Tubers and Bananas, Lima, Peru

Nora P. Castañeda-Álvarez
International Center for Tropical Agriculture, Cali, Colombia
School of Biosciences, University of Birmingham, Birmingham, UK

Harold A. Achicanoy
International Center for Tropical Agriculture, Cali, Colombia

Chrystian C. Sosa
International Center for Tropical Agriculture, Cali, Colombia

Richard E. Miller
Department of Biological Sciences, Southeastern Louisiana University, Hammond, LA, USA

Robert W. Scotland
Department of Plant Sciences, University of Oxford, Oxford, UK

John R. I. Wood
Department of Plant Sciences, University of Oxford, Oxford, UK

Genoveva Rossel
International Potato Center, CGIAR Research Program on Roots, Tubers and Bananas, Lima, Peru

Lauren A. Eserman
Department of Plant Biology, University of Georgia, Athens, GA, USA

Robert L. Jarret
Plant Genetic Resources Conservation Unit, United States Department of Agriculture – Agricultural Research Service, Griffin, GA, USA

G. C. Yencho
Department of Horticultural Science, North Carolina State University, Raleigh, NC, USA

Vivian Bernau
International Center for Tropical Agriculture, Cali, Colombia
Department of Horticulture and Crop Science, The Ohio State University, Columbus, OH, USA

Henry Juarez
International Potato Center, CGIAR Research Program on Roots, Tubers and Bananas, Lima, Peru

Steven Sotelo
International Center for Tropical Agriculture, Cali, Colombia

Stef de Haan
International Center for Tropical Agriculture, Cali, Colombia
International Potato Center, CGIAR Research Program on Roots, Tubers and Bananas, Lima, Peru

Paul C. Struik
Centre for Crop Systems Analysis, Wageningen University, Wageningen, Netherlands

Sognigbe N'Danikou
Horticulture and Genetics Unit, Faculty of Agronomic Sciences, University of Abomey-Calavi, BP: 2549 Abomey-Calavi, Republic of Benin
Bioversity International, West and Central Africa office, 08 BP 0932 Cotonou, Republic of Benin

Enoch G Achigan-Dako
Horticulture and Genetics Unit, Faculty of Agronomic Sciences, University of Abomey-Calavi, BP: 2549 Abomey-Calavi, Republic of Benin

Dedeou A Tchokponhoue
Horticulture and Genetics Unit, Faculty of Agronomic Sciences, University of Abomey-Calavi, BP: 2549 Abomey-Calavi, Republic of Benin

Chaldia OA Agossou
Horticulture and Genetics Unit, Faculty of Agronomic Sciences, University of Abomey-Calavi, BP: 2549 Abomey-Calavi, Republic of Benin

Carlos A Houdegbe
Horticulture and Genetics Unit, Faculty of Agronomic Sciences, University of Abomey-Calavi, BP: 2549 Abomey-Calavi, Republic of Benin

Raymond S Vodouhe
Bioversity International, West and Central Africa office, 08 BP 0932 Cotonou, Republic of Benin

Adam Ahanchede
Horticulture and Genetics Unit, Faculty of Agronomic Sciences, University of Abomey-Calavi, BP: 2549 Abomey-Calavi, Republic of Benin

F. Ngezahayo
Section Biologie-Chimie, Département des Sciences Naturelles, École Normale Supérieure, BP 6983, Bujumbura, Burundi

B. Liu
Key Laboratory of Molecular Epigenetics of the Ministry of Education (MOE), Northeast Normal University, Changchun 130024, China

Gea Galluzzi
Bioversity International, Regional Office for the Americas, c/o CIAT, km 17 Recta Cali-Palmira, Cali, Colombia

Isabel López Noriega
Bioversity International, Via dei Tre Denari, 472/a Maccarese, Rome 00057, Italy

Ganesh K. Jaganathan
Department of Biothermal Engineering, University of Shanghai for Science and Technology, Shanghai 20009, China

Bao-lin Liu
Department of Biothermal Engineering, University of Shanghai for Science and Technology, Shanghai 20009, China

Laura Zoratti
Department of Genetics and Physiology, University of Oulu, FI-90014 Oulu, Finland

Luisa Palmieri
Department of Food Quality and Nutrition, Research and Innovation Center, Fondazione Edmund Mach, Via E. Mach, 1-38010 San Michele all'Adige (TN), Italy

Laura Jaakola
Department of Arctic and Marine Biology, UiT The Arctic University of Norway, Climate Laboratory, 9037 Tromsø, Norway
Norwegian Institute of Bioeconomy Research, NIBIO Holt, 1431 A° s, Norway

Hely Ha¨ggman
Department of Genetics and Physiology, University of Oulu, FI-90014 Oulu, Finland

Murat Demir
Istanbul University, Faculty of Forestry, Forest Construction and Transportation Department 34473 Bahcekoy / Sariyer / Istanbul / Turkey

Ender Makineci
Istanbul University, Faculty of Forestry, Soil Science and Ecology Department 34473 Bahcekoy / Sariyer / Istanbul / Turkey

Beyza Sat Gungor
Istanbul University, Faculty of Forestry, Landscape Architecture Department 34473 Bahcekoy / Sariyer / Istanbul / Turkey

R. R. L. Atkinson
Department of Animal and Plant Sciences, University of Sheffield, Western Bank, Sheffield S10 2TN, UK

M. M. Burrell
Department of Animal and Plant Sciences, University of Sheffield, Western Bank, Sheffield S10 2TN, UK

K. E. Rose
Department of Animal and Plant Sciences, University of Sheffield, Western Bank, Sheffield S10 2TN, UK

C. P. Osborne
Department of Animal and Plant Sciences, University of Sheffield, Western Bank, Sheffield S10 2TN, UK

M. Rees
Department of Animal and Plant Sciences, University of Sheffield, Western Bank, Sheffield S10 2TN, UK

Preface

This book is an effort towards exploration of strategies for studying and protecting plants. We are at the stage in world history where we are facing widespread and devastating loss of plant biodiversity. The plant biodiversity at a particular location influences not only the local micro-climate, but in a broader horizon influences the types of animals, their habitat and water/soil resources for human population. Habitat destruction, over exploitation of resources and land, competition from cultivated plant species or varieties, environmental pollution and possible climate changes are all the human activities that have threatened and destroyed plant biodiversity. Swift action is needed to conserve genetic diversity of both cultivated and wild species. The ex situ conservation is used for all the major crops, where seeds can be dried and cooled and in situ conservation, literally in its "original" place, provided answer for crop diversity using traditional farming conditions. The book examines where and how ex situ, in situ and genetic conservation can be used for sustainable plant conservation and diversity maintenance.

Editor
Jaspreet Banga, Ph.D.

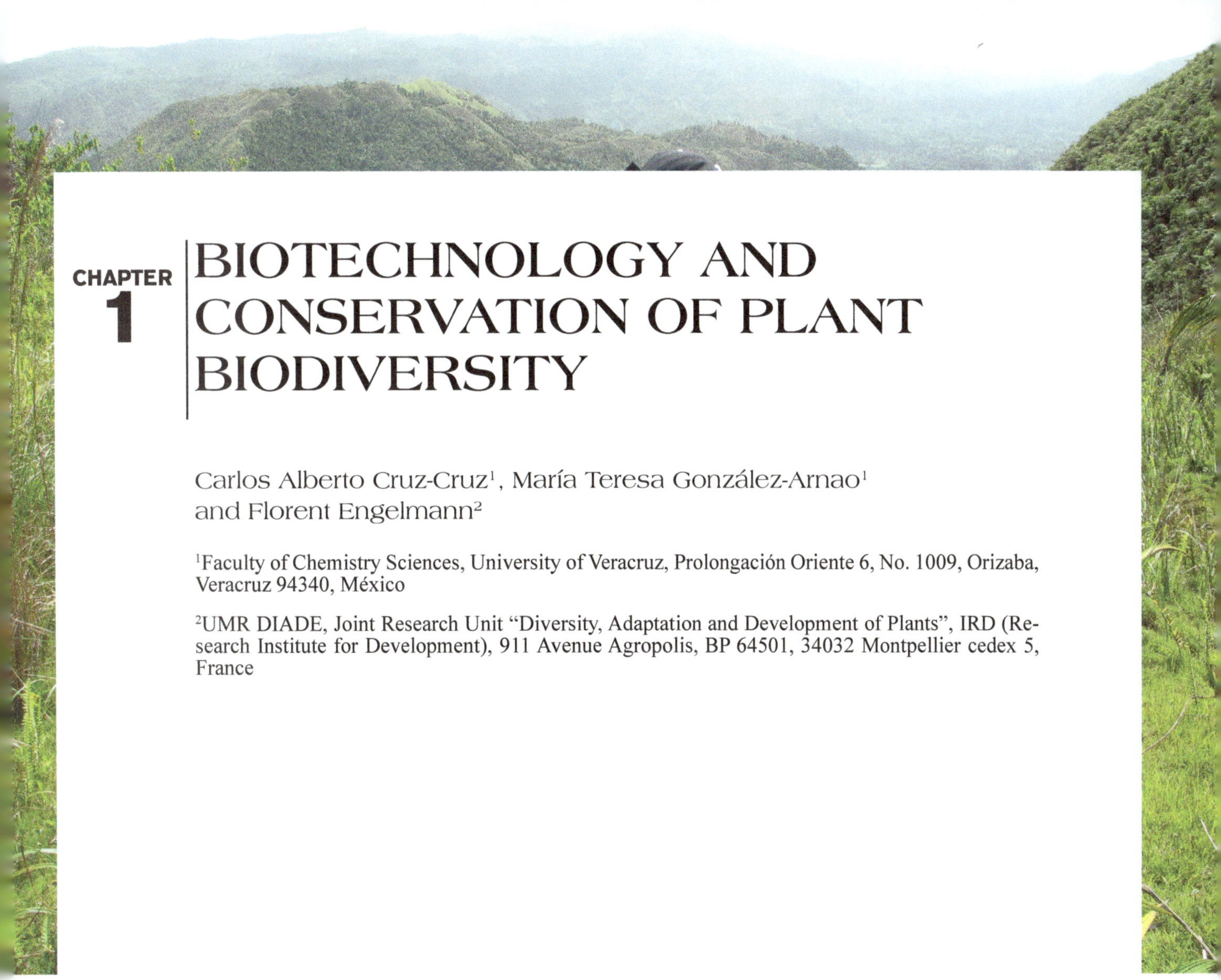

CHAPTER 1 BIOTECHNOLOGY AND CONSERVATION OF PLANT BIODIVERSITY

Carlos Alberto Cruz-Cruz[1], María Teresa González-Arnao[1] and Florent Engelmann[2]

[1]Faculty of Chemistry Sciences, University of Veracruz, Prolongación Oriente 6, No. 1009, Orizaba, Veracruz 94340, México

[2]UMR DIADE, Joint Research Unit "Diversity, Adaptation and Development of Plants", IRD (Research Institute for Development), 911 Avenue Agropolis, BP 64501, 34032 Montpellier cedex 5, France

ABSTRACT

Advances in plant biotechnology provide new options for collection, multiplication and short- to long-term conservation of plant biodiversity, using *in vitro* culture techniques. Significant progress has been made for conserving endangered, rare, crop ornamental, medicinal and forest species, especially for non-orthodox seed and vegetatively propagated plants of temperate and tropical origin. Cell and tissue culture techniques ensure the rapid multiplication and production of plant material under aseptic conditions. Medium-term conservation by means of *in vitro* slow growth storage allows extending subcultures from several months to several years, depending on the species. Cryopreservation (liquid nitrogen, −196 °C) is the only technique ensuring the safe and cost-effective long-term conservation of a wide range of plant species. Cryopreservation of shoot tips is also being applied to eradicate systemic plant pathogens, a process termed cryotherapy. Slow growth storage is routinely used in many laboratories for medium-conservation of numerous plant species. Today, the large-scale, routine application of cryopreservation is still restricted to a limited number of cases. However, the number of plant species for which cryopreservation techniques are established and validated on a large range of genetically diverse accessions is increasing steadily.

INTRODUCTION

The conservation of plant biodiversity is an important issue concerning the human population worldwide. The anthropogenic pressure, the introduction of alien species, as well as domesticated species and chronic weed infestation have dramatic effects on plant diversity, which is reflected in an increase in the number of threatened species. Plant biodiversity is a natural source of products to the medical and food industries. It provides different basic raw materials and contributes to supply new genetic information useful for breeding programs and for developing more productive crops and more resistant plants to biological and environmental stresses [1].

Conservation of plant biodiversity can be performed *in situ* or *ex situ.* The maintenance of plant species in their natural habitat, as well as the conservation of domesticated and cultivated species on the farm or in the surroundings where they have developed their distinctive characteristics represent the *in situ* strategies [2]. However, there is a heavy loss or decline of species, populations and ecosystem composition, which can lead to a loss of biodiversity, due to habitat destruction and

Citation: Carlos Alberto Cruz-Cruz, María Teresa González-Arnao and Florent Engelmann, Biotechnology and Conservation of Plant Biodiversity, doi: 10.3390/resources2020073.

the transformations of these natural environments; therefore, *in situ* methods alone are insufficient for saving endangered species. Additional approaches, like storage in seed banks, field gene collections, *in vitro* collections and botanical gardens, complement the preservation programs for plant biodiversity. They are classified as *ex situ* strategies, which means to maintain the biological material outside their natural habitats [2]. *Ex situ* conservation is a viable way for saving plants from extinction, and in some cases, it is the only possible strategy to conserve certain species [3]. *In situ* and *ex situ* methods are complementary and are not exclusive. They offer different alternatives for conservation, but selection of the appropriate strategy should be based on a number of criteria, including the biological nature of the species and the feasibility of applying the chosen methods [4].

Advances in plant biotechnology, especially those associated to *in vitro* culture and molecular biology, have also provided powerful tools to support and improve conservation and management of plant diversity [5]. At present, biotechnological methods have been used to conserve endangered, rare, crop ornamental, medicinal and forest species, allowing the conservation of pathogen-free material, elite plants and genetic diversity for short-, medium- and long-term. *In vitro*conservation is especially important for vegetatively propagated and for non-orthodox seed plant species [6]. Furthermore,*in vitro* techniques offer a safe mean to internationally exchange plant material, enable the establishment of extensive collections using minimum space, allow supply of valuable material for wild population recovery and facilitate molecular investigations and ecological studies [7].

This review briefly presents the *in vitro* techniques, which can be efficiently used to improve the conservation of plant biodiversity.

IN VITRO TECHNOLOGIES FOR COLLECTING PLANT BIODIVERSITY

Plant material collection is the first step to acquire plant germplasm. *In vitro* techniques can significantly increase collecting efficiency through the use of *in vitro* collecting, which is the process to initiate tissue cultures in the field [8]. For *ex situ* conservation, collecting cuttings of plants and seeds is generally the most cost-effective procedure. However, for some species, seeds are sterile or not available, or they have short longevity or viability [8], or they have unusual dormancy requirements and propagules may not be easily transported. In some cases, only few individuals of a given species still remain in specific areas; therefore, *in vitro* collecting of tissues would be less invasive than removing whole plants and will result in a more efficient method for sampling a large number of plants when seeds are not available [9].

Some species cannot be collected by traditional means, due to a seasonal pattern of development. Furthermore, some organs that are not strictly used for propagation, like shoots of trees, are more easily available for collecting at any time [10]. The deterioration of plant material, due to natural processes and microorganism attack, is another limiting factor affecting material integrity [10], and the excessive volume and weight of certain fruits can be a significant problem during the movement of the material collected [10]. Due to the limiting factors mentioned above, *in vitro* collecting broadens the possibilities for collecting living tissues. *In vitro* material can be dispatched internationally with fewer restrictions, even though it is still subject to import permits and phytosanitary certificates [8].

The material to be collected depends on each species. Due to cell totipotency, in theory, almost any part of the plant is sufficient to regenerate a whole organism under the appropriate growth conditions. For species producing orthodox seeds, the most common way to acquire plant material is through seed collection; nonetheless, different circumstances, such as seed absence or inadequate seed development, may hinder seed collection and for these cases, zygotic embryos or vegetative tissues, like budwoods, shoots, apices or leaves, can be collected [6]. For vegetatively propagated species, it is necessary to collect stakes, pieces of budwood, tubers or corms [6].

The different factors that must be considered during the *in vitro* collecting of plant tissue are: the appropriate tissue for *in vitro* collecting, the size of the tissue, soil residues and presence of diseased tissue, sterilization of plant tissue, removal of the disinfectant, nutrient medium and the conditions of storage, including light, temperature and humidity [10]. Since *in vitro* collecting is based on tissue culture techniques, its limitations are based on the recalcitrance of some species to regenerate or even to grow *in vitro* [11]. Furthermore, *in vitro* collecting may pose more challenges beyond those of normal tissue culture, as work is done in the field and culture exposure to air-borne contaminants may be unavoidable [12].

Microorganism removal is a critical factor that must be strictly controlled during *in vitro* collecting of plant material. Bacteria and fungi develop rapidly as saprophytes in culture media, and since their nutritional requirements are basically the same as plants, they compete with the plant for nutrients [12]; furthermore, microorganisms can produce phytotoxic metabolites that affect plant growth [13]. Different factors influence the level of explant contamination, like the age of tissues (older tissues are generally more infected than the younger ones), the localization of the tissues (in the air or underground) and the environment [12]. Surface sterilization is the first step in establishing aseptic cultures, which can be done at the collection site or in the laboratory after the tissue sample is placed on a transport medium [12,14]. Systemic antimicrobial agents must be added to the media to kill bacteria or fungi localized beneath the epidermis or in the intercellular spaces, so it is necessary to select the appropriate antibiotic depending on the target microorganism, antibiotic solubility, stability in light, its interactions with other media components and toxicity to humans. Several antibiotics and fungicides used for *in vitro* plant culture have been detailed and listed by Pence and Sandoval [12].

The first *in vitro* collecting systems were developed for cocoa (*Theobroma cacao* L.) and coconut (*Cocos nucifera* L.), generating two *in vitro* collecting methods that were used as a model to develop other protocols [10].

Cocoa seeds are highly recalcitrant, which represents a challenge for *ex situ* conservation; furthermore, the material generally used for propagation, mature seeds and cuttings, rapidly lose viability, and it is difficult to maintain alive the material over long distances [15]. In 1987, an *in vitro* collecting technique was developed for cocoa [16,17].

Collecting coconut seeds by conventional means is a costly and highly inefficient procedure, since seeds are bulky, heavy and highly recalcitrant [18,19]. *In vitro* collecting is based on the premise that the embryo is enough to grow and develop a coconut palm. The adaptation of *in vitro* culture techniques to collecting coconut embryos had two initial purposes: collecting plant material and the international exchange of coconut germplasm, avoiding the transmission of coconut diseases that are transferred by the nut, but not by the embryo [19]. The available coconut *in vitro* collecting techniques share some basic steps: the dehusking and cracking open of the nut, the extraction of a plug of endosperm containing the embryo, the dissection of the embryo from the endosperm and the inoculation of the embryo into culture [19]. Different protocols for*in vitro* collecting of coconut germplasm have been reported [14,20,21,22]. One of the most recent protocols involves storing the disinfected embryos in a KCl solution until they arrive to the laboratory; then, they are re-disinfected and inoculated under sterile conditions on semi-solid medium supplemented with sucrose and activated charcoal, placing them in the dark and then transferring cultures to light conditions once the shoots and roots start to develop [23]. Other representative examples of *in vitro* collecting techniques are presented in Table 1.

Table 1: Representative examples of *in vitro* collecting technique for selected species

Species	Explants/Tissue	Reference
Coffea arabica L. (coffee)	Single nodes with axillary buds from orthotropic stems	[24]
Musa L. sp.	Corms from sword shoots	[25]
Citrus L. sp.	Vegetative explants from straight twigs and seeds	[26]
Persea americana Miller (avocado) *Erythrina* L. sp. (flame tree) *Vanilla planifolia* Jackson *Pouteria* Aublet sp. (sapodilla)	Vegetative explants from straight twigs	[27]
Colocasia esculenta var. *esculenta* (Taro)	Corms from suckers	[28]
Gossypium hirsutum L. (Cotton)	Stem nodal	[29]

In vitro collecting represents an alternative for rare and endangered species, since usually this material is limited in supply and seed collection may be restricted. The removal of small amounts of appropriate tissue from the plant should not harm *in situ* populations [11]. It will be necessary to develop the appropriate protocol for *in vitro* collecting depending on the species. A good start will be to take guidance from literature on related species, and sometimes, educated judgments must be taken to develop a procedure for a species with limited amounts of material [11].

IN VITRO TECHNOLOGIES FOR PROPAGATION AND EXCHANGE OF PLANT BIODIVERSITY

The development of biotechnology has led to the production of a new category of germplasm, including clones obtained from elite genotypes, cell lines with special attributes and genetically transformed material [30]. This new germplasm is often of high added value and very difficult to produce. The development of efficient techniques to ensure its safe conservation is therefore of paramount importance.

Tissue culture techniques are of great interest for collecting, multiplication and storage of plant germplasm and are very useful for conserving plant biodiversity, including (a) genetic resources of recalcitrant seed and vegetatively propagated species; (b) rare and endangered plant species; and (c) biotechnology products, such as elite genotypes and genetically engineered material [6,31,32]. Tissue culture systems allow propagating plant material with high multiplication rates in an aseptic environment. Following two alternative morphogenic pathways, shoot organogenesis or somatic embryogenesis, tissue culture has been extensively developed and applied for propagation and regeneration of over 1000 different plant species [33], including numerous rare and endangered species [34,35]. Plant material generated by using *in vitro* culture techniques is "synchronized", miniaturized and relatively homogenous in terms of size, cellular composition and physiological state [6]. The first requirement for defining any conservation protocol *in vitro* is the establishment of fully operational tissues culture conditions for regeneration and multiplication of plant material. The factors that determine the response in plant regeneration are environmental, physical and genotypic. Tissue culture techniques should guarantee the generation of abundant material, the recovery of stored samples in high percentages and finally, the development of complete, true-to-type plants.

In vitro techniques have a clear role within *ex situ* conservation strategies, including for trees and endangered species, particularly where it is important to conserve specific genotypes or where normal propagules, such as recalcitrant seeds may not be suitable for long-term storage. These involve the use of conventional micropropagation systems, slow growth techniques and cryopreservation [36].

In vitro seed germination has been extensively employed for multiplication of a large number of orchid species [37] and could be a rapid mean for multiplying rare and endangered orchids. *In vitro* seed germination, micropropagation, somatic embryogenesis, zygotic embryo culture and callus culture systems have been developed successfully for a substantial number of native endangered Brazilian species [38]. These systems can be potentially used to further *in vitro* germplasm conservation studies. Somatic embryogenesis is an important method for mass production of tree species for forestry [39] and for the development of artificial seeds, making handling and direct planting easier [40]. Artificial seeds are encapsulated tissues, such as somatic embryos, shoot tips and axillary buds, which can be used for germplasm conservation. Artificial seeds are used for large scale clonal propagation, breeding of plants producing non-orthodox seeds or non-seed producing plants and facilitate the storage and transportation of samples [41].

Biodiversity hotspots around the globe are at risk and *in vitro* propagation methods have been used for rescuing and conserving endangered plants [42], in many countries [9], including Australia [43], Malaysia [44] and South Africa [45]. Although standard *in vitro* propagation methods are, in general, accessible, endangered species may have unusual growth requirements and, thus, may need modified procedures for *in vitro* culture. In addition, the limited amount of plant material available from rare and endangered species poses major challenges in the application of *in vitro* techniques [35].

It is already well known that micropropagation allows both rapid and massive clonal multiplication of plants; however, it does not ensure that material will be free of systemic agents, such as viruses, which can be present in tissues without manifesting symptoms and spread during the *in vitro* multiplication. However, among the *in vitro* techniques, shoot tip or meristem culture has been used for many decades to eliminate viruses in many species from vegetatively propagated plants [46,47]. This is based on the uneven distribution of viruses in the youngest tissues of the shoot apex, as their concentration tends to decrease progressively toward the apical meristem of the stem, where the cells are in constant and rapid division [48,49]. Since not all cells in a shoot apical meristem are infected with pathogens (e.g., virus, phytoplasmas and endophytic bacteria), it is possible to dissect out a non-infected region and manipulate this explant *in vitro* to produce virus-free plants [50,51]. As only the meristematic dome and the immediate covering (1st leaf primordia) are usually virus-free [50,51], the size of the meristem excised is critical. Therefore, excision and regeneration of tiny meristems might result in plants free of these pathogens. Regeneration ability is positively proportional to the size of the shoot tip, but pathogen eradication is more efficient using small shoot tips (0.2–0.4 mm). Hence, pathogen eradication using meristem culture is challenged by the difficulty of excising very small meristems mechanically to remove the infected tissues and of ensuring the survival and regeneration of the tiny meristems [47,49,52,53]. Meristem culture, in combination with thermotherapy, facilitates obtaining virus-free plants and ensures an easier production of disease-free stocks [48]. Then, *in vitro* culture techniques simplify the quarantine procedures for the international exchange of germplasm [6], because the sanitary status of the plants is safe and because it is easier to transport abundant amounts of a miniaturized material. These techniques have been successfully used for many years in virus eradication. Among woody plants, grapevine, apple and peach are the most frequent targets of sanitation protocols, because their health status is strictly regulated. Even when thermotherapy represents the preferred method for the host, viruses can also be eliminated with chemotherapy and tissue culture [54]. Tissue culture techniques have been used for virus elimination on woody, as well as herbaceous plants (Table 2).

Table 2: Representative examples of tissue culture technique used for virus elimination in selected woody and herbaceous plants

Species	Virus	Reference
Woody plants		
Grapevine	*Grapevine fanleaf virus* (GFLV) and *Grapevine leaf roll-associated virus-1* (GLRaV-1)	[55]
Banana	*Banana bract mosaic virus*	[56]
Citrus	*Citrus psorosis virus*	[57]
Cocoa	*Cocoa swollen shoot virus*	[58]
Rose	*Rose mosaic virus*	[59]
Herbaceous plants		
Sugarcane	*Sugarcane mosaic virus* (SCMV) *and sugarcane yellow leaf virus* (ScYLV)	[60]
Garlic	*Leek yellow stripe* (LYS) *and onion yellow dwarf virus* (OYDV)	[61]
Potato	*Potato leafroll virus* (PLRV) *and potato virus Y (PVY)*	[62]
Herbaceous plants		
Carnation	*Carnation latent virus (CLV)*	[63]
Chrysanthemum	*Cucumber mosaic and tomato aspermy virus*	[64]
Dahlia	*Dahlia mosaic virus*	[65]
Peanut	*Peanut mottle potyvirus (PMV) and peanut stripe potyvirus (PStV)*	[66]
Pumpkin	*Zucchini yellow mosaic virus, cucumber mosaic virus, alfalfa mosaic virus, bean yellow mosaic virus*	[67]

Collecting, conservation and utilization of plant genetic resources and their international distribution are essential components of international crop improvement programs. This is of special significance in plant movement, quarantine and to breeders who can import a large number of clones of choice, multiply them and distribute them among user groups. *In vitro* exchange is, therefore, catching on fast. It is important to use suitable, impact resistant and well-sealed culture containers. The packaging should provide adequate thermal insulation and protection against rough handling. The culture medium should be of a formation that includes a higher than usual concentration of gelling agent. The most rapid available shipping method should be preferred [68].

IN VITRO TECHNOLOGIES FOR CONSERVATION OF PLANT BIODIVERSITY

In vitro techniques used to achieve medium-term conservation allow the storage of biological material from several months to 2–3 years without subculture, depending on the technique used and on the plant material.

Growth reduction is generally attained by modifying the culture medium and/or the environmental conditions. Modifications of the culture medium can include dilution of mineral elements, reduction of sugar concentration, changes in the nature and/or concentration of growth regulators and addition of osmotically active compounds [6]. As regards the culture environment, it can be changed by reducing the temperature, combined or not, with a decrease in light intensity or by keeping cultures in complete darkness. The most frequently used combination of physical and chemical factors involves decrease of temperature, reduction of mineral elements and carbon source concentration in the medium and the use of low light intensity [69]. The temperatures reported for medium-term conservation are usually from 4 °C to room temperature [70]. However, tropical plant species are often cold-sensitive and have to be stored in the range of 15–20 °C or even higher, depending on their sensitivity [71]. Therefore, the procedure to enable extending subculture periods will mainly focus on modifying the chemical composition of culture medium.

Other additional parameters may also influence the efficiency of slow growth storage, such as the type of explants, their physiological state, as well as the type, volume and the kind of closure of culture vessels [31].

Standard *in vitro* culture conditions can also be used for medium-term storage when dealing with species that have a natural slow growing habit. Alternatively, the explants may be covered with paraffin, mineral oil or with liquid medium to reduce the growth rate. Modifications in gaseous environment, desiccation and/or encapsulation are other possible options [4]. Artificial seeds, which are produced by encapsulating plant propagules (shoot buds or somatic embryos) in a synthetic matrix, enable medium-term conservation of various plant species, like orchids, through encapsulation of protocorms [72].

Short- and medium-term conservation is routinely used in many laboratories in order to increase the intervals between subcultures required under the multiplication procedure. At the end of a storage period, cultures are transferred onto fresh medium and usually placed for a short period in optimal conditions to stimulate regrowth before entering the next storage cycle [4].

Slow growth has been successfully used for plant species of both temperate and tropical origin, including crops, forest trees, endangered species and medicinal plants [73,74]. Rare wild species, like *Gladiolus imbricatus*, which is an important resistance-gene pool in this genus by having resistance to abiotic and biotic stress, was stored up to one year on Murashige and Skoog medium [75], at low temperature and kept in the dark. After one year of storage, 25% of the plants could be successfully recovered [76].

Musa in vitro plantlets could be conserved at 15 °C without transfer for up to 15 months [77], while cassava shoot cultures demonstrated to be much more cold-sensitive, since they have to be conserved at temperatures higher than 20 °C [78]. The same authors reported that cassava shoot cultures could be stored for longer periods in a better condition by increasing the size of the storage containers. Another example related to the effect of culture vessels is the use of heat-sealable polypropylene bags instead of glass test tubes or plastic boxes, which was beneficial for the storage of several strawberry varieties [79].

The presence of a root system improved the storage capacities of coffee plantlets as observed by Kartha *et al.* [80]. The physiological properties of certain species could suggest that seedlings from recalcitrant seeds of some forest species could possibly be stored under *in vitro* conditions, which stimulate the under-canopy conditions that arrest development in the wild [36].

By using both mineral oil and silicone oil overlays, it was possible to restrict the growth of sweet potato shoot cultures maintained at 25 °C [81], and shoot cultures of several ginger species could be conserved for up to two years under mineral oil with high viability [82]. On the other hand, encapsulated grape shoot tips were stored for nine months at 23 °C [83], and encapsulated date palm somatic embryos were conserved for six months at 4 °C [84].

The advantage of slow growth techniques is that they use the same basic facilities used for plant micropropagation and that the storage regimes are based on modifying the conditions previously established for rapid multiplication. However, they do not alleviate the main problem associated with the high costs of labor and space requirements of any micropropagation system, in addition to the potential risks of somaclonal variation for some species [36].

LONG-TERM CONSERVATION THROUGH CRYOPRESERVATION

Cryopreservation is the maintenance of living cells, tissues organs and microorganisms at ultralow temperature (usually that of liquid nitrogen, −196 °C). Under cryogenic storage, the biological material can be conserved for extended durations, because at liquid nitrogen temperature, all metabolic activity and cell divisions are stopped and cells will not undergo genetic changes during storage, which may occur when they are maintained by serial subculturing. Furthermore, cryopreserved cells are stored in a small volume, requiring very limited maintenance (topping up storage containers with liquid nitrogen); samples are not continuously exposed to the risks of contamination and operator errors, due to frequent manipulations of the plant material [85].

Cryopreservation is the only technique that ensures the safe and cost-efficient long-term conservation of various categories of plants, including non-orthodox seed species, vegetatively propagated plants, rare and endangered species and biotechnology products [6].

In all cryopreservation processes, water removal plays a central role in preventing freezing injury and in maintaining post-thaw viability of cryopreserved material. There are two types of cryopreservation protocols that basically differ in their physical mechanisms: classical cryopreservation procedures, in which cooling is performed in the presence of ice; and the procedures based on vitrification, in which cooling normally takes place without ice formation [86].

Classical freezing procedures involve cryoprotection by using different cryoprotective solutions combined or not with pregrowth of material and followed by slow cooling (0.5–2.0 °C/min) to a determined prefreezing temperature (usually around −40 °C), rapid immersion of samples in liquid nitrogen, storage, rapid thawing and recovery. They are generally operationally complex, since they require the use of sophisticated and expensive programmable freezers.

Cryopreservation following classical protocols induces a freeze-dehydration process using a slow freezing regime. During the slow temperature decrease, ice is initially formed in the extracellular solution and this external crystallization promotes the efflux of water from the cytoplasm and vacuoles to the outside of the cells where it finally freezes. Therefore, cell dehydration will depend on the cooling rate and the prefreezing temperature set up before immersion of samples to liquid nitrogen [86].

Classical cryopreservation techniques have been successfully applied to undifferentiated culture systems of different plant species, such as cell suspensions and calluses [87,88]. They have also been employed with apices of cold-tolerant plants [89]. Successful cryopreservation of apices from tropical species, such as cassava (*Manihot esculenta*), is an exceptional example [90,91].

By contrast, the vitrification-based procedures involve cell dehydration prior to cooling by exposure of samples to highly concentrated cryoprotective media (usually called plant vitrification solutions, PVS) and/or by air desiccation. Cooling rate may be rapid or ultra-rapid, depending on how samples are immersed into liquid nitrogen. Vitrification *per se* is a physical process, defined as the transition of the liquid phase to an amorphous glassy solid at the glass transition (Tg) temperature [92]. This glass may contribute to preventing tissue collapse, solute concentration and pH alterations during dehydration. Therefore, the freeze-induced dehydration step characteristic of classical procedures is eliminated and the slow freezing regime is replaced by a rapid or ultra-rapid cooling process, characteristic of the vitrification-based protocols.

Several innovations have been implemented in order to increase the cooling rate from the temperature at, which cryoprotective treatments are performed (0 or 25 °C) and the final storage temperature (−196 °C). In that way, diverse new plant vitrification-based protocols have been developed and are systematically improved. Indeed, most of them are derived from two cryogenic strategies developed since 1990: vitrification [93,94,95] and encapsulation-dehydration [96,97].

Several vitrification-based procedures are well known at present: pregrowth; dehydration; pregrowth-dehydration; encapsulation-dehydration; vitrification; encapsulation-vitrification; droplet-vitrification [86], and more recently, a new procedure called Cryo-plate is emerging [98].

The pregrowth technique consists of cultivating samples in the presence of cryoprotectants, followed by rapid immersion in liquid nitrogen [85]. Dehydration consists of dehydrating explants usually by desiccation in the air current of a laminar airflow cabinet or with silica gel and, then, direct immersion in liquid nitrogen. Pregrowth-dehydration is the combination of the both previously mentioned methods. These techniques are mainly used for cryopreserving meristematic cultures, small size seeds, polyembryonic cultures, zygotic embryos or embryonic axes extracted from seeds, respectively [86].

The encapsulation-dehydration is based on the technology developed for the production of artificial seeds. Explants are encapsulated in alginate beads, pregrown in liquid medium enriched with sucrose for one to seven days, partially desiccated in the air current of a laminar air flow cabinet or with silica gel to a water content around 20% (fresh weight basis) and then immersed rapidly in liquid nitrogen. Encapsulation of explants allows the application of subsequent drastic dehydration processes prior to cryopreservation, which would otherwise be highly damaging or lethal to non-encapsulated samples. Cryopreservation using the encapsulation-dehydration procedure has been very effective for freezing apices of different plant species from temperate and tropical origin [99].

Vitrification involves treatment of samples with cryoprotective substances (loading), dehydration with highly concentrated plant vitrification solutions (PVS), rapid cooling and rewarming, removal of cryoprotectants and recovery. This procedure has been developed for apices, cell suspensions and somatic embryos of numerous different species [100,101].

Encapsulation-vitrification is a combination of encapsulation-dehydration and vitrification procedures, in which samples are encapsulated in alginate beads and then treated and cooled as under vitrification conditions [100].

Droplet-vitrification is a protocol derived from the combination of the vitrification procedure with the droplet-freezing technique developed by Kartha *et al.* [102] for cassava shoot tips [86]. Samples are treated with loading and vitrification solutions and then placed on an aluminum foil in minute droplets of vitrification solution or just in one small drop and the aluminum foil strip is directly immersed with the samples in liquid nitrogen [100].

The most recent cryogenic procedure developed, termed cryo-plate [98], combines the encapsulation-dehydration and droplet-vitrification techniques. In this method, shoot tips are attached with a thin calcium alginate layer to an aluminum cryo-plate, loaded, treated with PVS and then cooled by direct immersion of cryo-plates in liquid nitrogen [98,103].

The latest two cryopreservation techniques (droplet-vitrification and cryo-plate) have the common characteristic of providing higher cooling and warming rates compared to other vitrification-based procedures, since samples placed on aluminum foils (with a very high thermal conductivity), are plunged directly into liquid nitrogen for cooling and immersed in a sucrose supplemented medium at room temperature for warming. This significantly increases the probability of obtaining a vitrified state during cooling and of avoiding devitrification during warming [104].

In general, vitrification-based protocols have been very useful for cryopreserving complex organs like shoot-tips and somatic embryos that could not be effectively frozen following classical protocols [105,106,107,108]. In addition, the number of species to which they have been successfully applied is increasing steadily, and they are considerably contributing to improve the cryopreservation of tropical plant germplasm [86,100].

Cryopreservation of seeds is a very valuable strategy for the long-term conservation of tropical and subtropical forest species biodiversity, as it avoids problems related to embryo isolation and *in vitro* handling. Even for orthodox and intermediate seeds, cryostorage offers the advantages of seed longevity, so long as liquid nitrogen levels are maintained. Storage of dry seeds at −20 °C in seed banks for long periods might also lead to physiological and genetic damages in the very long-term [109].

It is important to realize that the use of cryopreservation should not be restricted to vegetatively propagated plants and non-orthodox seed species. Indeed, recent research findings have shown the necessity to also employ cryopreservation for long-term storage of orthodox seed species. Over the last 30 years, relatively widespread evidence has emerged of less than expected longevity at conventional seed bank temperatures [110]. These authors highlighted the observation that across nearly 200 species, those originating from drier (total rainfall) and warmer temperature (mean annual) locations tended to have greater seed P50 (time taken in storage for viability to fall to 50%) under accelerated ageing conditions, than species from cool and wet conditions [111]. Moreover, species P50 values were correlated with the proportion of accessions (not necessarily the same species) in that family, which significantly lost viability after 20 years under conditions for long term seed storage, that is, seeds pre-equilibrated with 15% relative humidity air and then stored at −20 °C [112]. Such relative underperformance at −20 °C was observed in 26% of the accessions [111]. In another study, it has been estimated that half-lives for the seeds of 276 species stored for an average of 38 years under cool (−5 °C) and cold (25 years at −18 °C) temperatures was >100 years only for 61 (22%) of the species [113]. Nonetheless, cryogenic storage did prolong the shelf life of lettuce (*Lactuca*) seeds with projected half-lives in the vapor and liquid phases of liquid nitrogen of 500 and 3400 years, respectively [114], up to 20-times greater than that predicted for that species in a conventional seed bank at −20 °C [115,116].

Therefore, cryopreservation could be integrated as an extra insurance policy for storing plant biodiversity and could also be considered appropriate for all orthodox seeds. One sub-sample of any accession might be systematically stored in liquid nitrogen, in addition to the samples stored under classical gene bank conditions [117].

There are several examples to illustrate the application of cryopreservation techniques above described to preserve rare and endangered species for various higher plants [32], including orchids [118], bryophytes and ferns [119].

Cryopreservation of seeds of 68 native Western Australian species out of the 90 species tested was successful using a classical protocol involving treatment with dimethyl sulfoxide (DMSO) [120]. Desiccation has been employed for freezing seeds of rare temperate orchids by direct immersion in liquid nitrogen [121]. Several authors have used the desiccation technique for freezing seeds of endangered, rare, ancient and wild *Citrus* species [73,122,123]. Encapsulation-dehydration has been used for cryopreserving protocorms of *Celisostoma areitinum*, a rare Thai orchid [124], and shoot-tips of the endemic endangered plant *Centaurium rigualii* [125]. Turner *et al.* [126] and Tanaka *et al.* [127] reported successful cryopreservation of shoot-tips of endangered Australian and Japanese species, respectively, using the vitrification technique. The droplet-vitrification procedure has been used for cryopreserving shoot-tips of wild potatoes [128] and of wild relatives of *Diospyros*[129].

For vegetatively propagated species, cryopreservation has a major applicability both in terms of species coverage, since protocols have been successfully established for root and tubers, fruit trees, ornamentals and plantation crops, both from temperate and tropical origin; and in terms of numbers of genotypes/varieties within a given species [91,130]. In comparison to the results obtained with vegetatively propagated species, the situation for recalcitrant seeds is still at a very preliminary stage. The desiccation technique is mainly employed for freezing embryos and embryonic axes [44]; however, survival is extremely variable and regeneration is frequently restricted to callusing or to incomplete development of plants.

At present, there is a growing number of gene banks and botanic gardens where cryogenic techniques are already employed for different types of materials. Some representative examples of large-scale application of cryopreservation are presented in Table 3.

Table 3: Large-scale application of cryopreservation techniques to different plant germplasm

Plant material	Gene bank/country	Reference
Seeds of 1200 accessions from 50 different species mainly of endangered medicinal plants	The National Bureau for Plant Genetic Resources (NBPGR), New Delhi, India	[131]
Seeds of more than 110 accessions of rare or threatened species	Kings Park and Botanic Garden, Perth, Australia	[132]
Seeds of coffee involving 450 accessions	IRD Montpellier, France	[133]
Dormant buds of apple involving 2200 accessions	National Center for Genetic Resources (CNGR), Fort Collins, USA	[134]
Dormant buds of mulberry involving 420 accessions	National Institute of Agrobiological Resources (NIAR), Yamagata, Japan	[135]
Shoot-tips of banana involving 630 accessions	INIBAP International Transit Center, Catholic University of Leuven, Belgium	[136]
Shoot-tips of cassava involving 540 accessions	International Center for Tropical Agriculture (CIAT), Cali, Colombia	[86]
Pollen of 13 pear cultivars and 24 *Pyrus* species	National Center for Genetic Resources (CNGR), Fort Collins, USA	[137]
Pollen of more than 700 accessions of traditional Chinese flower species	College of Landscape Architecture, Beijing Forestry University, Beijing , China	[138]
More than 1000 callus strains of species of pharmaceutical interest	Phytera, Sheffields, UK	[139]
Several thousand conifer embryogenic cell lines for large-scale clonal planting programs	Sylvagen, Vancouver, Canada	[140]
Embryogenic cell lines of coffee and cacao	Biotechnology Laboratory of the Nestlé Company, Notre Dame d'Oé, France	[141]

OTHER USES OF CRYOPRESERVATION

Besides germplasm conservation, cryopreservation may be employed to rejuvenate cell and callus cultures, cryoselect plant material with special properties and eliminate viruses from infected plants through apex conservation through a process known as cryotherapy [85].

Galerne *et al.* [142] mentioned that regrowth of cryopreserved Norway spruce callus was higher compared to that of non-cryopreserved controls. Aguilar *et al.* [143] showed that production of somatic embryos from cryopreserved *Citrus*embryogenic callus was higher compared to that of non-cryopreserved cultures. This is due to selection performed by liquid nitrogen exposure between different cell types. In cryopreserved samples, liquid nitrogen exposure kills more differentiated cells, which do not participate in culture growth, whereas more meristematic cells, which are responsible for growth, remain alive and reinitiate growth very rapidly and intensely after rewarming. By contrast, non-cryopreserved cell and callus cultures grow more slowly, because of the large proportion of differentiated cells, which do not proliferate actively [85].

Kendall *et al.* [144] could select freezing tolerant callus of *Triticum aestivum* by repeating their exposure to liquid nitrogen, and the cryoselected callus regenerated plants with enhanced cold hardiness. The seed progeny of some of the lines tested exhibited significantly enhanced tolerance to −12 °C, indicating that cryoselection appears to involve, at least in part, selection for genetic rather than epigenetic variants [31]. This could be an interesting strategy to induce cold tolerance in sensitive species.

A novel application of plant cryopreservation techniques is the high frequency pathogen eradication by means of cryotherapy of shoot tips, as first demonstrated by Brison *et al.* [145]. It eliminates systemic plant pathogens, such as viruses, phytoplasmas and bacteria, by treating shoot tips with liquid nitrogen using cryopreservation protocols. The uneven distribution of viruses and obligate vasculature-limited microbes in shoot tips allows elimination of the infected cells by injuring them with the cryo-treatment and regeneration of healthy shoots from the surviving pathogen-free meristematic cells. Healthy plants are regenerated from the surviving pathogen-free meristematic tissue. It allows treatment of large numbers of samples and results in a high frequency of pathogen-free regenerants. Therefore, it has the potential to replace more traditional methods like meristem culture, chemo- and thermo-therapy [49,146].

To date, severe pathogens have been eradicated in banana (*Musa* spp.) [147], *Citrus* spp. [148], grapevine (*Vitis vinifera*) [149], *Prunus* spp. [145], raspberry (*Rubus idaeus*) [150], potato (*Solanum tuberosum*) [151] and sweet potato (*Ipomoea batatas*) [52,53,152] using cryotherapy. These pathogens include nine viruses (banana streak virus, cucumber mosaic virus, grapevine virus A, plum pox virus, potato leaf roll virus, potato virus Y, raspberry bushy dwarf virus, sweet potato feathery mottle virus and sweet potato chlorotic stunt virus), sweet potato little leaf phytoplasma and Huanglongbing bacterium causing "citrus greening" [49,146].

CONCLUSIONS

In this paper, we have presented the new possibilities offered by biotechnology for improving *ex situ* conservation of plant biodiversity, through the development of efficient collection, exchange, multiplication, pathogen eradication and conservation methods. These methods are of particular interest for conserving non-orthodox seed and vegetatively propagated species, rare and endangered species, as well as biotechnology products. In recent years, progress has been especially important in the area of cryopreservation, with the development of vitrification-based protocols and with its new application for pathogen eradication by means of cryotherapy.

Optimized conservation strategies should rely on the complementary use of *in situ* and *ex situ* techniques. Until recently,*ex situ* conservation of plant genetic resources has been mostly based on the conservation of seeds in cold chambers and, to a lesser extent, on the maintenance of whole plants in field collections. The new biotechnological conservation methods, including *in vitro* slow growth storage and cryopreservation, need to be systematically integrated in conservation strategies, and the current *ex situ* conservation concepts should be modified accordingly to accommodate these technological advances. Various criteria should be considered to select the most appropriate methods for conserving a given gene pool, including, notably, the storage characteristics of the species involved, the applicability of the methods chosen in the storage environment, which will vary depending on the available infrastructures, as well as their cost-effectiveness. In many cases, research may still be needed to optimize the methods and to validate them on a range of genetically diverse accessions.

In conclusion, it should be emphasized that the new biotechnology-based conservation techniques developed are not meant to replace conventional *ex situ* conservation methods. They should be seen as additional tools provided to gene bank and botanic garden curators for optimizing the germplasm collections placed under their responsibility.

REFERENCES

1. Rao, N.K. Plant genetic resources: Advancing conservation and use through biotechnology. *African J. Biotech.* 2004, *3*, 136–145.
2. United Nations Conference on Environment and Development (UNCED), *Convention on Biological Diversity*; UNCED: Geneva, Switzerland, 1992.
3. Ramsay, M.M.; Jacskon, A.D.; Porley, R.D. A Pilot Study for *ex Situ* Conservation of UK Bryophytes. In Proceedings of EuroGard 2000—II European Botanic Garden Congress, Las Palmas de Gran Canaria, Spain, 10–15 April 2000; Botanic Garden Conservation International: Richmond, UK, 2000; pp. 52–57.
4. Engelmann, F. Germplasm Collection, Storage and Conservation. In *Plant Biotechnology and Agriculture*; Altman, A., Hasegawa, P.M., Eds.; Academic Press: Oxford, UK, 2012; pp. 255–268.
5. Withers, L.A. Collecting *in Vitro* for Genetic Resources Conservation. In *Collecting Plant Genetic Diversity*; Guarino, L., Rao, R., Reid, R., Eds.; Centre for Agricultural Bioscience International: Wallingford, UK,

1995; pp. 511–515.

6. Engelmann, F. Use of biotechnologies for the conservation of plant biodiversity. *In Vitro Cell. Dev. Biol. Plant* 2011, *47*, 5–16.
7. Tandon, P.; Kumaria, S. Prospects of Plant Conservation Biotechnology in India with Special Reference to Northeastern Region. In *Biodiversity: Status and Prospects*; Tandon, P., Kumaria, S., Eds.; Norasa Publishing House: New Delhi, India, 2005; pp. 79–91.
8. Pence, V.C. *In vitro* collecting (IVC). I. The effect of collecting method and antimicrobial agents on contamination in temperate and tropical collections. *In Vitro Cell. Dev. Biol. Plant* 2005, *41*, 324–332.
9. Reed, B.M.; Sarasan, V.; Kane, M.; Bunn, E.; Pence, V.C. Biodiversity conservation and conservation biotechnology tools. *In Vitro Cell. Dev. Biol. Plant* 2011, *47*, 1–4.
10. Withers, L.A. *In vitro* Collecting-Concept and Background. In *In Vitro Collecting Techniques for Germplasm Conservation*; Pence, V.C., Sandoval, J.A., Villalobos, V.M., Engelman, F., Eds.; International Plant Genetic Resources Institute: Rome, Italy, 2002; pp. 16–25.
11. Pence, V.C. *In vitro* Collecting—A Tool for Wild or Endangered Species Conservation. In *In Vitro Collecting Techniques for Germplasm Conservation*; Pence, V.C., Sandoval, J.A., Villalobos, V.M., Engelman, F., Eds.; International Plant Genetic Resources Institute: Rome, Italy, 2002; pp. 26–29.
12. Pence, V.C.; Sandoval, J.A. Controlling Contamination during *in Vitro* Collecting. In *In Vitro Collecting Techniques for Germplasm Conservation*; Pence, V.C., Sandoval, J.A., Villalobos, V.M., Engelman, F., Eds.; International Plant Genetic Resources Institute: Rome, Italy, 2002; pp. 30–40.
13. Falkiner, F. The criteria for choosing an antibiotic for control of bacteria in plant tissue culture. *IAPTC Newsl.* 1990, *60*, 13–23.
14. Rillo, E.P.; Paloma, M.B. Storage and transport of zygotic embryos of *Cocos nucifera* L. for *in vitro* culture. *Plant Gen. Res. Newsl.* 1991, *86*, 1–4.
15. Alvarenga, V.S.; Bianchetti, L.B.; López, P.E.; Sandoval, O.E.; Zacher, M.B. Cacao. In *In Vitro Collecting Techniques for Germplasm Conservation*; Pence, V.C., Sandoval, J.A., Villalobos, V.M., Engelman, F., Eds.; International Plant Genetic Resources Institute: Rome, Italy, 2002; pp. 47–51.
16. Yidana, J.A.; Withers, L.A.; Ivins, J. Development of a simple method for collecting and propagating cocoa germplasm*in vitro*. *Acta Hortic.* 1987, *212*, 95–98.
17. Yidana, J.A. The Development of *in Vitro* Collecting and Isozyme Characterization of Cocoa Germplasm. Ph.D. Thesis, University Nottingham, Nottingham, UK, March, 1988.
18. Engelmann, F.; Assy-Bah, B.; Bagniol, S.; Dumet, D.; Michaux-Ferriere, N. Cryopreservation of Date Palm, Oil Palm, and Coconut. In *Biotecnology in Agriculture and Forestry*; Bajaj, Y.P.S., Ed.; Springer-Verlag: Berlin, Germany, 1995; pp. 148–167.
19. Engelmann, F. Coconut. In *In Vitro Collecting Techniques for Germplasm Conservation*; Pence, V.C., Sandoval, J.A., Villalobos, V.M., Engelman, F., Eds.; International Plant Genetic Resources Institute: Rome, Italy, 2002; pp. 68–71.
20. Ashburner, G.R.; Faure, M.G.; Tomlinson, D.R.; Thompson, W.K. Collection of coconut (*Cocos nucifera*) embryos from remote locations. *Seed Sci. Technol.* 1996, *24*, 159–169.
21. Samosir, Y.M.S.; Godwin, I.D.; Adkins, S.W. A new technique for coconut (*Cocos nucifera*) germplasm collection from remote sites: Culturability of embryos following low temperature incubation. *Aust. J. Bot.* 1999, *47*, 69–75.
22. Assy-Bah, B.; Durand-Gasselin, T.; Pannetier, C. Use of zygotic embryo culture to collect germplasm of coconut (*Cocos nucifera* L.). *Plant Genet. Resour. Newsl.* 1987, *71*, 4–10.
23. Engelmann, F.; Malaurie, B.; N'Nan, O. *In Vitro* Culture of Coconut (*Cocos nucifera* L.) Zygotic Embryos. In *Plant Embryo Culture: Methods and Protocols, Methods in Molecular Biology Series*; Thorpe, T., Yeung, E., Eds.; Humana Press: Totowa, NJ, USA, 2011; pp. 63–74.
24. Lozoya-Saldaña, H.; Oicatá, M.; Borbor-Ponce, M.M.; Calderón-Díaz, J.H. Coffe. In *In Vitro Collecting Techniques for Germplasm Conservation*; Pence, V.C., Sandoval, J.A., Villalobos, V.M., Engelman, F., Eds.; International Plant Genetic Resources Institute: Rome, Italy, 2002; pp. 42–46.
25. Montoya-Henao, L.M.; Tapia, C.; Espadas, F.L.; Sandoval, G.; Sandoval, J.A. Musa. In *In Vitro Collecting Techniques for Germplasm Conservation*; Pence, V.C., Sandoval, J.A., Villalobos, V.M., Engelman, F., Eds.; International Plant Genetic Resources Institute: Rome, Italy, 2002; pp. 52–55.
26. Brenes-Hines, A.; García-Tapia, V.; Velasco-Urquizo, E. Citrus. In *In Vitro Collecting Techniques for Germplasm Conservation*; Pence, V.C., Sandoval, J.A., Villalobos, V.M., Engelman, F., Eds.; International Plant Genetic Resources Institute: Rome, Italy, 2002; pp. 56–60.
27. Sandoval, J.A.; Villalobos, V.M. Avocado. In *In Vitro Collecting Techniques for Germplasm Conservation*; Pence, V.C., Sandoval, J.A., Villalobos, V.M., Engelman, F., Eds.; International Plant Genetic Resources Institute: Rome, Italy, 2002; pp. 61–64.
28. Taylor, M. Taro. In *In Vitro Collecting Techniques for Germplasm Conservation*; Pence, V.C., Sandoval, J.A., Villalobos, V.M., Engelman, F., Eds.; International Plant Genetic Resources Institute: Rome, Italy, 2002; pp. 65–67.
29. Altman, D.W.; Fryxell, P.A.; Koch, S.D.; Howell, C.R. *Gossypium* germplasm conservation augmented by tissue culture techniques for field collecting. *Econ. Bot.* 1990, *44*, 106–113.
30. Engelmann, F. Cryopreservation of Embryos. In *Reproductive Biology and Plant Breeding*; Dattée, Y., Dumas, C., Gallais, A., Eds.; Springer Verlag: Berlin, Germany, 1992; pp. 281–290.

31. Engelmann, F. *In vitro* conservation of tropical plant germplasm—A review. *Euphytica* 1991, *57*, 227–243.

32. Bunn, E.; Turner, S.R.; Panaia, M.; Dixon, K.W. The contribution of *in vitro* technology and cryogenic storage to conservation of indigenous plants. *Austr. J. Bot.* 2007, *55*, 345–355.

33. Villalobos, V.M.; Engelmann, F. *Ex situ* conservation of plant germplasm using biotechnology. *World J. Microb. Biotechnol.* 1995, *11*, 375–382.

34. Fay, M.F. Conservation of rare and endangered plants using *in vitro* methods. *In Vitro Cell. Dev. Biol. Plant* 1992, *28*, 1–4.

35. Sarasan, V.; Cripps, R.; Ramsay, M.M.; Atherton, C.; McMichen, M.; Prendergast, G.; Rowntree, J.K. Conservation *in vitro* of threatened plants—Progress in the past decade. *In Vitro Cell. Dev. Biol. Plant* 2006, *42*, 206–214.

36. Blakesley, D.; Pask, N.; Henshaw, G.G.; Fay, M.F. Biotechnology and the conservation of forest genetic resources: *In vitro* strategies and cryopreservation. *Plant Growth Regul.* 1996, *20*, 11–16.

37. Gangaprasad, A.; Decruse, S.W.; Seeni, S.; Menon, V.S. Micropropagation and restoration of endangered Malabar daffodil orchid. Ipsea malabarica. *Lindleyana* 1999, *14*, 47–56.

38. Pilatti, F.K.; Aguiar, T.; Simões, T.; Benson, E.E.; Viana, A.M. *In vitro* and cryogenic preservation of plant biodiversity in Brazil. *In Vitro Cell. Dev. Biol. Plant* 2011, *47*, 82–98.

39. Kriebel, H.B. Introduction. In *Somatic Embryogenesis in Woody Plants*; Jain, S.M., Gupta, P.K., Newton, R.J., Eds.; Kluwer Academic: Dordrecht, The Netherlands, 1995; Volume 1, pp. 1–8.

40. Carlson, W.C.; Hartle, J.E. Manufactured Seed of Woody Plants. In *Somatic Embryogenesis in Woody Plants*; Jain, S., Gupta, P., Newton, R., Eds.; Kluwer Academic Press: London, UK, 1995; Volume 1, pp. 253–263.

41. Ravi, D.; Anand, P. Production and applications of artificial seeds: A review. *Int. Res. Biol. Sci.* 2012, *1*, 74–78.

42. Pence, V.C. *In vitro* Collection. In *A Colour Atlas of Plant Propagation and Conservation*; Bowes, B.G., Ed.; Manson Publishing: London, UK, 1999; pp. 87–96.

43. Ashmore, S.E.; Hamilton, K.N.; Offord, C.A. Conservation technologies for safeguarding and restoring threatened flora: Case studies from Eastern Australia. *In Vitro Cell. Dev. Biol. Plant* 2011, *47*, 99–109.

44. Normah, M.N.; Makeen, A.M. Cryopreservation of Excised Embryos and Embryonic Axes. In *Plant Cryopreservation: A Practical Guide*; Reed, B., Ed.; Springer: New York, NY, USA, 2008; pp. 211–240.

45. Berjak, P.; Bartels, P.; Benson, E.; Harding, K.; Mycock, D.; Pammenter, N.W.; Sershen; Wesley-Smith, J. Cryoconservation of South African plant genetic diversity. *In Vitro Cell. Dev. Biol. Plant* 2011, *47*, 65–81.

46. Ashmore, S. *Status Report on the Development and Application of In Vitro Techniques for the Conservation and Use of Plant Genetic Resources*; Engelmann, F., Ed.; International Plant Genetic Resources Institute: Rome, Italy, 1997.

47. Faccioli, V.C.; Marani, F. Virus Elimination by Meristem Tip Culture and Tip Micrografting. In *Plant Virus Disease Control*; Hadidi, A., Khetarpal, R.K., Koganezawa, H., Eds.; APS Press: St. Paul, MN, USA, 1998; pp. 346–380.

48. Abdelnour-Esquivel, A.; Bermudez, L.C.; Alvarenga, S.; Rivera, C. Cultivo de meristemas, termo y quimioterapia en chayote (*Sechium edule* Jacq. Sw.) para la erradicación del virus del mosaico del chayote (ChMV) [in Spanish]. *Manejo Integr. Plag. Agroecol. Costa Rica* 2006, *77*, 17–23.

49. Wang, Q.C.; Valkonen, J.P.T. Cryotherapy of shoot tips: Novel pathogen eradication method. *Trends Plant Sci.* 2009, *14*, 199–122.

50. Grout, B.W.W. Meristem-Tip Culture for Propagation and Virus Elimination. In *Plant Cell Culture Protocols*; Hall, R.D., Ed.; Humana Press Inc.: Totowa, NJ, USA, 1990; pp. 115–125.

51. Kane, M.E. Shoot Culture Procedures. In *Plant Development and Biotechnology*; Triagiano, R.N., Gray, D.G., Eds.; CRC Press LLC: Boca Raton, FL, USA, 2005; pp. 145–157.

52. Wang, Q.C.; Valkonen, J.P.T. Efficient elimination of sweetpotato little leaf phytoplasma from sweetpotato by cryotherapy of shoot tips. *Plant Pathol.* 2008, *57*, 338–347.

53. Wang, Q.C.; Valkonen, J.P.T. Eradication of two synergistically interacting viruses from sweetpotato using shoot tip culture and cryotherapy of shoot tips. *J. Virol. Methods* 2008, *154*, 135–145.

54. Panattoni, A.; Luvisi, A.; Triolo, E. Review. Elimination of viruses in plants: Twenty years of progress. *Span. J. Agric. Res.* 2013, *11*, 173–188.

55. Youssef, S.A.; Al-Dhaher, M.M.A.; Shalaby, A.A. Elimination of *Grapevine fanleaf virus* (GFLV) and *Grapevine leaf roll-associated virus-1* (GLRaV-1) from infected grapevine plants using meristem tip culture. *Int. J. Virol.* 2009, *5*, 89–99.

56. Ramos, C.S.; Zamora, A.B. Isolation of banana plantlets cv. señorita freed of *Banana bract mosaic virus*. *Philipp. Agric. Sci.* 1999, *82*, 25–34.

57. D'Onghia, A.M.; Carimi, F.; de Pasquale, F.; Djelouah, K.; Martelli, G.P. Elimination of *Citrus psorosis* virus by somatic embryogenesis from stigma and style cultures. *Plant Pathol.* 2001, *50*, 266–269.

58. Quainoo, A.K.; Wetten, A.C.; Allainguillaume, J. The effectiveness of somatic embryogenesis in eliminating the *Cocoa swollen shoot virus* from infected cocoa trees. *J. Virol. Methods* 2008, *149*, 91–96.

59. Golino, D.A.; Sim, S.T.; Lee, J.; Rowhani, A. Elimination of rose mosaic viruses using microshoot tip tissue culture.*Acta Hort.* 2007, *751*, 237–240.

60. Ramgareeb, S.; Snyman, S.J.; Van-Antwerpen, T.; Rutherford, R.S. Elimination of virus and rapid propagation of diseasefree sugarcane (*Saccharum* spp. cultivar NCo376) using apical meristem culture. *Plant Cell Tissue Org.* 2010, *100*, 175–181.

61. Ramírez-Malagón, R.; Pérez-Moreno, L.; Borodanenko, A.; Salinas-González, G.J.; Ochoa-Alejo, N. Differential organ infection studies, potyvirus elimination, and field performance of virus-free garlic plants produced by tissue culture.*Plant Cell Tissue Org.* 2006, *86*, 103–110.

62. Fang, Y.; Dhital, S.P.; Li, K.; Khu, D.; Kim, H.; Song, Y.; Lim, H. Utilization of single nodal cuttings and therapies for eradicating double-infected potato virus (PLRV, PVY) from *in vitro* plantlets of potato (*Solanum tuberosum*). *J. K. Soc. Hort. Sci.* 2005, *48*, 119–125.

63. Mangal, M.; Bhardwaj, S.V.; Mangal, A.K.; Kaur, R. Production of *Carnation latent virus*-free stock of carnations.*Phytomorphology* 2002, *52*, 1–6.

64. Kumar, S.; Khan, M.S.; Raj, S.K.; Sharma, A.K. Elimination of mixed infection of *Cucumber mosaic* and *Tomato aspermy virus* from Chrysanthemum morifolium Ramat. cv. Pooja by shoot meristem culture. *Sci. Hortic.* 2009, *119*, 108–112.

65. Sedivá, J.; Novák, P.; Laxa, J.; KaÀka, J. Micropropagation, detection and elimination of DMV in the Czech collection of Dahlia. *Acta Hort.* 2006, *725*, 495–498.

66. Morris, J.B.; Dunn, S.; Pinnow, D.L.; Hopkins, M.S.; Pittman, R.N. Meristem culture for virus elimination and peanut interspecific hybrid preservation. *Crop Sci.* 1997, *37*, 591–594.

67. Kabir, A.H.; Pal, S.P.; Sarker, K.K.; Sharmin, S.A.; Alam, M.F. Virus elimination and pathogen-free plantlets regeneration in *Cucurbita pepo* L. *Arch. Phytopathol. PFL* 2010, *43*, 527–537.

68. Sahijram, L.; Rajasekharan, P.E. Tissue Culture Strategies Applicable to in Vitro Conservation of Tropical Fruit Crops. In *Tropical Fruits in Asia: Conservation and Use*; IBPGR Progress Report; International Board for Plant Genetic Resources: Rome, Italy, 1996; pp. 113–119.

69. Holobiuc, I.; Paunescu, A.; Blindu, R. *Ex situ* conservation using *in vitro* methods in some Caryophyllaceae plant species from the Red List of vascular plants in Romania. *Rom. J. Biol. Plant Biol.* 2007, *49*, 3–16.

70. Tassy, C.; Feuillet, C.; Barret, P. A method for the medium-term storage of plant tissue samples at room temperature and successive cycles of DNA extraction. *Plant Mol. Biol. Rep.* 2006, *24*, 247–248.

71. Paunescu, A. Biotechnology for endangered plant conservation: A critical overview. *Rom. Biotech. Lett.* 2009, *14*, 4095–4103.

72. Devi, J.; Ray, B.K.; Chetia, S.; Deka, P.C. Regeneration of low temperature stored encapsulated protocorms of orchids.*J. Orchid. Soc. India* 1998, *12*, 39–41.

73. Lambardi, M.; Halmagyi, A.; Benelli, C.; de Carlo, A.; Vettori, C. Seed cryopreservation for conservation of ancient*Citrus* germplasm. *Adv. Hortic. Sci.* 2007, *21*, 198–202.

74. Keshavachandran, R.; Smitha, M.K.; Vijayaraghavan, R. *In Vitro* Conservation of Endangered Medicinal Plants. In Recent Trends in Horticultural Biotechnology, Volumes I and II, Proceedings of Indian Council of Agricultural Research National Symposium on Biotechnological Interventions for Improvement of Horticultural Crops: Issues and Strategies, Vellanikkara, India, 10–12 January 2005; Keshavachandran, R., Nazeem, P., Girija, D., John, P.S., Peter, K.V., Eds.; Indian Council of Agricultural Research: New Delhi, India, 2005; pp. 439–442.

75. Murashige, T.; Skoog, F. A revised medium for rapid growth and bioassays with tobacco tissue cultures. *Physiol. Plant*1962, *15*, 473–497.

76. Rakosy-Tican, E.; Bors, B.; Szatmari, A.M. *In vitro* culture and medium-term conservation of the rare wild species*Gladiolus imbricatus*. *African J. Biotech.* 2012, *11*, 14703–14712.

77. Banerjee, N.; de Langhe, E. A tissue culture technique for rapid clonal propagation and storage under minimal growth conditions of *Musa* (banana and plantain). *Plant Cell Rep.* 1985, *4*, 351–354.

78. Roca, W.M.; Reyes, R.; Beltran, J. Effect of Various Factors on Minimal Growth in Tissue Culture Storage of Cassava Germplasm. In Proceedings of the Sixth Symposium of the International Society for Tropical Root Crops, Lima, Peru, 21–26 February 1984; pp. 441–446.

79. Reed, B.M. Application of gas-permeable bags for *in vitro* cold storage of strawberry germplasm. *Plant Cell Rep.* 1991,*10*, 431–434.

80. Kartha, K.K.; Mroginski, N.L.; Pahl, K.; Leung, N.L. Germplasm preservation of coffee (*Coffea arabica* L.) by *in vitro*culture of shoot apical meristems. *Plant Sci. Lett.* 1981, *22*, 301–307.

81. De Goes, M. Studies on the Conservation of Sweet Potato [*Ipomoea batatas* (L) LAM] Germplasm. Ph.D. Thesis, University of Bath, Bath, UK, 1993.

82. Dekkers, A.J.; Rao, A.N.; Goh, C.J. *In vitro* storage of multiple shoot cultures of gingers at ambient temperature. *Sci. Hortic.* 1991, *47*, 157–167.

83. Guanino, T.; Silvanini, A.; Benelli, C.; Beghe, D.; Fabbri, A. Synthetic seed production and conservation of Kober 5BB grapevine rootstock. *Ital. Hort.* 2009, *16*, 267–270.

84. Hamed, A.M.; Gomma, A.H.; Aly, A.S.; Ibrahim, I.A. *In vitro* storage of encapsuylated somatic embryos of *Phenix dactylifera* L. *Bull. Fac. Agric. Cairo Univ.* 2003, *54*, 585–598.

85. Engelmann, F. Plant cryopreservation: Progress and prospects. *In Vitro Cell. Dev. Biol. Plant* 2004, *40*, 427–433.

86. Gonzalez-Arnao, M.T.; Panta, A.; Roca, W.M.; Escobar, R.H.; Engelmann, F. Development and large scale application of cryopreservation techniques for shoot and somatic embryo cultures of tropical crops. *Plant Cell Tissue Org. Cult.*2008, *92*, 1–13.

87. Kartha, K.K.; Engelmann, F. Cryopreservation and Germplasm Storage. In *Plant Cell and Tissue Culture*; Vasil, I.K., Thorpe, T.A., Eds.; Kluwer: Dordrecht, Germany, 1994; pp. 195–230.

88. Withers, L.A.; Engelmann, F. *In Vitro* Conservatipon of Plant Genetic Resources. In *Biotechnology in*

Agriculture; Altman, A., Ed.; Marcel Dekker Inc.: New York, NY, USA, 1998; pp. 57–88.

89. Reed, B.M.; Uchendu, E. Controlled Rate Cooling. In *Plant Cryopreservation: A Practical Guide*; Reed, B.M., Ed.; Springer Verlag: Berlin, Germany, 2008; pp. 77–92.

90. Escobar, R.H.; Mafla, G.; Roca, W.M. A methodology for recovering cassava plants from shoot tips maintained in liquid nitrogen. *Plant Cell Rep.* 1997, *16*, 474–478.

91. Engelmann, F. Importance of Cryopreservation for the Conservation of Plant Genetic Resources. In *Cryopreservation of Tropical Plant Germplasm: Current Research Progress and Application*; Engelmann, F., Takagi, H., Eds.; Japan International Research Center for Agricultura: Tsukuba, Japan; International Plant Genetic Resources Institute: Rome, Italy, 2000; pp. 8–20.

92. Fahy, G.M.; MacFarlane, D.R.; Angell, C.A.; Meryman, H.T. Vitrification as an approach to cryopreservation. *Cryobiology* 1984, *21*, 407–426.

93. Langis, R.; Schnabel, B.; Earle, B.J.; Steponkus, P.L. Cryopreservation of carnation shoot tips by vitrification. *Cryobiology* 1990, *276*, 658–659.

94. Sakai, A.; Kobayashi, S.; Oiyama, I. Cryopreservation of nucellar cells of navel orange (*Citrus sinensis* Osb. var.*Brasiliensis* Tanaka) by vitrification. *Plant Cell Rep.* 1990, *9*, 30–33.

95. Towill, L.E. Cryopreservation of isolated mint shoot tips by vitrification. *Plant Cell Rep.* 1990, *9*, 178–180.

96. Dereuddre, J.; Scottez, C.; Arnaud, Y.; Duron, M. Resistance of alginate-coated axillary shoot tips of pear tree (*Pyrus communis* L. cv Beurre Hardy) *in vitro* plantlets to dehydration and subsequent freezing in liquid nitrogen. *C. R. Acad. Sci. Paris* 1990, *310*, 317–323.

97. Fabre, J.; Dereuddre, J. Encapsulation-dehydration: A new approach to cryopreservation of *Solanum* shoot tips. *CryoLetters* 1990, *11*, 413–426.

98. Yamamoto, S.; Rafique, T.; Priyantha, W.S.; Fukui, K.; Matsumoto, T.; Niino, T. Development of a cryopreservation procedure using aluminium cryo-plates. *CryoLetters* 2011, *32*, 256–265.

99. Gonzalez-Arnao, M.T.; Engelmann, F. Cryopreservation of plant germplasm using the encapsulation-dehydration technique: Review and case study on sugarcane. *CryoLetters* 2006, *27*, 155–168.

100. Sakai, A.; Engelmann, F. Vitrification, encapsulation-vitrification and droplet-vitrification: A review. *CryoLetters* 2007,*28*, 151–172.

101. Sakai, A.; Hirai, D.; Niino, T. Development of PVS-based Vitrification and Encapsulation-Vitrification Protocols. In*Plant Cryopreservation: A Practical Guide*; Reed, B.M., Ed.; Springer Verlag: Berlin, Germany, 2008; pp. 33–58.

102. Kartha, K.K.; Leung, N.L.; Mroginski, L.A. *In-vitro* growth responses and plant regeneration from cryopreserved meristems of cassava (*Manihot esculenta* Crantz). *Z. Pflanzenphysiol.* 1982, *107*, 133–140.

103. Yamamoto, S.; Rafique, T.; Fukui, K.; Sekizawa, K.; Niino, T. V-cryo-plate procedure as an effective protocol for cryobanks: Case study of mint cryopreservation. *CryoLetters* 2012, *33*, 12–23.

104. Panis, B.; Piette, B.; Swennen, R. Droplet vitrification of apical meristems: A cryopreservation protocol applicable to all Musaceae. *Plant Sci.* 2005, *168*, 45–55.

105. Panis, B. Cryopreservation of Banana (*Musa* spp.) Germplasm. Ph.D. Thesis, Catholoc University of Leuven, Leuven, Belgium, February 1995.

106. Gonzalez-Arnao, M.T.; Engelmann, F.; Huet, C.; Urra, C. Cryopreservation of encapsulated apices of sugarcane: Effect of freezing procedure and histology. *CryoLetters* 1993, *14*, 303–308.

107. Gonzalez-Arnao, M.T.; Ravelo, M.M.; Urra-Villavicencio, C.; Montero, M.M.; Engelmann, F. Cryopreservation of pineapple (*Ananas comosus*) apices. *CryoLetters* 1998, *19*, 375–382.

108. Gonzalez-Arnao, M.T.; Juarez, J.; Ortega, C.; Navarro, L.; Duran-Vila, N. Cryopreservation of ovules and somatic embryos of citrus using the encapsulation-dehydration technique. *CryoLetters* 2003, *24*, 85–94.

109. Stanwood, P.C.; Bass, L.N. Seed germplasm preservation using liquid nitrogen. *Seed Sci. Technol.* 1981, *9*, 423–437.

110. Li, D.Z.; Pritchard, H.W. The science and economics of *ex situ* plant conservation. *Trends Plant Sci.* 2009, *14*, 614–621.

111. Probert, R.J.; Daws, M.I.; Hay, F. Ecological correlates of *ex situ* seed longevity: A comparative study on 195 species.*Ann. Bot.* 2009, *104*, 57–69.

112. Food and Agriculture Organization of the United Nations, International Plant Genetic Resources Institute; *Genebank Standards*; Food and Agriculture Organization of the United Nations: Rome, Italy; International Plant Genetic Resources Institute: Rome, Italy, 1994; pp. 1–17.

113. Walters, C.; Wheeler, L.J.; Grotenhuis, J.M. Longevity of seeds stored in a genebank: Species characteristics. *Seed Sci. Res.* 2005, *15*, 1–20.

114. Walters, C.; Wheeler, L.J.; Stanwood, P.C. Longevity of cryogenically stored seeds. *Cryobiology* 2004, *48*, 229–244.

115. Roberts, E.H.; Ellis, R.H. Water and seed survival. *Ann. Bot.* 1989, *63*, 39–52.

116. Dickie, J.B.; Ellis, E.H.; Kraak, H.L.; Ryder, K.; Tompsett, P.B. Temperature and seed storage longevity. *Ann. Bot.* 1990,*65*, 197–204.

117. Pritchard, H.W.; Ashmore, S.; Berjak, P.; Engelmann, F.; González-Benito, M.E.; Li, D.Z.; Nadarajan, J.; Panis, B.; Pence, V.; Walters, C. Storage Stability and the Biophysics of Preservation. In Proceedings of the Plant Conservation for the Next Decade: A Celebration of Kew's 250th Anniversary, Royal Botanic Garden Kew, London, UK, 12–16 October 2009; p. 11.

118. Hirano, T.; Ishikawa, K.; Mii, M. Advances in Orchid Cryopreservation. In *Floriculture, Ornamental and Plant Biotechnology, Advances and Topical Issues*; Teixeira da Silva, J.A., Ed.; Global Science Books: Kenobe, Japan, 2006; Volume 2, pp. 410–414.

119. Pence, V.C. Cryopreservation of Bryophytes and Ferns.

In *Plant Cryopreservation: A Practical Guide*; Reed, B.M., Ed.; Springer Verlag: Berlin, Germany, 2008; pp. 117–140.

120. Touchell, D.H.; Dixon, K.W. Cryopreservation of seed of Western Australia native species. *Biodiv. Cons.* 1993, *2*, 594–602.

121. Nikishina, T.V.; Popova, E.V.; Vakhrameeva, M.G.; Varlygina, T.I.; Kolomeitseva, G.L.; Burov, A.V.; Popovich, E.A.; Shirokov, A.I.; Shumilov, V.; Yu, V.; *et al.* Cryopreservation of seeds and protocorms of rare temperate orchids. *Russ. J. Plant Physiol.* 2007, *54*, 121–127.

122. Malik, S.K.; Chaudhury, R. The cryopreservation of embryonic axes of two wild and endangered *Citrus* species. *Plant Genet. Res. Charact. Util.* 2006, *4*, 204–209.

123. Hamilton, K.N.; Ashmore, S.E.; Pritchard, H.W. Thermal analysis and cryopreservation of seeds of Australian wild*Citrus* species (Rutaceae): *Citrus australasica, C. inodora* and *C. garrawayi. CryoLetters* 2009, *30*, 268–279.

124. Maneerattanarungroj, P.; Bunnag, S.; Monthatong, M. *In vitro* conservation of *Cleisostoma areitinum* (Rchb. f.) Garay, rare Thai orchid species by an encapsulation-dehydration method. *Asian J. Plant Sci.* 2007, *6*, 1235–1240.

125. Gonzalez-Benito, M.E.; Perez, C. Cryopreservation of nodal explants of an endangered plant species (*Centaurium rigualii* Esteve) using the encapsulation-dehydration method. *Biodiv. Cons.* 1997, *6*, 583–590.

126. Turner, S.R.; Senaratna, T.; Bunn, E.; Tan, B.; Dixon, K.W.; Touchell, D.H. Cryopreservation of shoot tips from six endangered Australian species using a modified vitrification protocol. *Ann. Bot.* 2001, *87*, 371–378.

127. Tanaka, D.; Niino, T.; Tsuchiya, Y.; Shirata, K.; Uemura, M. Cryopreservation of shoot tips of endangered Hayachine-usuyukiso [*Leontopodium hayachinense* (Takeda) Hara et Kitam.] using a vitrification protocol. *Plant Genet. Res. Charact. Util.* 2009, *6*, 164–166.

128. Yoon, J.W.; Kim, H.H.; Cho, E.G.; Ko, H.C.; Hwang, H.S.; Park, Y.E.; Engelmann, F. Cryopreservation of cultivated and wild potato varieties by droplet vitrification procedure. *Acta Hortic.* 2007, *760*, 203–208.

129. Niu, Y.L.; Luo, Z.R.; Zhang, Y.F. Studies on cryopreservation of two *Diospyros* spp. germplasm by modified droplet vitrification. *J. Wuhan Bot. Res.* 2009, *27*, 451–454.

130. Benelli, C.; de Carlo, A.; Engelmann, F. Recent advances in the cryopreservation of shoot-derived germplasm of economically important fruit trees of *Actinidia, Diospyros, Malus, Olea, Prunus, Pyrus* and *Vitis. Biotech. Adv.* 2013, *31*, 175–185.

131. Mandal, B.B. Cryopreservation Research in India: Current Status and Future Perspectives. In *Cryopreservation of Tropical Plant Germplasm—Current Research Progress and Applications*; Engelmann, F., Takagi, H., Eds.; Japan International Research Center for Agricultura: Tsukuba, Japan; International Plant Genetic Resources Institute: Rome, Italy, 2000; pp. 282–286.

132. Touchell, D.H.; Dixon, K.W. Cryopreservation for seedbanking of Australian species. *Ann. Bot.* 1994, *40*, 541–546.

133. Dussert, S.; Couturon, E.; Bertrand, B.; Nuissier, F.; Joët, T.; Pavis, C.; Engelmann, F. Progress in the Implementation of the French Coffee Cryobank. In Proceedings of the 2nd International Symposium on Plant Cryopreservation, Fort Collins, CO, USA, 11–14 August 2013. in press.

134. Forsline, P.L.; McFerson, J.R.; Lamboy, W.F.; Towill, L.E. Development of base and active collections of *Malus*germplasm with cryopreserved dormant buds. *Acta Hort.* 1999, *484*, 75–78.

135. Niino, T. Cryopreservation of Germplasm of Mulberry. In *Biotechnology in Agriculture and Forestry, Volume 32, Cryopreservation of Plant Germplasm I*; Bajaj, Y.P.S., Ed.; Springer Verlag: Berlin, Germany, 1995; pp. 102–113.

136. Panis, B.; van den Houwe, I.; Piette, B.; Swennen, R. Cryopreservation of the Banana Germplasm Collection at the ITC (INIBAP Transit Centre). In Proceedings of the 1st Meeting of COST 871 Working Group 2: Technology, Application and Validation of Plant Cryopreservation, Florence, Italy, 10–13 May 2007; pp. 34–35.

137. Reed, B.M.; DeNoma, J.; Chang, Y. Application of Cryopreservation Protocols at a Clonal Genebank. In*Cryopreservation of Tropical Plant Germplasm—Current Research Progress and Applications*; Engelmann, F., Takagi, H., Eds.; Japan International Research Center for Agricultura: Tsukuba, Japan; International Plant Genetic Resources Institute: Rome, Italy, 2000; pp. 246–249.

138. Li, B.L.; Zhang, Y.L.; Wang, H.; Song, C.H.; Liu, Y. Pollen cryo-bank establishment and application of traditional chinese flowers. *Cryobiology* 2009, *59*, 414.

139. Spencer, M. The challenges of developing cryopreservation strategies to suit the requirements of a large industrial *in vitro* plant cell collection. *Cryobiology* 1999, *39*, 325.

140. Cyr, D.R. Cryopreservation: Roles in Clonal Propagation and Germplasm Conservation of Conifers. In*Cryopreservation of Tropical Plant Germplasm—Current Research Progress and Applications*; Engelmann, F., Takagi, H., Eds.; Japan International Research Center for Agricultura: Tsukuba, Japan; IPGRI: Rome, Italy, 2000; pp. 261–268.

141. Florin, B.; Brulard, E.; Lepage, B. Establishment of a cryopreserved coffee germplasm bank. *Cryobiology* 1999, *39*, 309.

142. Galerne, M.; Bercetche, J.; Dereuddre, J. Cryoconservation of embryogenic cellus of Norway spruce (*Picea abies* L. Karst)—Influence of different factors on callus recovery and on embryos and plantlet production. *Bull. Soc. Fr. Lett. Bot.* 1992, *4–5*, 331–344.

143. Aguilar, M.E.; Engelmann, F.; Michaux-Ferrière, N. Cryopreservation of embryogenic cell suspensions of *Citrus deliciosa* Tan. and histological study. *CryoLetters* 1993, *14*, 217–228.

144. Kendall, E.J.; Qureshi, J.A.; Kartha, K.K.; Leung, N.;

Chevrier, N.; Caswell, K.; Chen, T.H.H. Regeneration of freezing-tolerant spring wheat (*Triticum aestivum* L.) plants from cryoselected callus. *Plant Physiol.* 1990, *94*, 1756–1762.

145. Brison, M.; de Boucaud, M.T.; Pierronnet, A.; Dosba, F. Effect of cryopreservation on the sanitary state of a cv of*Prunus* rootstock experimentally contaminated with Plum Pox Potyvirus. *Plant Sci.* 1997, *123*, 189–196.

146. Wang, Q.C.; Panis, B.; Engelmann, F.; Lambardi, M.; Valkonen, J.T. Cryotherapy of shoot tips: A technique for pathogen eradication to produce healthy planting materials and prepare healthy plant genetic resources for cryopreservation. *Ann. Appl. Biol.* 2009, *154*, 351–363.

147. Helliot, B.; Panis, B.; Poumay, Y.; Swennen, R.; Lepoivre, P.; Frison, E. Cryopreservation for the elimination of cucumber mosaic and banana streak viruses from banana (*Musa* spp.). *Plant Cell Rep.* 2002, *20*, 1117–1122.

148. Ding, F.; Jin, S.; Hong, N.; Zhong, Y. Vitrification-cryopreservation, an efficient method for eliminating *Candidatus Liberobacter asiaticus*, the citrus Huanglongbing pathogen, from *in vitro* adult shoot tips. *Plant Cell Rep.* 2008, *27*, 241–250.

149. Wang, Q.C.; Mawassi, M.; Li, P.; Gafny, R.; Sela, I.; Tanne, E. Elimination of grapevine virus A (GVA) by cryopreservation of *in vitro*-grown shoot tips of *Vitis vinifera* L. *Plant Sci.* 2003, *165*, 321–327.

150. Wang, Q.C.; Cuellar, W.J.; Rajamäki, M.L.; Hirata, Y.; Valkonen, J.P.T. Combined thermotherapy and cryotherapy for virus eradication: Relation of virus distribution, subcellular changes, cell survival and viral RNA degradation in shoot tips to efficient production of virus-free plants. *Mol. Plant Pathol.* 2008, *9*, 237–250.

151. Wang, Q.C.; Liu, Y.; Xie, Y.; You, M. Cryotherapy of potato shoot tips for efficient elimination of potato leaf roll virus (PLRV) and potato virus Y (PVY). *Potato Res.* 2006, *49*, 119–129.

152. Feng, C.; Yin, Z.; Ma, Y.; Zhang, Z.; Chen, L.; Wang, B.; Li, B.; Huang, Y.; Wang, Q. Cryopreservation of sweetpotato (*Ipomoea batatas*) and its pathogen eradication by cryotherapy. *Biotech. Adv.* 2011, *29*, 84–93.

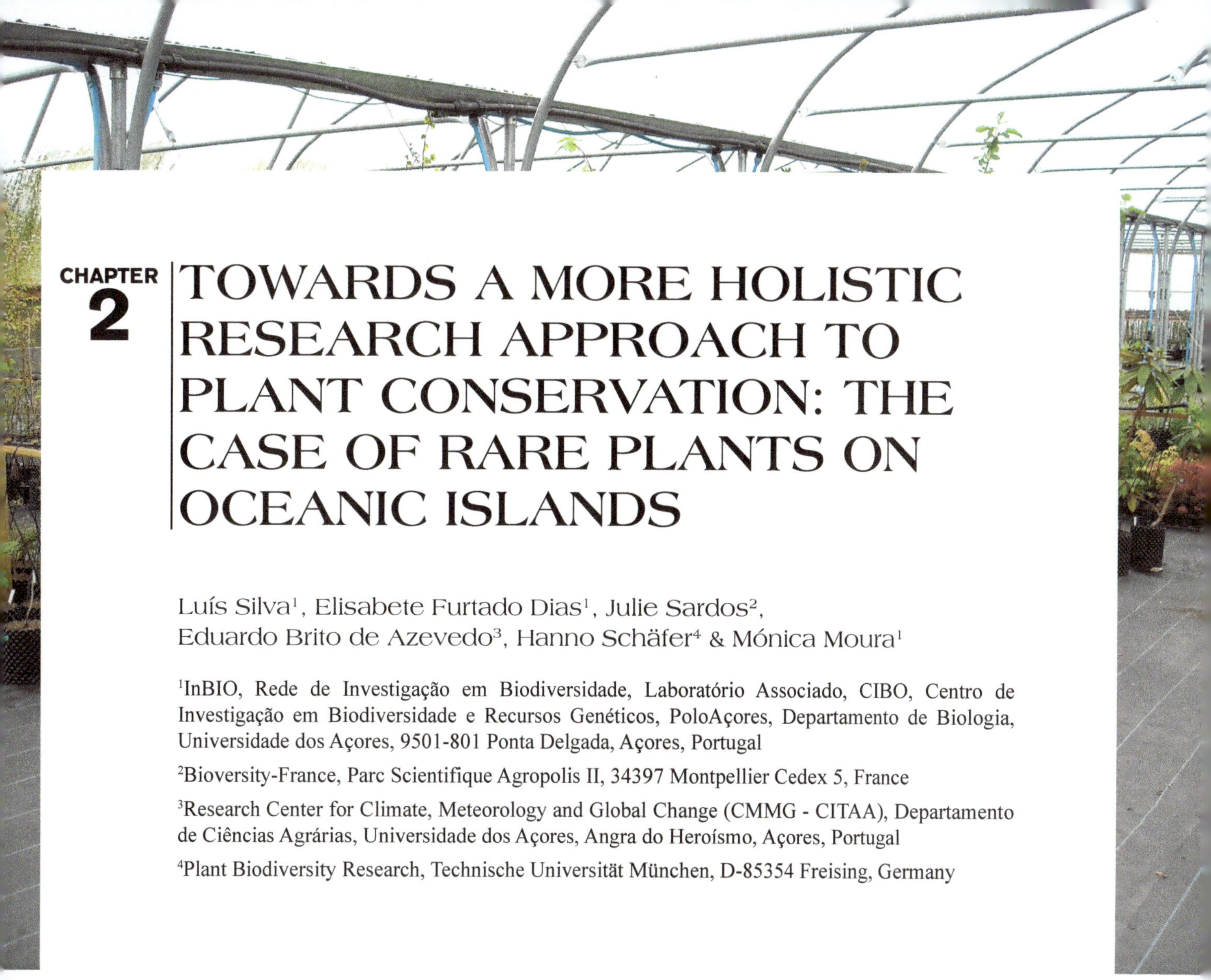

CHAPTER 2

TOWARDS A MORE HOLISTIC RESEARCH APPROACH TO PLANT CONSERVATION: THE CASE OF RARE PLANTS ON OCEANIC ISLANDS

Luís Silva[1], Elisabete Furtado Dias[1], Julie Sardos[2], Eduardo Brito de Azevedo[3], Hanno Schäfer[4] & Mónica Moura[1]

[1]InBIO, Rede de Investigação em Biodiversidade, Laboratório Associado, CIBO, Centro de Investigação em Biodiversidade e Recursos Genéticos, PoloAçores, Departamento de Biologia, Universidade dos Açores, 9501-801 Ponta Delgada, Açores, Portugal

[2]Bioversity-France, Parc Scientifique Agropolis II, 34397 Montpellier Cedex 5, France

[3]Research Center for Climate, Meteorology and Global Change (CMMG - CITAA), Departamento de Ciências Agrárias, Universidade dos Açores, Angra do Heroísmo, Açores, Portugal

[4]Plant Biodiversity Research, Technische Universität München, D-85354 Freising, Germany

ABSTRACT

Research dedicated to rare endemic plants is usually focused on one given aspect. However, holistic studies, addressing several key issues, might be more useful, supporting management programs, while unravelling basic knowledge about ecological and population level processes. A more comprehensive approach to research is proposed, encompassing: phylogenetics/systematics, pollination biology and seed dispersal, propagation, population genetics, species distribution models (SDMs), threats and monitoring. We present a holistic study dedicated to *Veronica dabneyi*, an endangered chamaephyte endemic to the Azores. *Veronica dabneyi* was mainly found associated with other endemic taxa, however invasive plants were also present and together with introduced cattle, goats and rabbits are a major threat. Most populations grow at somewhat rocky and steep locations that appeared to work as refuges. Seed set in the wild was generally high and recruitment of young plants from seed seemed to be frequent. In the laboratory, it was possible to germinate and fully develop *V. dabneyi* seedlings, which were planted at their site of origin. No dormancy was detected and time for 50% of germination was affected by incubation temperature. Eight new microsatellite markers were applied to 72 individuals from 7 sites. A considerable degree of admixture was found between samples from the two islands Flores and Corvo, with 98% of the genetic variability allocated within populations. Levels of heterozygosity were high and no evidence of inbreeding was found. SDMs based on climatic and topographic variables allowed the estimation of the potential distribution of *V. dabneyi* on Flores and Corvo using ENFA and Maxent. The inclusion of land use variables only slightly increased the information explained by the models. Projection of the expected habitat in Faial largely coincided with the only historic record of *V. dabneyi* in that island. This research could be the basis for the design of a recovery plan, showing the pertinence of more holistic research approaches to plant conservation.

INTRODUCTION

Island endemic plants are among the most threatened group of organisms worldwide (Caujapé-Castells et al. 2010). Research dedicated to rare endemic plants is usually focused on one or few aspects, such as conservation genetics, propagation or distribution (e.g. Dubuis et al. 2013; Evans et al. 2014; Mir et al. 2014). Multidisciplinary studies, addressing several

Citation: Luís Silva, Elisabete Furtado Dias, Julie Sardos, Eduardo Brito de Azevedo, Hanno Schäfer & Mónica Moura, Towards a more holistic research approach to plant conservation: the case of rare plants on oceanic islands, doi: 10.1093/aobpla/plv066.

key issues, are much more useful but still the exception (e.g. Menges 1990; Halbur et al. 2014). They have the potential to provide science-based evidence to management or recovery programs while at the same time unravelling basic knowledge about the population processes and the ecology of endangered species. We propose a general framework towards more holistic conservation research, particularly when devoted to rare plants on oceanic islands. We suggest that such approach should include the following critical areas: (1) phylogenetics/ systematics (i.e. DNA sequences and morphology) to determine origin and close relatives as well as the existence of unaccounted taxa (Bateman et al. 2013); (2) population genetics to estimate genetic structure and diversity (Dias et al. 2014), to identify possible cases of inbreeding depression (Li et al. 2012), and ensure adequate provenance of propagation material (Silva et al. 2011; Hancock and Hughes 2014); (3) germination biology and propagation methods, to identify possible biological constraints (Beaunea et al. 2013) and support the species recovery (Pence 2013); (4) pollination biology and research on dispersal mechanisms, to identify possible biological/ecological constraints (Ollerton et al. 2011; Rodríguez et al. 2015); (5) the identification of threats, including invasive plants (Foxcroft et al. 2013) and animals, particularly herbivores (Donlan et al. 2003; Garzón-Machado et al. 2010, Barrios-Garcia et al. 2013); (6) species distribution models to determine environmental constraints and potentially favorable areas, possible impacts of climate change and of other anthropogenic alterations (Costa et al. 2013; Marcer et al. 2013); and (7) long-term monitoring (minimum 10 years, depending on the duration of the species' generation time) to evaluate population fluctuations and the effect of management actions (Godefroid et al. 2011). Besides addressing those biological issues, conservation research should also integrate possible societal aspects (i.e. the different stakeholders that will have a direct or indirect role in the conservation process).

Here we present an example of a multidisciplinary study, using Veronica dabneyi Hochst. (Plantaginaceae Juss.), known as "veronica" or "Azorean speedwell", a rare chamaephyte (subshrub) endemic to the Azores islands. The genus Veronica L. is the largest of Plantaginaceae with about 450 species. It is distributed worldwide, with a large range of life forms from diverse habitats (Albach et al. 2005). Based on morphological affinities and preliminary molecular data, V. dabneyi is a close relative of V. officinalis L., belonging to the subgenus Veronica L., a clade also including V. alpina L. and V. montana L. (Albach and Meudt 2010).

Veronica dabneyi was first described by Karl C.F. Hochstetter in 1838, after a visit to Faial Island. Since the plant was cultivated as ornamental in the garden of the American consul, Charles William Dabney in Horta (Faial), Hochstetter attributed the specific name presently used. Seubert (1844) published the description with drawings in his Flora Azorica. Later, Watson (1870) reported V. dabneyi for São Miguel, Faial, and Corvo Islands, without specifying the exact places where the species was found. Cunha and Sobrinho (1939) collected V. dabneyi in Faial, at the inner side of the Caldeira summit, the only record published during the 20th century (specimen in LISU, seen by HS). The species was later cited as extinct by Catarino et al. (2001). However, it turned out that the species was still extant in inaccessible parts of the western islands (Pereira et al. 2002). Those authors observed a number of mature plants on Flores and Corvo, but noticed that young plants were rare and seed production was low, probably due to predation by goats and rabbits, and to the occurrence of environmental disturbance (e.g. landslides, trampling by cattle). The species was found to be associated with a vegetation type described by Sjögren (1973) as Festucetum jubatae, dominated by Festuca francoi Fern.Prieto, C.Aguiar, E.Dias & M.I.Gut (Poaceae Barnhart). The conservation status was reevaluated according to the IUCN criteria and it was classified as Extinct in Faial and as Critically Endangered in Flores and Corvo (Pereira et al. 2002), based on an area of occurrence of 63 km^2, with sixteen subpopulations in Flores, and one in Corvo. In general, the populations only included a few individuals, while no individuals were found in the wild at Faial. Veronica dabneyi was thus classified as Extinct in Faial and as Critically Endangered in Flores and Corvo. In a global analysis of the conservation status of Azorean indigenous species, Silva et al. (2009) suggested possible natural threats (storms, strong wind, landslides), biological limitations (isolation of populations), as well as threats of human origin (expansion of invasive plants and introduced herbivores, changes in land use) as causes of species decline.

Two aspects that should be addressed as important contributions to the species long-term survival are estimates of genetic diversity and population genetic structure. These are crucial for any sort of recovery plan for endangered plants, particularly in islands, where genetic diversity has been often expected to be lower than in mainland populations (e.g. Caujapé-Castells et al. 2008; Silva et al. 2011; Martins et al. 2013; Moreira et al. 2013; Moura et al. 2013; Dias et al. 2014).

Propagation measures, particularly those allowing the maintenance of the genetic variability, such as seed germination, are also critical in the recovery of endangered plants (e.g. Moura et al. 2010; Martins et al. 2012; Moreira et al. 2012).

Modelling is nowadays a common approach for predicting species distributions. This is based on statistically or theoretically derived response surfaces that link the known distribution of a species to the pertinent environmental descriptors, allowing to estimate its potential distribution, as well as to determine the environmental factors limiting its range (Guisan and Zimmermann 2000). Species distribution modeling has been used in a wide range of applications (Elith and Leathwick 2009), including the evaluation and management of endangered species (e.g. Engler et al. 2004; Marcer et al. 2013).

Integrating the above aspects in the design of a multidisciplinary research program is an example of a more holistic approach to be applied in plant conservation, as a basis for recovery plans. Here, we reanalyze available data for V. dabneyi, and provide new information on germination rate and population genetic structure and diversity. We also analyze species distribution in order to understand what ecological factors might be constraining it. Due to the small size of the populations, and to the degree of isolation and fragmentation, we expect to find reduced levels of seed set, comparably low levels of genetic diversity, and some degree of differentiation between the studied populations. Because it is rare, we expect the species to have a relatively restricted ecological niche and high levels of ecological specialization.

METHODS

Sampling

Veronica dabneyi populations

Veronica dabneyi was searched in areas with potentially suitable habitat on the islands of Corvo, Flores and Faial (Figure 1), between the years 2000- 2014. The search was not successful on Faial but on Flores and Corvo a total of seven (sub)populations were found. All cited locations here are recorded in the Atlantis database of the "Azorean Biodiversity Portal" (Borges et al. 2010), including data from Pereira et al. (2002) and Schäfer (2003).

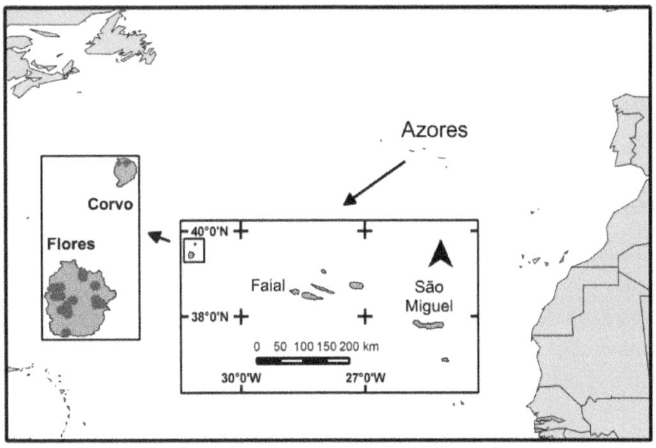

Figure 1: Location of the Azores Archipelago in the North Atlantic Ocean, and recorded presence of Veronica dabneyi in Flores and Corvo islands.

Demographic data

Three populations were studied in detail: Miradouro Craveiro Lopes and Tapada da Forcada (Flores), and Madeira Seca (Corvo). Variables measured included: individual number and size, stem length, number of inflorescences, and inflorescence height (measured in loco), number of fruits per inflorescence, number of seeds per fruit, seed diameter and weight, using a digital caliper and an electronic scale (measured/counted at the laboratory). Fruits were sampled in July 2008 from Miradouro Craveiro Lopes, and in June 2010 form Madeira Seca and Tapada da Forcada.

Associated flora and invasive species

Characterization of the associated flora was based on the work by Pereira et al. (2002) and on data collected by the authors in 2010, using a 1 square meter plot centered on V. dabneyi individuals or groups of individuals. Presence of plant invaders (see Silva et al. 2008) was recorded from the immediate surroundings of V. dabneyi populations. Presence of cattle, goats and rabbits was confirmed through direct visual observation of the animals or by the presence of characteristic feces.

Plant material for genetic analysis

In 2008 and 2010, a collection of leaf material was carried out to complement the samples already available at the DNA bank collection of the AZB herbarium (Biology Department, Azores University). Depending on leaf size, one or two leaves per individual were collected and immediately stored in a plastic bag with silica gel. After drying, the leaves were vacuum sealed and stored in folders. The number of individuals sampled per site varied from 10 to 30 individuals depending on population size. In total, 72 individuals from 7 different sites were sampled (Figure 1, Table 1). Plant material from obtained from the F1 of Miradouro Craveiro Lopes (i.e. obtained from the germinated seedlings) was also used for comparison with the mother population.

Table 1: Location and number of samples of Veronica dabneyi for population genetic analysis in the Islands of Flores and Corvo (Azores). Voucher code for material deposited at AZB (Herbarium Ruy Telles Palhinha), population code used in the study, local designation for each site, elevation (m a.s.l.), UTM coordinates (X, Y; WGS84 25S), and number of samples per site (N). Three individuals resulting from seed germination were also included (i.e. F1 Miradouro Craveiro Lopes)

Island	Voucher	Code	Site	Elevation	X	Y	N
Flores	VD-FLML-001	VDFLML	Miradouro Craveiro Lopes	482	651717	4365683	26
Flores	VD-FLTP-001	VDFLTP	Tapada da Forcada	490	654964	4373203	12
Corvo	VD-COMS-005	VDCOMS	Madeira Seca	442	662989	4398153	18
Corvo	VD-COMC-002	VDCOMC	Caldeirão I	600	662653	4397918	4
Corvo	VDCO1A	VDCO1	Caldeirão II	569	662640	4397900	3
Corvo	VDCO3A	VDCO3	Arribas do Caldeirão	255	661310	4397900	3
Corvo	VDCO2A	VDCO2	Arribas	255	661330	4397900	3

Propagation

Seed germination

Fresh seeds collected in 2008 and 2010, as well as 2-year-old seeds stored at room temperature were used for germination tests. Germination tests were done in Petri dishes using growth chambers with automatic temperature control (error margin of ca. 1°C) and a light period of 12 hours per day, provided by six fluorescent lamps with a photosynthetic photon flux density (PPFD) of 19- 22 μmol m^{-2} s^{-1} . The chambers were set to the incubation temperatures of 25/20°C, 20/15°C, 15/10°C and 10/5°C, and the highest temperature coincided with the 12 hours of the photophase. Germination proceeded in light or in darkness (Petri dish covered with aluminum foil) with three replicates per treatment and 17, 18 or 56 seed per replicate, depending on the total number of seeds sampled at each site. Seeds were monitored daily and considered to be germinated when the radicle extruded. Seeds from the dark treatments were observed under a green light. Accumulated germination curves were adjusted to a Gompertz model to allow the calculation of the time, in days, necessary for 50% seed germination (T50). The latter model has been successfully used to describe accumulated germination while allowing a biological interpretation (see Latera & Bazzalo 1999; Moura & Silva 2010).

Seedling growth and establishment

Germinated seedlings were planted in Jiffy® peat pellets in a growth chamber with regulated temperature (20°C) and photoperiod (12 hours of photophase) for two months, and regularly watered to avoid substrate desiccation. The seedlings were transported to Flores Island and planted near the mother plants at Miradouro Craveiro Lopes, and were followed up every six months during two years.

Genetic analysis

General DNA extraction

DNA was extracted from dry leaves using a modified CTAB extraction method (Doyle and Dickson 1987) without the final ethanol wash. DNA was then precipitated by adding 450 µl of isopropanol and re-suspended in 50 µl of pure water. DNA quality and quantity were measured using a Nanodrop 2000 (Thermo Fisher Scientific) spectrophotometer. Samples were conserved at -20°C until use.

Microsatellite development

Total DNA from fresh leaves of one individual of V. dabneyi was sent to the Savannah River Ecology Laboratory (University of Georgia, USA), where the enrichment procedure described in Glenn and Schable (2005), with the exceptions described in Lance et al. (2010), was followed for microsatellite isolation. CAP3 (Huag and Madan 1999) was used to assemble sequences at 98 % sequence identity using a minimal overlap of 75 bp. Search for microsatellite DNA loci was conducted using the program MSATCOMMANDER version 0.8.1 (Faircloth 2008) and primers designed with Primer3 (Rozen and Skaletsky 2000). One primer from each pair was extended on the 5'-end with an engineered sequence (M13R tag 5'-GGAAACAGCTATGACCAT-3') to enable the use of a third primer identical to the M13R (Schuelke 2000), and a GTTT "pigtail" was added to the 50-end of the untagged primer to facilitate accurate genotyping (Brownstein et al. 1996). Out of the 202 sequences of primer pairs provided by the Savannah River Ecology Lab, we selected 24 primer pairs, 12 with expected PCR products ranging between 100 and 200 bp (A series) and 12 exhibiting expected PCR products ranging between 200 and 300 bp (B series) to allow later the multiloading of PCR products. All the primer pairs (with the tag sequence included) were selected on criteria of non-complementarities within and between primers, low secondary structures and 3'-end instability (Rychlik 1995).

Microsatellite selection and full-scale genotyping

All 24 primer pairs were tested on eight samples of V. dabneyi using a unlabelled tag primer (M13R) in a final volume of 25 µl consisting of 25 ng of DNA, 75 µg/ml BSA, 1 x NH_4 Buffer, 2 mM $MgCl_2$, 0.4 mM of untagged primer, 0.08 mM of tagged primer, 0.36 mM of Universal dyed M13R, 200mM of dNTPs, 1 U of Immolase (Bioline) and using a Biometra TGradient thermocycler. Touchdown thermal cycling programs (Don et al. 1991) encompassing a 10°C span of annealing temperatures ranging between 63 and 53°C were used for all loci. The PCR program included the following steps: 95°C for 7 min (hot start); 96°C for 3 min; 20 cycles of 95°C for 30 s, the highest annealing temperature of 63°C (decreased by 0.5°C per cycle) for 30 s, and 72°C for 30 s; 20 cycles of 95°C for 30 s, 53°C for 30 s, and 72°C for 30 s; and finally 72°C for 10 min for the final extension of the PCR products. Five µl of PCR products was then run on a 3.5 % agarose gel, stained with SafeViewTM Classic Nucleic Acid Stain (ABM Inc.) and visualized under UV to check for amplification, polymorphism and scorability of the bands. Ten primer pairs exhibited scorable amplified products of the expected length range and with at least two alleles. After analysis of the quality of the PCR products obtained with the universal primer M13R, ten primers with acceptable to high scorability were selected to run the complete study (Table 2). After optimization, the amplifications for the whole sample were performed using the protocol indicated above with the alterations presented in Table 3, and the M13R labeled either with PET, FAM, NED or VIC. Amplification products were diluted, multiloaded, run on an ABI- 3130xl Genetic Analyzer and sized with LIZ500 size standard. The genotypes obtained were scored using the software GeneMarker V.1.97 Demo version (Softgenetics).

Table 2: Set of SSR primers that showed scorable and polymorphic products for the analysis of Veronica dabneyi population genetic structure in the Azores

Primer	motif	Forward	Reverse
Vd2B04	AC(9)	GTTTAGTGACGAGGACATTGATTG	GGAAACAGCTATGACCATCCTTCTAACATCGCAAACTG
Vd2B09	AC(10)	GTTTGCACACTGAAGGGTATCAAC	GGAAACAGCTATGACCATAAATCGGTGAATGTTTGATC
Vd3A01	AAG(9)	GTTTGTGTTCAGCTTGGAATTGAG	GGAAACAGCTATGACCATCTCTTCGACCAAATTCTTG
Vd3A03	ATC(18)	GGAAACAGCTATGACCATAAGTTCTTGCTCTGCTTGTC	GTTTCTTGTAGCCCAGATTGAAAC
Vd3A05	ATC(14)	GGAAACAGCTATGACCATCTAAACTCCCTTTCACTGG	GTTTGCGTCGAAGTACAAGAACAG
Vd3B07	AAG(15)	GTTTAGCTCGGAAACTTTGTAATG	GGAAACAGCTATGACCATGCAATAAAGTGATTAAGTGG
Vd4A01	AATG(6)	GTTTCCCACTATCCAACCATAATC	GGAAACAGCTATGACCATAACTCAGCTCAGCGTGAC
Vd4B01	AAAC(11)	GGAAACAGCTATGACCATAACCACATCACTCCAAACAG	GTTTGACTGGGCTAGAGTTGTC
Vd4B04	ACAT(20)	GTTTAATCCATTGTGTGCAGTCTC	GGAAACAGCTATGACCATCACCTCCCACACTTAATC
Vd4A03	AAAC(6)	GGAAACAGCTATGACCATGCTTTAATTTGTGCGTATC	GTTTCCTATCCCTTAACCTTTCTTC

Table 3: Optimization conditions used in the full-scale genotyping of Veronica dabneyi

Primer	Optimization
Vd2B04	0.36 μM VIC, 0.5 U Immolase
Vd2B09	0.36 μM NED, 0.5 U Immolase
Vd3A01	0.36 μM FAM, 0.5 U Immolase
Vd3A03	0.04 μM tagged primer, 0.36 μM FAM, 0.5 U Immolase
Vd3A05	0.04 μM tagged primer, 0.36 μM VIC, 0.5 U Immolase
Vd3B07	0.04 μM tagged primer, 0.36 μM NED, 0.5 U Immolase, 50 ng DNA
Vd4A01	0.04 μM tagged primer, 0.36 μM NED, 0.5 U Immolase
Vd4B01	0.36 μM VIC, 0.5 U Immolase
Vd4B04	0.2 μM tagged primer, 0.2 μM FAM, 0.75 U Immolase, 50 ng DNA
Vd4A03	0.36 μM PET, 0.5 U Immolase

Analysis of genetic data

Population structure was analyzed with GenAlEx 6.5 (Peakall and Smouse 2012), to obtain mean values per population of the total number of alleles, the number of alleles with a minimum allele frequency of 5%, the number of effective alleles, the Shannon's Information Index, the number of private alleles, the expected heterozygosity, R_{st} and the estimation of gene flow. A Principal Coordinates Analysis (PCoA), and an AMOVA were also performed. Furthermore, we used a Bayesian approach to estimate the number of genetic clusters present in the whole sample. This model-based analysis was run with the software STRUCTURE version 2.3.3 (Pritchard et al. 2000), using a batchoriented web program package for construction of super matrices ready for phylogenomic analyses (Kumar et al. 2009). We ran 10 replicates for each K value ranging from 1 to 10 with a burn-in length of 50,000 followed by 500,000 iterations of each chain using the admixture model along with the assumption of correlated allele frequencies between groups (Falush et al. 2003).

STRUCTURE then partitioned individuals of the sample according to the membership coefficient Q, that ranges from 0 (lowest affinity to the group) to 1 (highest affinity to a group), across K groups. Estimation of the best K value was conducted with STRUCTURE Harvester (Earl and von Holdt 2012) following Evanno et al. (2005) method. The optimal K repetitions were permuted in Clumpp version 1.1.2 (Jakobsson and Rosenberg 2007), using the Greedy algorithm, with results graphically represented using Distruct version 1.1 (Rosenberg 2004). The population matrix is available at DEMIURGE (http://www.demiurge-project.org/) with digest code D-NMICR-98.

Species distribution modelling

Modelling approaches

Since it is likely that the current distribution range on the studied islands is much reduced as a consequence of human activities (e.g. changes in land use and biological invasions), true absences are not available in this case. We thus opted to use modelling methods based on presences only, namely Ecological Niche Factor Analysis (ENFA) and maximum entropy modeling.

Such an approach allows not only to estimate the potential distribution and the habitat suitability for the species but also to identify the macroecological factors that might affect species distribution (e.g. altitude, climate, land use). The ENFA (Hirzel et al. 2002, 2006, 2007; Martinez et al. 2006; Hirzel & Le Lay 2008) provides smooth responses to environmental factors (Vaclavík & Meentemeyer 2012). This is desirable for modelling potential distributions, as models fitting complex responses may not accurately predict the distribution of species that are not at equilibrium. This approach was used successfully to model Azorean plant species, both invasive and native (Costa et al. 2012, 2013a; Moreira et al.

Distribution data

The species presences (Figure 1) recorded at the Atlantis data base (shape file format) were transformed into raster format at the same resolution as the ecogeographical variables for input in Biomapper (Idrisi raster format) and in Maxent (ASCI format),

Ecogeographical variables

As the number of presences was relatively low (< 100), we followed Lomba et al. (2010) regarding the number of ecogeographical variables (EGVs) that should be used. We used three EGVs categories: climate, topography and land cover. Climatic variables were selected from the CIELO Model (Azevedo 1996; Azevedo et al. 1999), a raster GIS environment with 100 m spatial resolution which models local scale climate variables relying on limited available data from synoptic coastal meteorological stations, and based on physical models that simulate the movement of air masses and their interaction with island topography (for more information see http://www.climaat.angra.uac.pt or Azevedo (2003)). We used the annual average of minimum, maximum, mean and range values of temperature (TMIN, TMAX, TM, TRAG), relative humidity (RHMIN, RHMAX, RHM, RHRAG) and precipitation (PMIN, PMAX, PM, PRAG). In addition, these climatic variables were submitted to a principal component analysis (PCA) as most of them were highly correlated. The principal components explaining more than 90% of variance in the original variables were held and used alternatively.

Those components, used as five alternative EGVs, corresponded to the first two components extracted from temperature (TPC1-2) and relative humidity variables (RHPC1-2), and to the first component extracted from precipitation variables (PCP). The topographic and land cover EGVs were acquired from the supporting data available in the CIELO Model database, which matches the same spatial resolution of 100 m. To characterize the topography, we used the elevation (ELE) and the slope (SLP). Land-cover was defined in six classes: (1) forest, (2) natural vegetation, (3) pasture, (4) agriculture, (5) barren/bare areas and (6) urban/ industrial areas. We tested two different approaches for land cover data. The land cover classes were sorted in the foregoing order to define an ordinal land cover variable (OLC), from "like forest" (forest) to "unlike forest" (urban/industrial areas). Moreover, distance variables were calculated for each land cover class (DLC1-6). Distance variables express the distance between the focal cell and the closest cell belonging to a given land cover class. In total, 26 EGVs were tested. A global model was calculated in Maxent including all available variables. Only variables providing more information to the model were kept for each of the following descriptors: elevation, temperature, relative humidity, and rainfall. These four variable groups were differently combined with other physiographic descriptors: aspect, slope, flow accumulation, and hill shade.

Modelling

Models were run with Maxent, and the best model selected, based on the analysis of jackknife permutations and AUC (Area Under the Curve). The same data was used to run the model in Biomapper. An R script was used to calculate the Boyce index and the Boyce curve, as well as to compare the habitat suitability maps obtained in the two methods, using the McNemar Test (for further details on model validation and comparison see Costa et al., 2012, 2013a, and references within). The best model was then changed to include one or two variables describing the distance to different types of land use. The best model was selected and compared as previously. The projection of the potential distribution of V. dabneyi in Faial Island was obtained using Maxent.

RESULTS

Demographic analysis

Veronica dabneyi individuals usually grow linearly and horizontally, according to the procumbent nature of the stems. At Miradouro Craveiro Lopes we found 15 plants, with a stem length of 16 to 70 cm. Only 4 (27%) showed inflorescences, ranging from 1 to 2, with 7.5 to 18 cm in height and 6 to 23 fruits per inflorescence (mean = 14.6, sd = 6.9). At Tapada da Forcada we found 12 plants, with a stem length of 8 to 65 cm, and 5 seedlings. Only 6 (50%) showed inflorescences, ranging from 1 to 15, with 8 to 21 cm in height and 8 to 31 fruits per inflorescence (mean = 19.1, sd = 5.8). At Madeira Seca (Corvo) we found 10 plants, with a stem length of 6 to 35 cm, and 1 seedling. Of those, 9 (90%) showed inflorescences, ranging from 1 to 24, with 3 to 13.5 cm in height and 4 to 20 fruits per inflorescence (mean = 8.8, sd = 3.6).

In Flores the number of seeds per fruit ranged from 1 to 24 (mean = 10.4, sd = 6.3) while in Corvo, a larger variation was found (1 to 31 seeds, mean = 15.4, sd = 7.2). Seed diameter was similar on both islands (Flores: mean = 1.20 mm, sd = 0.13 mm; Corvo: mean = 1.09 mm, sd = 0.17 mm). However, the weight of 100 seeds ranged from 0.00049 g in Flores, to up to 0.00074 g in Corvo. Based on those values, we estimated seed production as 5870 ± 309.9 (mean ± se) seeds at Madeira Seca, 4969 ± 500.4 seeds at Tapada da Forcada and only 760 ± 76.6 at Miradouro Craveiro Lopes.

Associated flora and invasive species

The flora closely associated with the presence of V. dabneyi included mostly other endemic taxa (Table 4). However, surrounding V. dabneyi populations, Hydrangea macrophylla (Thunb.) Ser. (Hydrangeaceae Dumort.), one of the most problematic invasive species in Flores and Corvo Islands (Silva et al., 2008), was found. In Flores the largest populations were located along steep road sides. In Corvo, the population of Madeira Seca is established on vertical walls of a volcanic chimney. At this location, the main threats identified were the presence of feral goats (Figure 3A) and the proximity of H. macrophylla clumps (Figure 3B).

Figure 3: Two of the most relevant threats affecting Veronica dabneyi populations in the Azores: A) feral goats free roaming and grazing very close to Madeira Seca in Corvo Island; B) Hydrangea macrophylla clumps very close to a rock outcrop in Madeira Seca.

Table 4: Vascular plants associated with Veronica dabneyi, observed by Pereira et al. (2002) and by the authors

Taxa	Family	Life form	Origin	Pereira *et al.*	Authors
Blechnum spicant	Blechnaceae	Hemicryptophyte	Native	X	X
Calluna vulgaris	Ericaceae	Chamaephyte	Native		X
Centaurium scilloides	Gentianaceae	Chamaephyte	Endemic		X
Deschampsia foliosa	Poaceae	Hemicryptophyte	Endemic	X	X
Euphrasia azorica	Orobanchaceae	Chamaephyte	Endemic		X
Festuca francoi	Poaceae	Hemicryptophyte	Endemic	X	X
Frangula azorica	Rhamnaceae	Phanerophyte	Endemic		X
Hedera azorica	Araliaceae	Phanerophyte (scandent)	Endemic		X
Holcus rigidus	Poaceae	Hemicryptophyte	Endemic	X	X
Hypericum foliosum	Hypericaceae	Phanerophyte	Endemic		X
Ilex azorica	Aquifoliaceae	Phanerophyte	Endemic		X
Juniperus brevifolia	Cupressaceae	Phanerophyte	Endemic		X
Lotus pedunculatus	Fabaceae	Hemicryptophyte	Introduced		X
Luzula purpureosplendens	Juncaceae	Hemicryptophyte	Endemic		X
Lysimachia azorica	Primulaceae	Chamaephyte	Endemic	X	X
Myosotis azorica	Boraginaceae	Chamaephyte	Endemic		X
Picconia azorica	Oleaceae	Phanerophyte	Endemic		X
Rubia agostinhoi	Rubiaceae	Chamaephyte	Endemic		X
Rubus hochstetterorum	Rosaceae	Phanerophyte (scandent)	Endemic		X
Scabiosa nitens	Caprifoliaceae	Hemicryptophyte	Endemic	X	X
Selaginella kraussiana	Selaginellaceae	Hemicryptophyte	Native	X	X
Viburnum treleasei	Adoxaceae	Phanerophyte	Endemic		X
Woodwardia radicans	Blechnaceae	Hemicryptophyte	Native		X

Seed germination and seedling establishment

The germination percentage of the seeds collected at Craveiro Lopes in July 2008 was high for all temperature regimes (90-100%), with faster germination occurring at the higher ones (Figure 2). However, seed batches lost viability after 2 years of storage at room temperature, with no germination in

2010. Seeds collected in June 2010 at Tapada da Forcada and at Madeira Seca did not germinate, however they were most likely not mature at the time of collection.

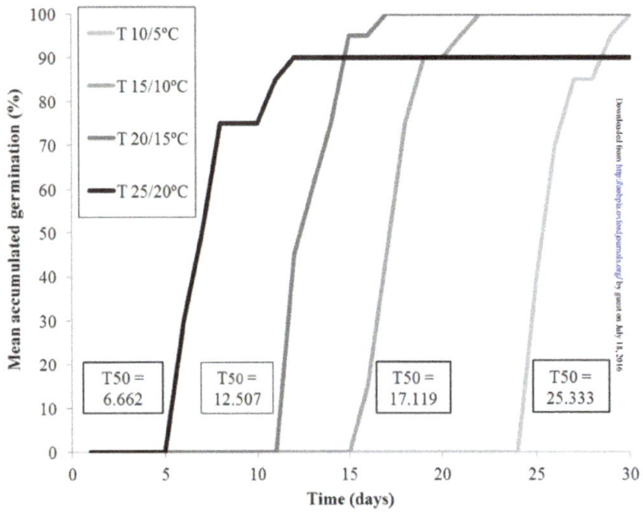

Figure 2: Veronica dabneyi accumulated seed germination curves obtained under 12 hours of light and four alternating temperature regimes in Petri dishes in acclimatized chambers, after 30 days. Time for 50% germination in days is shown for each treatment, based on the Gompertz model that was adjusted to each curve (all models with a $R^2 > 0.9$).

Population genetics

A total of 72 samples from 7 populations was analyzed at 10 microsatellite loci. The results of the AMOVA indicated that the majority of the genetic variation was found within the populations (98%), and only a small portion among populations (2%). Permutation test showed that the R_{st} value (0.033) was not significant (p=0.160) with an estimated gene flow of 7.249. Also, R_{it} value (-0.451) was significant and negative (Prand≥data, 1), revealing a lack of genetic population structure. The inbreeding coefficient, R_{is} (-0.501) was significant and negative (Prand≥data, 1). Diversity patterns were similar across the sampled populations (Figure 4). The average number of alleles ranged from 2.5 to 5, with the average number of private alleles per population below 1 (Figure 4). Expected heterozygosity was somewhat homogeneous across the sampled populations, ranging from 0.46 to 0.59 (Figure 4). The results obtained with the PCoA and STRUCTURE (Figures 5 and 6) showed a considerable degree of admixture, with three genetic clusters identified and represented at various degrees in all the sampled populations. Miradouro Craveiro Lopes individuals seem to encompass all the genetic variability found in the other populations of both Flores and Corvo (Figures 5 and 6). The F1 generation (i.e. resulting from seed germination) obtained from Miradouro Craveiro Lopes population showed genetic patterns compatible with those obtained in the source population (Figure 6). The Corvo populations showed slightly different genetic patterns, at the exception of the Arribas population that is similar to those found in Flores (Figure 6).

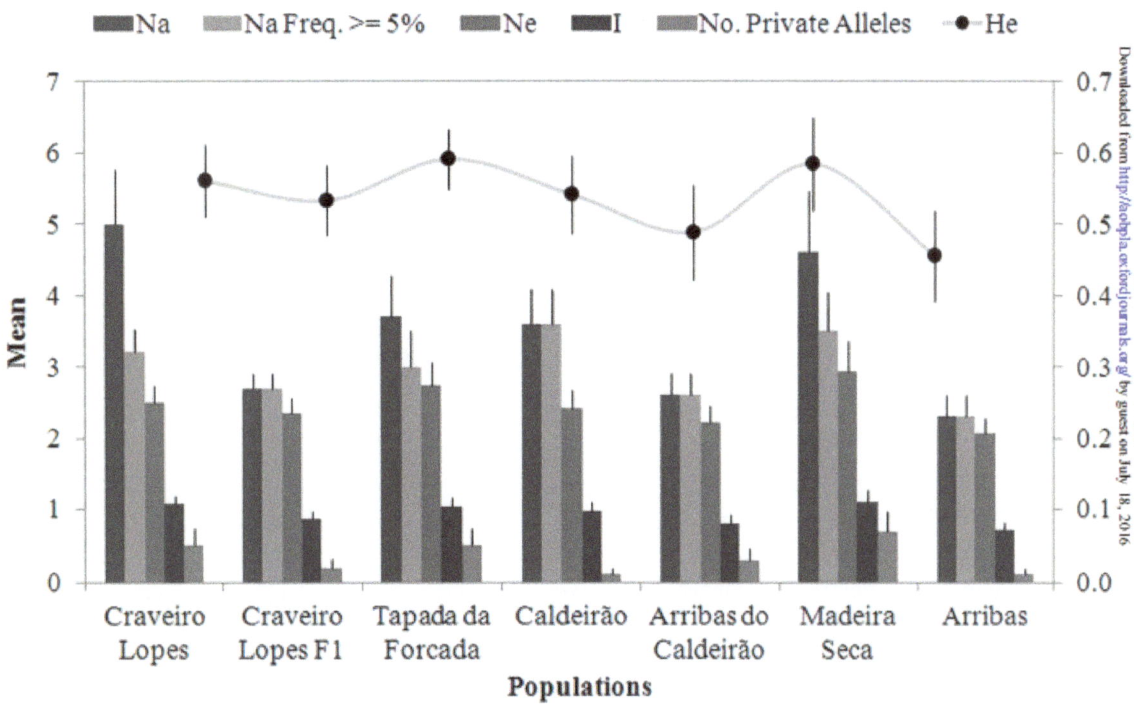

Figure 4: Mean allelic patterns across populations. Results of an analysis using GenAlEx on a total of 72 samples from 7 populations of Veronica dabneyi from Flores and Corvo islands (Azores) analyzed with 10 microsatellite loci. Na = total number of alleles; Na (Freq >= 5%) = number of alleles with a frequency >= 5%; Ne = number of effective alleles; I = Shannon's Information Index; No. Private Alleles = number of alleles unique to a single population. The line represents expected heterozygosity.

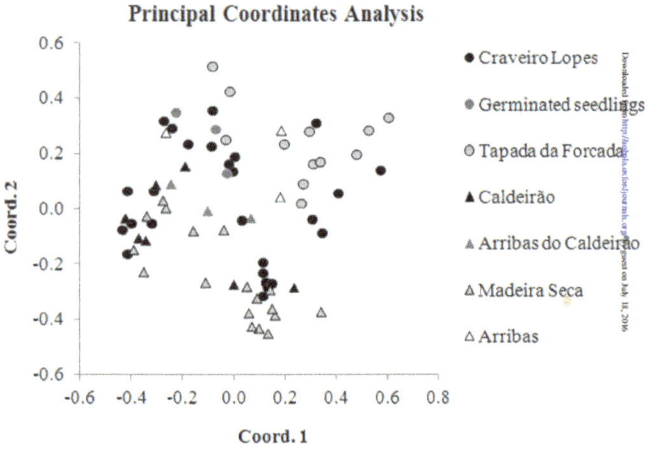

Figure 5: Graphic representation of the two first Principal Coordinates Analysis axes which explained 63.8% of the detected genetic variation. Results obtained with GenAlEx for 10 microsatellite loci applied to 72 samples of Veronica dabneyi collected at six sites in Flores and Corvo islands, Azores. Germinated seedlings were derived from seeds obtained from Miradouro de Craveiro Lopes.

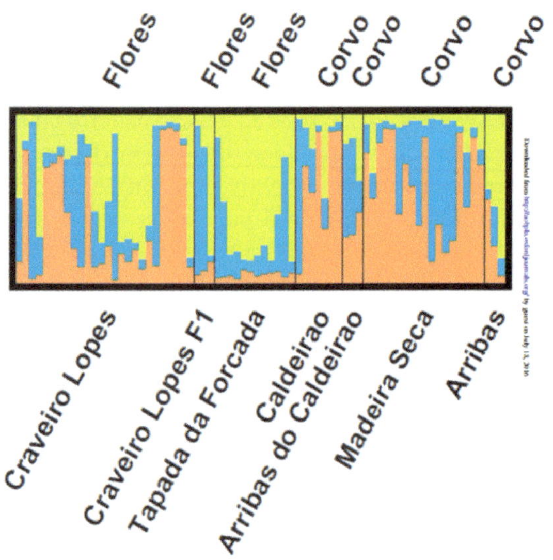

Figure 6: Results of the analysis of genetic clusters using STRUCTURE and DISTRUCT on a total of 72 samples from 7 populations of Veronica dabneyi from Flores and Corvo islands (Azores) analyzed with 10 microsatellite loci. The number of genetic clusters was estimated at 3, using a model including admixture.

Species distribution modelling

The best distribution model found in Maxent (AUC = 0.840) also corresponded to a good model in Biomapper (total information explained = 0.852, for the first three factors; Boyce Index =0.990) (Figure 7A). Although there was a slight increase in the AUC (0.865) and in the total information explained (0.871 for the first three factors; Boyce Index = 0.953), according to the shape of the Boyce curve the best model including land use information was not as good as the model based on physiographic and climatic data alone (Table 5, Figure 7B). When considering only the cells with habitat suitability above the third quartile, the McNemar test did not show significant differences between the results obtained with Maxent and Biomapper (model without land use data, c ^{2}=0.264, p=0.607; model with land use data, χ^2=0.250, p=0.617; Figure 8). Likewise, marginality and specialization gave similar results for both models (marginality 0.426 and 0.349; specialization 1.108 and 1.144). The low marginality indicated that the habitat actually occupied by the species is similar to the average conditions of the available habitat. The relatively low specialization suggested that the conditions for species occurrence were not narrow. The analysis of the score matrix for the best models showed a positive link with elevation and slope, which affect the niche of the species, a negative link with the temperature, a positive link with rainfall and a negative association with high relative humidity ranges. It also demonstrates a negative association with the distance to uncultivated areas and a positive association with the distance to cultivated areas (Table 5).

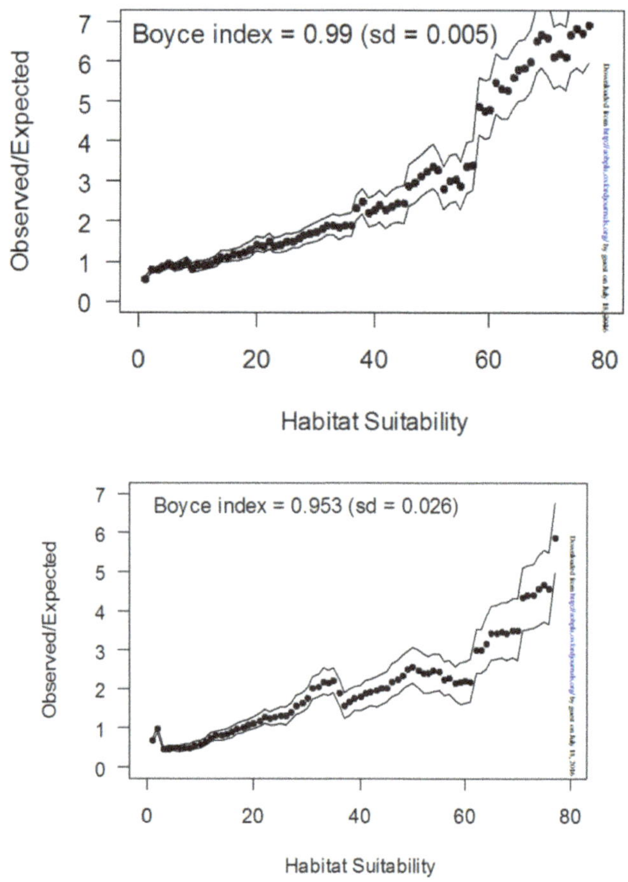

Figure 7: Boyce curve and Boyce Index for the best distribution models for Veronica dabneyi, calculated using six ecogeographical variables in Biomapper: A) model with climatic and physiographic variables only (elevation, hillshade summer, annual rain fall, relative humidity annual range, annual mean temperature, slope); B) Model including land use information (elevation, annual rain fall, annual mean temperature, slope, distance to the nearest cell with agriculture, distance to the nearest cell with abandoned/unused land).

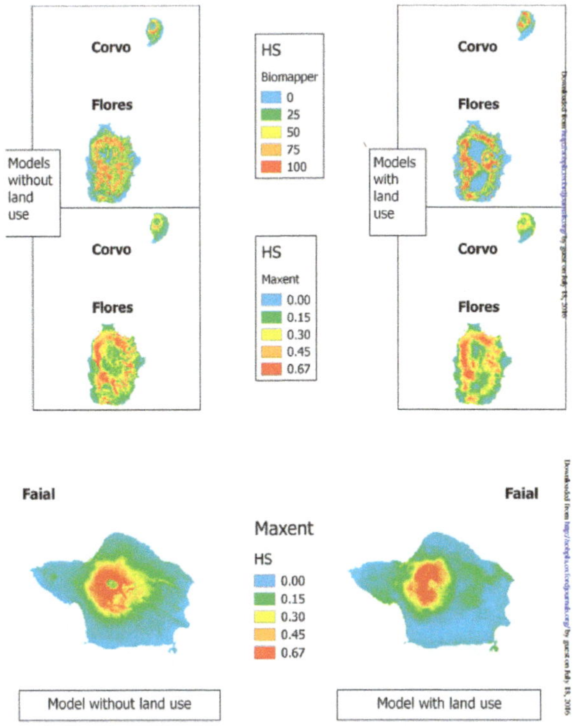

Figure 8: Habitat suitability maps for Veronica dabneyi in Corvo, Flores and Faial islands, calculated using six ecogeographical variables. A) Models derived using Biomapper and Maxent for Flores and Corvo islands, either using physiographic and climatic variables only, or including land use information also. B) Projection of the potential habitat on Faial Island calculated using Maxent. Habitat suitability values were approximately divided according to the respective quartiles.

Table 5: Results of modeling Veronica dabneyi distribution in Flores and Corvo islands using ENFA (Biomapper). Score matrix of each variable for the first three extracted factors for best models including or not, information regarding land use

EGV	Factors 1	2	3
Model without land use			
Elevation	0.50	0.58	-0.04
Hillshade Summer	-0.32	0.18	-0.57
Slope	0.28	0.25	-0.23
Mean annual temperature	-0.47	-0.17	-0.46
Annual rainfall	0.28	0.00	-0.57
Relative humidity annual range	-0.52	0.74	0.30
Model with land use			
Elevation	0.61	0.30	0.64
Mean annual temperature	-0.57	0.38	0.50
Annual rainfall	0.34	-0.01	0.08
Distance to agricultural land	0.21	0.45	-0.52
Distance to uncultivated land	-0.13	0.74	0.24
Slope	0.34	0.12	0.02

DISCUSSION

Demographic analysis

The heterogeneity of plant sizes in the three study populations suggests that there is frequent recruitment of new individuals from seed. This is more evident at Tapada da Forcada (Flores), a population with an almost uniform distribution of life stages, including 30% seedlings and 35% seed producing individuals. At Madeira Seca (Corvo) more than 80% of the individuals were reproducing, whereas at Miradouro Craveiro Lopes individuals were usually large but produced fewer inflorescences, resulting in a much lower total seed set. The existence of larger individuals with low number of inflorescences might be associated with regular brush cutting (see below). The shortest plant with an inflorescence measured 6 cm, although fruit production was more common in individuals with a length ≥ 12 cm. This might imply that seed producing plants can be relatively young. Despite the considerable variation in plant size and number of inflorescences, the estimated seed set for two of the studied populations was relatively high. The very small size/weight of the seeds, rapid germination, absence of dormancy and relatively fast decline on viability suggested that persistence as seed bank should be low (Yu et al. 2007), and this was confirmed in our study by the absence of germination following two years of storage at room temperature.

Habitat and threats

Veronica dabneyi was found to be mainly associated with other endemic species. The conservation of native vegetation cover is thus of the utmost importance for the preservation of this species. It grows at sites with low vegetation cover mostly dominated by Festuca francoi and Deschampsia foliosa Hack. (Poaceae), generally found in forest openings and at steep locations such as waterfalls or road side slopes (Sjögren 1973). Therefore, the conservation of this type of vegetation, by avoiding changes in steep areas will also be necessary. While populations of V. dabneyi were generally recorded on steep locations, the largest ones were found along steep road sides. On Flores, the population found at Miradouro Craveiro Lopes is located in front of a viewpoint, where brush cutting is regular. Changes in the road or at the top of the road side slope might affect this population. A similar situation was found at Tapada da Forcada, where the existing population occurs in a very steep slope, along the road. At some sites, the presence of H. macrophylla is a threat that should not be ignored, since this plant invader develops pure stands, and can outcompete all other plant species (Silva et al. 2008). In the Azores (Costa et al. 2013b), as in similar island groups and in many regions worldwide, plant invaders are commonly found within protected areas (see Foxcroft et al. 2013) and hydrangeas are even actively planted by Azorean farmers and the forest department as "green fences". The hydrangea shrub can form impenetrable stands and stop cattle from falling down cliffs. These cliff-top hydrangea plantations spread throughout the cliff sides, outcompeting endemic species that survive in these inaccessible refugia. Hydrangea clumps are very close to several V. dabneyi populations both in Flores and Corvo and might overgrow those populations in the near future. Thus, H. macrophylla clumps that are detected near V. dabneyi occurrences should be considered as priority targets for removal/control measures, and its plantation on cliff tops should be banned.

Instead, barbed wire fences should be preferred to keep cattle away from dangerous areas with no significant impact on natural vegetation. Other important herbaceous invasive species like Hedychium gardnerianum Ker Gawl. (Zingiberaceae Martinov) (Silva et al. 2008) should also be monitored at V. dabneyi sites.

Although direct evidence of V. dabneyi consumption by goats, cattle and rabbits is rare, its distribution at Caldeirão in Corvo Island, mostly on steep, rocky outcrops, suggests a possible retreat from areas more fully accessible to herbivores (Milchunas and Noy-Meir 2002). During our visits we found that cattle were free to roam at Caldeirão, even on sensitive vegetation like peat bogs that are important systems for water absorption and retention, and that goats were generally feral or unherded. This poses a big threat not only to V. dabneyi but also to other herbaceous endemic species (Houston et al. 1994; Silva et al. 2008) like Tolpis azorica (Nutt.) P.Silva (Asteraceae Bercht. & J.Presl) or Euphrasia azorica H.C.Watson (Scrophulariaceae Juss.), which are eaten in all accessible places. In the Azores it was not yet possible to use herbivore-exclusion plots, but this approach was used in the Canary Islands, showing that herbivores exert a strong negative effect on plant establishment, demanding the implementation of conservation measures, such as large fenced areas, control activities and eradication (Garzón-Machado et al. 2010).

Meanwhile, the necessary changes in herbivore management clearly include a societal component, involving stakeholders that are not directly linked to conservation (e.g. hunters, farmers, the agriculture services).

Seed germination and seedling establishment

The fastest germination treatment occurred under the 25/20°C temperature regime (T50=6.7 days), germination reaching 90%. The rate of 100% germination was obtained with a temperature regime of 20/15°C, given the second best T50 (12.6 days). Similarly, high germination percentages were also obtained for the related Veronica arvensis L. (King 1975). For conservation purposes, seeds should be collected after full development of the fruits which generally occurs by mid July, and promptly sown. It was possible to grow seedlings obtained from germination in the laboratory and to plant them at the site of origin. For the first two years the plants were monitored every six months and it was found that the seedlings planted next to the mother population survived at rate of about 50%. Some early mortality due to transportation/establishment, and later mortality associated with roadside maintenance activities was found. Furthermore, the genetic analysis of the germinated seedlings showed no considerable reduction in the genetic variability. Thus, the use of seeds might be a valuable option in a future recovery plan, in cases where population reinforcement might be preferred.

Population genetics

Contrary to our initial expectations that were based on the existence of small and possibly isolated populations, the levels of genetic diversity found in the populations were relatively high, as was the level of genetic admixture, with no evidence of inbreeding. Expected heterozygosity was similar to that found for other Azorean herbaceous endemic taxa (Dias et al. 2014), which are much more common than V. dabneyi. The concentration of most of the genetic variation within populations was similar to patterns found for some of the endemic trees of the archipelago (Martins et al. 2013; Moreira et al. 2013). In a study using AFLP markers with Veronica hederifolia L., a European species invasive in China, high levels of genetic diversity were also found, and most of the total variance was attributed to that within (76%) rather than between the populations (24%) (Wu et al. 2010). Regarding Hebe speciosa (R.Cunn. ex A.Cunn.) Andersen (Plantaginaceae), a threatened endemic New Zealand shrub, using AFLP markers it was found that there is negligible contemporary gene flow, and that some of the populations exhibited extremely low genetic diversity (Armstrong and De Lange 2005). In V. dabneyi the high levels of gene flow and genetic admixture among the sampled populations presently impede clear population differentiation, even between Flores and Corvo. It should be noted that flower morphology of V. dabneyi suggests an entomophily syndrome (Garnock-Jonesa 1976): many-flowered inflorescence well above the level of the leaves; the background color of the corolla is generally lavender with darker guide marks radiating from this ring to the surrounding corolla lobes, especially the posterior lobe. The related Veronica chamaedrys L. is mainly pollinated by hoverflies and short-tongued bees. Other possible pollinators include the Ichneumonidae (Garnock-Jonesa 1976). We frequently observed a range of Diptera species on V. dabneyi flowers, especially syrphid flies, dung flies and small, unidentified dipterans. Even though they probably do not regularly cross the channel between Flores and Corvo, they could easily be blown between islands in stormy weather, thus transferring pollen from island to island. Alternatively, the present population genetic structure might be the result of a bottleneck effect, consequence of the fragmentation of a previously wider species distribution range that was reduced due to land use changes (Arenas et al. 2012). In fact, our modelling results do suggest that the potential distribution of V. dabneyi could have been wider in the past. Still another possibility is that the population at the Miradouro Craveiro Lopes, which shares most of the sampled genetic diversity, could be the result of a human translocation of plant material, although written records are not available. In this regard, although seed germination might be a good option to reinforce the most depauperate populations, we do not support translocation of plants between different populations and, particularly, between different islands. The species seems to comprise one global meta-population with gene flow among populations, but there is no reason for translocation, which would be an artificial intervention in the natural gene flow patterns. Meanwhile, the occurrence of natural gene flow might be viewed as a positive factor for the conservation of V. dabneyi, ensuring transfer of genetic information among populations, and avoiding extreme cases of inbreeding depression (Li et al. 2012).

Species distribution modelling

The macroecological factors modeled showed that, at this scale, V. dabneyi is neither a marginal nor a highly specialized species. Both modelling approaches used, ENFA and Maxent, provided similar results, showing that its potential distribution largely coincides with the intermediate elevation zone in Flores and with the Caldeirão zone in Corvo. While ENFA had already been used in the Azores to model invasive and native trees (Costa et al. 2012, 2013a; Moreira et al. 2014; Martins et al. 2015), this is a first result devoted to modeling the distribution of a rare plant in the Azores, further supported by the agreement obtained with Maxent. Veronica dabneyi was shown to prefer intermediate elevations, high slopes, relatively low temperature, high rainfall, and small variations in relative humidity. Land use data did not increase model quality in a sensible way but suggested the existence of a negative correlation with agricultural land. This largely coincides with the type of habitat known for V. dabneyi, suggesting that microenvironmental factors like vegetation cover at a specific location or the presence of a rock outcrop might also be relevant for its establishment (Batik et al. 1992; Svenning 2001; Crain et al. 2014). It should also be stressed that, as stated above, V. dabneyi is mostly found associated with other native and endemic plants at sites with relatively low stature vegetation, which is frequently found at steep locations (e.g. volcanic craters, road side slopes). However, even those sites are not completely protected from herbivores (e.g. goats, rabbits) or human disturbance (e.g. roadside maintenance). Interestingly, the model based on Flores and Corvo occurrence data correctly predicted that V. dabneyi would have adequate habitat at the Faial Caldeira, in agreement with the previous record (Cunha and Sobrinho 1939).

Conservation measures

Since the species shows a considerable degree of genetic diversity, high seed production, and high germination rate, conservation measures should be devoted to: (1) monitoring of natural populations to detect possible changes associated with human impacts; (2) effective management of herbivores, especially feral goats, in Corvo and Flores Island Natural Parks, delimiting areas from which cattle should be prohibited and feral goats removed; and (3) the use of population circumscribed seed germination and seedling growth to recover the most depauperate populations. Eventually, more field work should be directed to Faial Island, particularly in the whole Caldeira area, in order to detect a potentially still existing V. dabneyi population on that island. We hope that this study will stimulate the development of a scientifically based recovery plan for this species, while serving as a model to similar studies devoted to other rare or endangered endemic plant species worldwide.

Towards a more holistic approach to research in plant conservation

Although it is a common approach to dedicate attention to specific areas involved in plant conservation, we consider that a more holistic approach, devoted to multidisciplinary studies of endangered plants, should serve as basis for designing management or recovery plans. The latter are an important gap in the Azores where conservation efforts dedicated to endemic plants do not follow integrated recovery plans, contrary to the situation, for example, in the Canary Islands (BOC 2012). In fact, more than 500000 native plants are produced annually in the Azorean nurseries but their use does not follow approved recovery plans or strategies.

Why should conservation research follow a more holistic approach? First, the absence of phylogenetic/systematic revues dedicated to endemic species might lead to erroneous conservation decisions or to the lack of action where it is needed. It was recently found that the most endangered plant in the Azores, known only from one location, is in fact an introduced species (Schäfer et al. 2011). In contrast, for several native genera the number of taxa present in the islands is still unclear and often underestimated as shown by two recent studies that discovered overlooked endemic taxa with specific distribution patterns (Bateman et al. 2013; Moura et al. 2015). At another level, accumulated evidence for the population genetics of endemic trees in the Azores shows that levels of genetic diversity, and patterns of population structure, vary considerably among the evaluated taxa (Silva et al. 2011; Moreira et al. 2013; Martins et al. 2013; Moura et al. 2013), demanding a detailed study per taxon. Moreover, the absence of long-term monitoring data will not only preclude the evaluation of recovery programs (Godefroid et al. 2011), but also impede a sound evaluation of conservation status, which has then to be determined based on distribution areas and not on observed population trends (Moreira et al. 2014; Martins et al. 2015). Monitoring is also linked to other relevant factors such as the detection of high mortality rates associated with herbivore pressure, making the propagation of high numbers of individuals an almost complete loss if no measures are taken to control predation (Donlan et al. 2003; Garzón-Machado et al. 2010). In the future, other aspects such as climate change will have to be integrated in long-term management or recovery programs, making modelling approaches fundamental tools to support decision making (Fordham et al. 2012).

This more holistic approach can be accomplished by evaluating conservation status and possible management actions, based on a series of previous publications devoted to the target species (e.g. Moreira et al. 2014; Martins et al. 2015), or by developing multidisciplinary projects from the onset of the research program, like it was done in the present paper. Undoubtedly the need to address different aspects involved in the assessment and restoration of endangered plants arises directly from the Global Strategy for Plant Conservation, namely from its Objective 1 (Plant diversity is well understood, documented and recognized, CBD 2012). In our view, the different targets of this strategy will only be implemented if holistic approaches to research in plant conservation are effectively implemented in the near future.

Multidisciplinary studies like the one presented here, although sometimes longer, would increase the success of recovery and long-term maintenance of rare species, therefore improving the outcome of conservation investment, besides being a more powerful tool to halt plant extinctions.

CONTRIBUTIONS BY THE AUTHORS

LS participated in sampling, performed species distribution modelling, conceived the idea for the paper and wrote the first draft; MM, EFD and JS were involved in the genetic analysis; MM performed the germination tests; EBA provided climatic data; HS participated in field work. All authors helped revise the manuscript.

ACKNOWLEDGEMENTS

We are grateful to Hugo Costa for help in managing ecogeographical variables, to Selícia Cordeiro for help in the lab and to Raquel Alcaria, Graciete Maciel, José Martins, Orlanda Moreira, Mark Carine and Fred Rumsey for help in field work. We thank two anonymous reviewers who helped to improve the manuscript quality.

REFERENCES

1. Albach DC, Meudt HM, Oxelman B. 2005. Piecing together the 'new' Plantaginaceae. American Journal of Botany 92:297-315.
2. Albach DC, Meudt HM. 2010. Phylogeny of Veronica in the Southern and Northern Hemispheres based on plastid, nuclear ribosomal and nuclear low-copy DNA. Molecular Phylogenetics and Evolution 54:457-471.
3. Arenas M, Ray N, Currat M, Excoffier L. 2012. Consequences of range contractions and range shifts on molecular diversity. Molecular Biology and Evolution 29: 207-218.
4. Armstrong TTJ, De Lange PJ. 2005. Conservation genetics of Hebe speciosa (Plantaginaceae) an endangered New Zealand shrub. Botanical Journal of the Linnean Society 149:229-239.
5. Azevedo EB, Pereira LS, Itier B. 1999. Modelling the local climate in island environments: water balance applications. Agriculture Water Management 40:393-403.
6. Azevedo EB. 1996. Modelação do Clima Insular à Escala Local – Modelo CIELO Aplicado à Ilha Terceira. PhD Thesis, Universidade dos Açores, Angra do Heroísmo, Portugal.
7. Azevedo EB. 2003. Projecto CLIMAAT – Clima e Meteorologia dos Arquipélagos Atlânticos. PIC Interreg_IIIB-Mac2, 3/A3.
8. Barrios-Garcia MN, Classen AT, Simberloff D. 2014. Disparate responses of above- and belowground properties to soil disturbance by an invasive mammal. Ecosphere 5(4): art44 (http://dx.doi.org/10.1890/ES13-00290.1).
9. Bateman RM, Rudall PJ, Moura M. 2013. Systematic revision of Platanthera in the Azorean archipelago: not one but three species, including one of Europe's rarest orchids. PeerJ. DOI: 10.7717/peerj.218.
10. Batik SK, Pandey HN, Tripathi RS, Rao P. 1992. Microenvironmental variability and species diversity in treefall gaps in a sub-tropical broadleaved forest. Vegetatio 103:31-40.
11. Beaunea D, Bollachec L, Muganzaad MD, Bretagnollebe F, Hohmanna G, Frutha B. 2013. Artificial germination activation of Dialium corbisieri by imitation of ecological process. Journal of Sustainable Forestry 32(6):565- 575
12. BOC. 2012. Decreto 329/2011, de 22 de diciembre, por el que se aprueba el Plan de Recuperación de la especie vegetal "flor de mayo leñosa" (Pericallis hadrosoma). Boletín Oficial de Canarias núm. 2, Martes 3 de enero de 2012:71-84.
13. Borges PAV, Gabriel R, Arroz A, Costa A, Cunha R, Silva L, Mendonça E, Martins AF, Reis F, Cardoso P. 2010. The Azorean Biodiversity Portal: an internet database for regional biodiversity outreach. Systematics and Biodiversity 8:423-434.
14. Brownstein MJ, Carpten JD, Smith JR. 1996. Modulation of non templated nucleotide addition by Taq DNA polymerase: primer modifications that facilitate genotyping. Biotechniques 20:1004-1010.
15. Catarino FM, Carvalho JA, Dias E, Fernandes F, Fontinha S, Jardim R, Roselló-Graell A. 2001. Acções de conservação da flora em Portugal. In Fundación Ramón Areces, ed. Conservación de Especies Vegetales Amenazadas En la Región Mediterránea Occidental. Madrid: Editorial Centro de Estudios Ramón Areces, S.A., 63-92.
16. Caujapé-Castells J, Marrero-Rodríguez A, Baccarani-Rosas M, Cabrera-García N, Vilches-Navarrete B. 2008. Population genetics of the endangered Canarian endemic Atractylis arbuscula (Asteraceae): implications for taxonomy and conservation. Plant Systematics and Evolution 274:99-109.
17. Caujapé-Castells J, Tye A, Crawford D.J, Santos-Guerra A, Sakai A, Beaver K, Lobin W, Vincent Florens FB, Moura M, Jardim R, Gomes I, Kueffer C. 2010. Conservation of oceanic island floras: Present and future global challenges. Perspectives in Plant Ecology, Evolution and Systematics 12: 107-129.
18. CBD. 2012. Convention on Biological Diversity. Global Strategy for Plant Conservation: 2011-2020. Botanic Gardens Conservation International, Richmond, UK.
19. Costa H, Aranda SC, Lourenço P, Medeiros V, Azevedo EB, Silva L. 2012. Predicting successful replacement of forest invaders by native species using species distribution models: The case of Pittosporum undulatum and Morella faya in the Azores. Forest Ecology and Management 279: 90-96.
20. Costa H, Bettencourt MJ, Silva CMN, Teodósio J, Gil A, Silva L. 2013b. Invasive alien plants in the Azorean protected areas: invasion status and mitigation actions. In: Foxcroft LC, Pyšek P, Richardson DM, Genovesi P, eds. Plant Invasions in Protected Areas: Patterns, Problems and Challenges. Springer: Dordrecht, 375-394.
21. Costa H, Medeiros V, Azevedo EB, Silva L. 2013a. Evaluating the ecologicalniche factor analysis as a modelling tool for environmental weed management in island systems. Weed Research 53:221-230.
22. Crain BJ, Sánchez-Cuervo AM, White JW, Steinberg SJ. 2014. Conservation ecology of rare plants within complex local habitat networks. Oryx (DOI: http://dx.doi.org/10.1017/S0030605313001245).
23. Cunha AG, Sobrinho LG. 1939. Estudos Botânicos no Arquipélago dos Açôres. Revista da Faculdade de Ciências, Universidade de Lisboa 1:177-220.
24. Dias EF, Sardos J, L Silva, Maciel MGB, Moura M. 2014. Microsatellite markers unravel the population genetic structure of the Azorean Leontodon: implications in conservation. Plant Systematics and Evolution 300:987-1001.
25. Don RH, Cox PT, Wainwright BJ, Baker K, Mattick JS. 1991. 'Touchdown' PCR to circumvent spurious priming during gene amplification. Nucleic Acids Research 19:4008.
26. Donlan CJ, Croll DA, Tershy BR. 2003. Islands, exotic herbivores, and invasive plants: Their roles in Coastal California restoration. Restoration Ecology 11 (4):524-530.

27. Doyle JJ, Dickson EE. 1987. Preservation of plant samples for DNA restriction endonuclease analysis. Taxon 36:715-722.

28. Dubuis A, Giovanettina S, Pellissier L, Pottier J, Vittoz P, Guisan A. 2013. Improving the prediction of plant species distribution and community composition by adding edaphic to topo-climatic variables. Journal of Vegetation Science 24:593-606.

29. Earl DA, von Holdt BM. 2012. STRUCTURE HARVESTER: a website and program for visualizing STRUCTURE output and implementing the Evanno method. Conservation Genetics Resources 4:359-361.

30. Elith J, Leathwick JR. 2009. Species distribution models: Ecological explanation and prediction across space and time. The Annual Review of Ecology, Evolution, and Systematics 40:677-97.

31. Engler R, Guisan A, Rechsteiner L. 2004. An improved approach for predicting the distribution of rare and endangered species from occurrence and pseudo-absence data. Journal of Applied Ecology 41:263-274.

32. Evanno G, Regnault S, Goudet J. 2005. Detecting the number of clusters of individuals using the software STRUCTURE: a simulation study. Molecular Ecology 14:2611-2620.

33. Evans SM, Sinclair EA, Poore AGB, Steinberg PD, Kendrick GA, Vergés A. 2014. Genetic diversity in threatened Posidonia australis seagrass meadows. Conservation Genetics 15:717-728.

34. Faircloth BC. 2008. MSATCOMMANDER: detection of microsatellite repeat arrays and automated, locus-specific primer design. Molecular Ecology Resources 8:92-94.

35. Falush D, Stephens M, Pritchard JK. 2003. Inference of population structure using multilocus genotype data: linked loci and correlated allele frequencies. Genetics 164:1567-1587.

36. Fordham DA, Resit Akçakaya H, Araújo MB, Elith J, Keith DA, Pearson R, Auld TD, Mellin C, Morgan JW, Regan TJ, Tozer M, Watts MJ, White M, Wintle BA, Yates C, Brook BW. 2012. Plant extinction risk under climate change: are forecast range shifts alone a good indicator of species vulnerability to global warming? Global Change Biology 18:1357-1371.

37. Foxcroft LC, Pyšek P, Richardson DM, Genovesi P. 2013. Plant Invasions in Protected Areas: Patterns, Problems and Challenges. Springer: Dordrecht.

38. Garnock-Jonesa PI. 1976. Breeding systems and pollination in New Zealand Parahebe (Scrophulariaceae). New Zealand Journal of Botany 14:291- 298.

39. Garzón-Machado V, González-Mancebo JM, Palomares-Martínez A, AcevedoRodríguez A, Fernández-Palacios JM, Del-Arco-Aguilar M, Pérez-de-Paz PL. 2010. Strong negative effect of alien herbivores on endemic legumes of the Canary pine forest. Biological Conservation 143:2685-2694.

40. Glenn TC, Schable NA. 2005. Isolating microsatellite DNA loci. Methods in Enzymology 395:202-222.

41. Godefroid S, Piazza C, Rossi G, Buord S, Stevens A-D, Aguraiuja R, Cowell C, Weekley CW, Vogg G, Iriondo JM, Johnson I, Dixon B, Gordon D, Magnanon S, Valentin B, Bjureke K, Koopman R, Vicens M, Virevaire M, Vanderborght T. 2011. How successful are plant species reintroductions? Biological Conservation 144(2):672-682.

42. Guisan A, Zimmermann NE. 2000. Predictive habitat distribution models in ecology. Ecological Modelling 135:147-186.

43. Halbur MM, Sloop CM, Zanis MJ, Emery NC. 2014. The population biology of mitigation: impacts of habitat creation on an endangered plant species. Conservation Genetics 15:679-695.

44. Hancock N, Hughes L. 2014. Turning up the heat on the provenance debate: Testing the 'local is best' paradigm under heat wave conditions. Austral Ecology 3:600-611.

45. Hirzel AH, Hausser J, Chessel D, Perrin N. 2002. Ecological-niche factor analysis: how to compute habitat suitability maps without absence data? Ecology 83: 2027-2036.

46. Hirzel AH, Hausser J, Perrin N. 2007. Biomapper 4.0. Laboratory for Conservation Biology, University of Lausanne, Lausanne, Switzerland.

47. Hirzel AH, Le Lay G, Helfer V, Randin C, Guisan A. 2006. Evaluating the ability of habitat suitability models to predict species presences. Ecological Modelling 199: 142-152.

48. Hirzel AH, Le Lay G. 2008. Habitat suitability modelling and niche theory. Journal of Applied Ecology 45:1372-1381.

49. Houston DB, Schreiner EG, Moorhead BB. 1994. Mountain goats in Olympic National Park: biology and management of an introduced species. United States Department of the Interior, National Park Service, Olympic National Park, Port Angeles, Washington. Scientific Monograph NPS/NROLYM/NRSM-94/25.

50. Huang X, Madan A. 1999. CAP3: A DNA sequence assembly program. Genome Research 9:868-877.

51. Jakobsson M, Rosenberg NA. 2007. CLUMPP: a cluster matching and permutation program for dealing with label switching and multimodality in analysis of population structure. Bioinformatics 23:1801-1806.

52. King TJ. 1975. Inhibition of seed germination under leaf canopies in Arenaria serpyllifolia, Veronica arvensis and Cerastum holosteoides. New Phytologist 75: 87-90.

53. Kumar S, Skjæveland A, Orr RJS, Enger P, Ruden T, Mevik B, Burki F, Botnen A, Shalchian-Tabrizi K. 2009. AIR: a batch-oriented web program package for construction of supermatrices ready for phylogenomic analyses. BMC Bioinformatics 10:357.

54. Lance SL, Light JE, Jones KL, Hagen C, Hafner JC. 2010. Isolation and characterization of 17 polymorphic microsatellite loci in the kangaroo mouse, genus Microdipodops (Rodentia: Heteromyidae). Conservation Genetic Resources 2:139-141.

55. Latera P, Bazzalo ME. 1999. Seed-to-seed allelopathic effects between two invaders of burned pampa grasslands. Weed Research 39:297-308.

56. Li Y-Y, Guan S-M, Yang S-Z, Luo Y, Chen X-Y. 2012. Genetic decline and inbreeding depression in an extremely rare tree. Conservation Genetics 13(2):343-347.

57. Lomba A, Pellissier L, Randin C, Vicente J, Moreira F, Honrado J, Guisan A. 2010. Overcoming the rare species modelling paradox: A novel hierarchical framework applied to an Iberian endemic plant. Biological Conservation 143:2647-2657.

58. Marcer A, Sáez L, Molowny-Horas R, Pons X, Pino J. 2013. Using species distribution modelling to disentangle realised versus potential distributions for rare species conservation. Biological Conservation 166:221-230.

59. Martinez I, Carreño F, Escudero A, Rubio A. 2006. Are threatened lichen species well-protected in Spain? Effectiveness of a protected areas network. Biological Conservation 133:500-511.

60. Martins J, Costa H, Moreira O, EB de Azevedo, Moura M & Silva L. 2015. Distribution and conservation status of the endangered Azorean tree Picconia azorica. International Journal of Biological Sciences and Applications 2(1):1-9.

61. Martins JM, Moreira OCB, Rainha NFP, Baptista JAB, Silva L, Moura M. 2012. Morphophysiological dormancy and germination in seeds of the Azorean tree Picconia azorica. Seed Science & Technology 40, 163-176.

62. Martins JM, Moreira OCB, Sardos J, Maciel MG, Silva L, Moura M. 2013. Population genetics and conservation of the Azorean tree Picconia azorica. Biochemical Systematics and Ecology 49:135-143.

63. Menges ES. 1990. Population viability analysis for an endangered plant. Conservation Biology. 4:52-62.

64. Mir BA, Mir SA, Koul S. 2014. In vitro propagation and withaferin A production in Withania ashwagandha, a rare medicinal plant of India. Physiology and Molecular Biology Plants 20(3):357-64.

65. Moreira O, Costa H, Martins J, EB de Azevedo, Moura M, Silva L. 2014. Present and potential distribution of the endangered tree Prunus lusitanica subsp. azorica: Implications in conservation. International Journal of Biological Sciences and Applications 1(5):190-200.

66. Moreira O, Martins J, Silva L, Moura M. 2012. Seed germination and seedling growth of the endangered Azorean cherry Prunus azorica. HortScience 47:1222-1227.

67. Moreira OCB, Martins JM, Sardos J, Maciel MGB, Silva L, Moura M. 2013. Population genetic structure and conservation of the Azorean tree Prunus azorica. Rosaceae). Plant Systematics and Evolution 299:1737-1748.

68. Moura M, Silva L, Caujapé-Castells J. 2013. Population genetics in the conservation of the Azorean shrub Viburnum treleasei Gand. Plant Systematics and Evolution 299:1809-1817.

69. Moura M, Silva L, Dias EF, Schaefer H, Carine M. 2015. A revision of the genus Leontodon (Asteraceae) in the Azores based on morphological and molecular evidence. Phytotaxa 210(1):24-46.

70. Moura M, Silva L. 2010. Seed germination of Viburnum treleasei Gand., an Azorean endemic with high ornamental potential. Propagation of Ornamental Plants 10: 129-135.

71. Ollerton J, Winfree R, Tarrant S. 2011. How many flowering plants are pollinated by animals? Oikos 120(3):321-326.

72. Peakall R, Smouse PE. 2012. GenAlEx 6.5: genetic analysis in Excel. Population genetic software for teaching and research - an update. Bioinformatics 28:2537-2539.

73. Pence VC. 2013. In vitro methods and the challenge of exceptionals for Target 8 of the Global Strategy for Plant Conservation. Annals of the Missouri Botanical Garden 99(2):214-220.

74. Pereira JC, Schäfer H, Paiva J. 2002. New records of Veronica dabneyi (Hochst. Scrophulariaceae), an Azorean endemic plant not collected since 1938. Botanical Journal of the Linnean Society 139:311-315.

75. Phillips SJ, Anderson RP, Schapire RE. 2006. Maximum entropy modeling of species geographic distributions. Ecological Modelling 190:231-259.

76. Phillips SJ, Dudík M, Schapire RE. 2004. A maximum entropy approach to species distribution modeling. Proceedings of the 21st International Conference on Machine Learning, Banff, Canada.

77. Phillips SJ, Dudík M. 2008. Modeling of species distributions with Maxent: new extensions and a comprehensive evaluation. Ecography 31:161-175.

78. Pritchard JK, Stephens M, Donnelly P. 2000. Inference of population structure using multilocus genotype data. Genetics 155:945-959.

79. Rodríguez B, Siverio F, Siverio M, Barone R & Rodríguez A. 2015. Nectar and pollen of the invasive century plant Agave americana as a food resource for endemic birds. Bird Study (DOI:10.1080/00063657.2015.1015484).

80. Rosenberg N. 2004. Distruct: a program for the graphical display of population structure. Molecular Ecology Notes 4:137-138.

81. Rozen S, Skaletsky H. 2000. Primer3 on the WWW for general users and for biologist programmers. Methods in Molecular Biology 132:365-386.

82. Rychlik W. 1995. Selection of primers for polymerase chain reactions. Molecular Biotechnology 3:129-134.

83. Schäfer H, Carine MA, Rumsey FJ. 2011. From European priority species to invasive weed: Marsilea azorica (Marsileaceae) is a misidentified alien. Systematic Botany 36(4):845-853.

84. Schäfer H. 2003. Chorology and diversity of the Azorean flora. Dissertationes Botanicae 374:1-130.

85. Schuelke M. 2000. An economic method for the fluorescent labeling of PCR fragments: a poor man's approach to genotyping for research and highthroughput diagnostics. Nature Biotechnology 18:233-234.

86. Seubert M. 1844. Flora Azorica quam ex collectionibus

schedisque Hochstetteri patris et filii elaboravit et tabulis XV propria manu aeri incisis illustravit Mauritius Seubert. Adolphum Marcum, Bonnae.

87. Silva L, Elias RB, Moura M, Meimberg H, Dias E. 2011. Genetic variability and differentiation among populations of the Azorean endemic gymnosperm Juniperus brevifolia: Baseline information for a conservation and restoration perspective. Biochemical Genetics 49:715-734.

88. Silva L, Martins M, Maciel G, Moura M. 2009. Azorean Vascular Flora. Priorities in Conservation. Amigos dos Açores & CCPA: Ponta Delgada.

89. Silva L, Ojeda-Land E, Rodrıguez-Luengo J-L. 2008. Invasive terrestrial flora and fauna of Macaronesia. Top 100 in Azores, Madeira and Canaries. ARENA: Ponta Delgada.

90. Sjögren J. 1973. Recent changes in the vascular flora and vegetation of the Azores Islands. Memórias da Sociedade Broteriana 22:5-453.

91. Svenning J-C. 2001. On the role of microenvironmental heterogeneity in the ecology and diversification of neotropical rain-forest palms (Arecaceae). The Botanical Review 67:1-53.

92. Václavík T, Meentemeyer RK. 2012. Equilibrium or not? Modelling potential distribution of invasive species in different stages of invasion. Diversity and Distributions 18:73-83.

93. Watson HC. 1870. Botany of the Azores. In: Godman FC, ed. Natural history of the Azores or Western Islands. London: Taylor & Francis, 113-288.

94. Wu H, Qiang S, Peng G. 2010. Genetic diversity in Veronica hederifolia (Plantaginaceae), an invasive weed in China, assessed using AFLP markers. Annals Botanica Fennici 47:190-198.

95. Yu S, Sternberg M, Kutiel P, Chen H. 2007. Seed mass, shape, and persistence in the soil seed bank of Israeli coastal sand dune flora. Evolutionary Ecology Research 9:325-340.

CHAPTER 3

PLANT AND ANIMAL ENDEMISM IN THE EASTERN ANDEAN SLOPE: CHALLENGES TO CONSERVATION

Jennifer J Swenson[10,1], Bruce E Young[1], Stephan Beck[2], Pat Comer[1], Jesús H Córdova[3], Jessica Dyson[1,11], Dirk Embert[4], Filomeno Encarnación[5], Wanderley Ferreira[6], Irma Franke[3], Dennis Grossman[1,12], Pilar Hernandez[1,13], Sebastian K Herzog[7], Carmen Josse[1], Gonzalo Navarro[6], Víctor Pacheco[3], Bruce A Stein[1,14], Martín Timaná[1,15], Antonio Tovar[8], Carolina Tovar[8], Julieta Vargas[9] and Carlos M Zambrana-Torrelio[2,16]

[1] NatureServe, 4600 North Fairfax Drive, Floor 7, Arlington, VA 22203, USA

[2] Herbario Nacional de Bolivia, Universidad Mayor de San Andrés, La Paz, Bolivia

[3] Museo de Historia Natural, Universidad Nacional Mayor de San Marcos, Apartado 140434, Lima-14, Perú

[4] Fundación Amigos de la Naturaleza, km 7,5 Doble Vía la Guardia, Santa Cruz de la Sierra, Bolivia, Casilla 2241

[5] Instituto de Investigaciones de la Amazonía Peruana, Iquitos, Perú

[6] Rumbol, S.R.L. Av. Dorbigni 1608, Cochabamba, Bolivia

[7] Asociación Armonía, BirdLife Internacional, Avenida Lomas de Arena 400, Casilla 3566, Santa Cruz de la Sierra, Bolivia

[8] Centro de Datos para la Conservación, Departamento de Manejo Forestal, Facultad de Ciencias Forestales, Universidad Nacional Agraria La Molina, Apartado 456, Lima 100, Perú

[9] Museo Nacional de Historial Natural, Colección Boliviana de Fauna, Casilla 8706, La Paz, Bolivia

[10] Nicholas School of the Environment, Duke University, Box 90328, Durham, NC 27708, USA

[11] The Nature Conservancy, 99 Bedford St., 5th Floor, Boston MA 02111 USA

[12] The Nature Conservancy, 4245 Fairfax Drive, Arlington, VA 22203 USA

[13] Ontario Ministry of Natural Resources, 50 Bloomington Road W, Aurora, ON L4G 3G8

[14] National Wildlife Federation, 901 E Street, NW Suite 400, Washington DC, 20004 USA

[15] Departamento de Ciencias, Pontificia Universidad Católica del Perú, Av. Universitaria 1801, Lima 32, Peru

[16] EcoHealth Alliance - 460 W 34th Street, 17th Floor, New York, NY 10001, USA

ABSTRACT

Background

The Andes-Amazon basin of Peru and Bolivia is one of the most data-poor, biologically rich, and rapidly changing areas of the world. Conservation scientists agree that this area hosts extremely high endemism, perhaps the highest in the world, yet we know little about the geographic distributions of these species and ecosystems within country boundaries. To address this need, we have developed conservation data on endemic biodiversity (~800 species of birds, mammals,

amphibians, and plants) and terrestrial ecological systems (~90; groups of vegetation communities resulting from the action of ecological processes, substrates, and/or environmental gradients) with which we conduct a fine scale conservation prioritization across the Amazon watershed of Peru and Bolivia. We modelled the geographic distributions of 435 endemic plants and all 347 endemic vertebrate species, from existing museum and herbaria specimens at a regional conservation practitioner's scale (1:250,000-1:1,000,000), based on the best available tools and geographic data. We mapped ecological systems, endemic species concentrations, and irreplaceable areas with respect to national level protected areas.

Results

We found that sizes of endemic species distributions ranged widely (< 20 km^2 to > 200,000 km^2) across the study area. Bird and mammal endemic species richness was greatest within a narrow 2500-3000 m elevation band along the length of the Andes Mountains. Endemic amphibian richness was highest at 1000-1500 m elevation and concentrated in the southern half of the study area. Geographical distribution of plant endemism was highly taxon-dependent. Irreplaceable areas, defined as locations with the highest number of species with narrow ranges, overlapped slightly with areas of high endemism, yet generally exhibited unique patterns across the study area by species group. We found that many endemic species and ecological systems are lacking national-level protection; a third of endemic species have distributions completely outside of national protected areas. Protected areas cover only 20% of areas of high endemism and 20% of irreplaceable areas. Almost 40% of the 91 ecological systems are in serious need of protection (= < 2% of their ranges protected).

Conclusions

We identify for the first time, areas of high endemic species concentrations and high irreplaceability that have only been roughly indicated in the past at the continental scale. We conclude that new complementary protected areas are needed to safeguard these endemics and ecosystems. An expansion in protected areas will be challenged by geographically isolated micro-endemics, varied endemic patterns among taxa, increasing deforestation, resource extraction, and changes in climate. Relying on pre-existing collections, publically accessible datasets and tools, this working framework is exportable to other regions plagued by incomplete conservation data.

BACKGROUND

Numerous global conservation prioritization schemes have been developed that are centered on biodiversity, endemism and vulnerability (e.g. [1, 2, 3, 4, 5]). Characterizing global areas of high biodiversity under threat as "hotspots" [1] or "priority ecoregions" [6], for example, has identified priorities using a variety of weighting schemes (e.g. [3, 4]). However, the information that underlies these prioritizations in the best cases can consist of coarse scale species range maps, typically hand-drawn by knowledgeable researchers from available locality data [7, 8, 9, 10]. In less than ideal cases, lists of known species by large areal units such as ecoregions are used [11]. Although the range maps are convenient accompaniments for species accounts in field guides, they are too coarse for landscape-level conservation planning (Figure 1). There are often errors in the locality information that is used to generalize range maps, and they typically overestimate areas of occupancy because of the coarse scale at which they are drawn [12, 13]

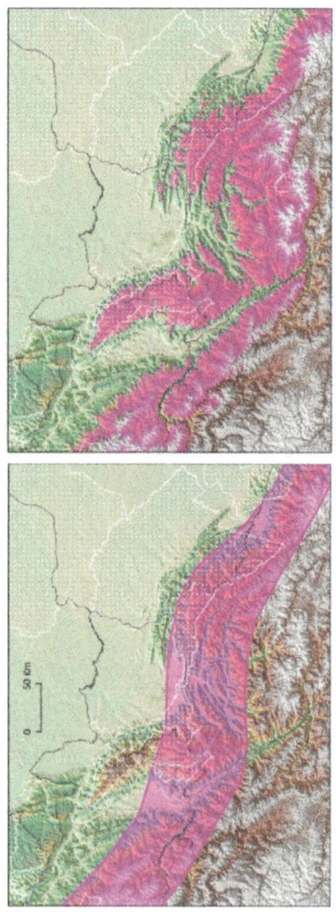

Figure 1: Comparison of hand drawn vs. modelled species distribution map. Hand-drawn range map (a) used in many continental studies with (b), a modeled species distribution for *Cycanolyca viridicyanus* in southwestern Peru (Vilcabamba). National protected areas (white), and department boundaries (black lines) and elevation as backdrop.

Citation: Jennifer J Swenson, Bruce E Young, Stephan Beck, Pat Comer, Jesús H Córdova, Jessica Dyson, Dirk Embert, Filomeno Encarnación, Wanderley Ferreira, Irma Franke, Dennis Grossman, Pilar Hernandez, Sebastian K Herzog, Carmen Josse, Gonzalo Navarro, Víctor Pacheco, Bruce A Stein, Martín Timaná, Antonio Tovar, Carolina Tovar, Julieta Vargas and Carlos M Zambrana-Torrelio, Plant and animal endemism in the eastern Andean slope: challenges to conservation, DOI: 10.1186/1472-6785-12-1.

Global prioritization areas themselves are typically too large to protect in their entirety (e.g. the Andean ‹hotspot› *sensu*[1], covers an area over four times the size of Germany and crosses over seven Andean countries), and are not practical nor intended for use in national or departmental planning. For many data-poor countries however, global datasets such as these are the only consistent estimates of biodiversity that are available. Effective on-the-ground conservation efforts and decisions require planning and biodiversity information at a much finer scale [14].

Endemic species are restricted to a particular geographic area-occurring nowhere else-and are important components in most global conservation prioritizations. A focus on endemic species richness can provide unique information about biodiversity patterns [3, 15] compared to all-encompassing species richness that is dominated by generalist (non-endemic) species [4], which are typically the lowest priority for conservation. Areas high in endemism are especially valuable because they may represent areas of high past speciation in evolutionary hotspots [16]. The forces that create areas of high species endemism and richness are still not well understood, which is one argument for their preservation for further study [17]. Another reason for preservation is that these areas may function as species refugia during future climate changes, as they may have in the past. Globally, areas of high endemism are currently underrepresented by the protected area network [2].

The Andes region of South America harbours one of the largest assemblages of endemic plant and animal species and is one of the most biodiverse and threatened areas of the world [1, 2, 3, 4, 5]. Explanations for such a concentration of endemics include past climate shifts, geotectonic events, modern ecological interactions, and limited dispersal. This area was historically isolated from the lowlands by the Andean uplift, which created a complex mosaic of high mountains and deep inter-Andean valleys. Researchers generally agree that this ancient uplift and isolation were important drivers in speciation, resulting in high concentrations of endemic birds [18, 19, 20, 21, 22], mammals [23], and plants [24, 25, 26, 27]. Analyses of Andean amphibians are limited but indicate similar drivers of environmental divergence [28, 29, 30] and colonization from different regions [31]. Recent climatic stability influenced by topography has created ideal conditions for high biodiversity (very humid areas) and endemism (dissected topography creating isolated dry valleys) [32].

Despite the agreement among scientists about the origins and existence of the extremely high endemic diversity of this region, it remains scientifically understudied [33]. We have very limited knowledge of current patterns of Andean species distributions and diversity within this globally prioritized area [14]. National-level efforts to prioritize conservation in Peru and Bolivia have previously explored gaps in protected area coverage, but have been hindered by the limited information available on species status and distribution [34,35]. The information available is primarily of bird diversity patterns rather than other taxon groups [36, 37, 38,39, 40]. Yet even the most recent endemism studies of birds were delimited by a 1/4° grid (~28 × 28 km) as the unit of analysis [36, 37]. Studies of the spatial pattern of Andean endemic mammal richness are lacking, possibly due to unstable taxonomy and incomplete knowledge about distributions [41]. A worldwide distributional analysis at a coarse scale with a 1° grid (~111 × 111-km) showed a relative concentration of endemic mammal species along the east side of the Andes in Peru and northern Bolivia [5]. As well, a regional study in Peru corroborated this pattern [42]. We are unaware of spatially explicit analyses of amphibian endemic patterns, although several authors have suggested that higher concentrations of endemics should be found in montane regions [43, 44, 45]. Knowledge of endemic plants in this region varies widely by taxonomic group. Analyses of a few better-known groups suggest peaks of diversity and endemism in the eastern Andes [17, 46, 47, 48, 49]. Vegetation and land cover maps of this region have variable coarse spatial and classification detail; different regions employ distinct classification schemes and methods that make joining maps along borders difficult.

The development of computer-aided models to predict species distributions presents an opportunity to develop distribution information at the scale necessary for in-country conservation planning [50, 51]. With the goal of producing relatively fine resolution species and ecosystem data within a repeatable framework of methods, we created geographic distributions of endemic birds, mammals, amphibians, plants, and mapped their ecosystems on the eastern slope of the Andes in Peru and Bolivia at a scale applicable to conservation planning (1 km^2 grid, less than < 1/60°, 1:250,000 - 1:1,000,000). This multiple taxon approach enables a broader characterization of diversity, given that one taxonomic group or species is not always representative of other taxa [15, 52, 53]. By geographically integrating this data, we identify areas of high endemic concentrations and irreplaceable areas (greatest number of narrowly distributed endemics) across the study area [54]. We characterize the ecological systems where endemic species reside and perform a gap analysis to identify species ranges, endemic concentrations and ecological systems currently located outside of established national-level protected areas. In addition to pinpointing candidate areas for future protection efforts, the results highlight several challenges to conservation in the region.

METHODS

In addition to the following descriptions of endemic distribution modelling, mapping of ecological systems and geographical analysis of all the overlapping datasets, the Supporting Information Additional Files 1, 2, 3, 4, 5, 6, contain further method details.

Study Area

Our study focused on the Amazon basin of Peru and Bolivia, from treeline in the eastern Andes (~3500 m), downslope to the Amazon lowlands and extending to the Brazilian border (Figure 2). The southern limit extends to the edge of the southern subtropical uplands where the biogeographic province of the *chiquitanía*begins. The area hosts a wide range of ecosystems from the wetlands of the Beni savanna and the Iquitos*várzea*, to xeric habitats of inter-Andean valleys and humid montane forests along much of the eastern Andean slope. Many areas are difficult to access because of lack of transportation infrastructure, entrance restrictions into indigenous lands and patrolling of illegal crops [36]. The study area extends from 5°23’ to 18° 15’ S latitude and from 60° 23’ to 79° 26’ W longitude and covers 1,249,282 km^2.

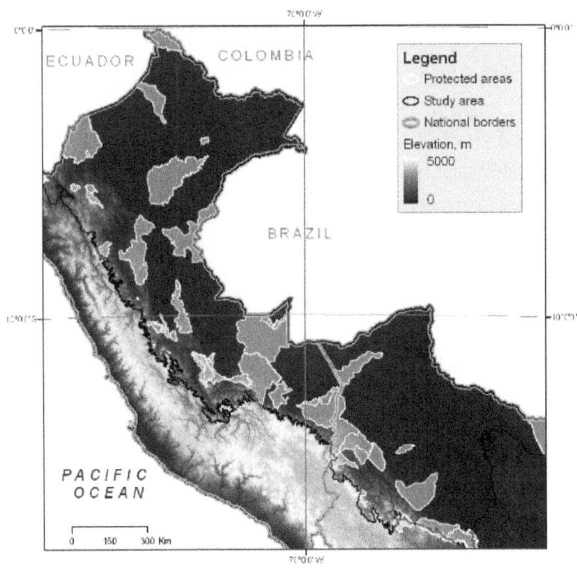

Figure 2: Study area. The Amazon basin of Peru and Bolivia with current protected areas. Protected areas information from INRENA-Peru, and FAN-Bolivia, elevation from Shuttle Radar Topography Mission.

Endemic Species and Locality Data

More than a century of collecting in South America has yielded large numbers of plant and animal specimens that provide locality data for species geographic distribution predictions. To represent a diverse suite of species, we modelled the geographic distributions of all bird, mammal, and amphibian species that are endemic to our study area [7, 8, 9] (Table 1). We identified which species were endemic based on pre-existing hand drawn range maps [8, 9, 55] and consultation with regional experts. We also modelled distributions of endemic plants but limited our analysis to 15 representative focal groups (families or genera) generally well known, and relatively well sampled in both countries (details on criteria for inclusion can be found in Additional File 1): Acanthaceae, Anacardiaceae, Aquifoliaceae, Brunelliaceae, Campanulaceae, Chrysobalanaceae, Cyatheaceae, Ericaceae, *Inga* (Fabaceae), *Mimosa* (Fabaceae), Loasaceae, Malpighiaceae, Marcgraviaceae, *Fuchsia* (Onagraceae), and Passiflora (Passifloraceae). As with the vertebrates, we modelled distributions for all species in these groups that are endemic to the study area.

Table 1: Summary of endemic species groups and modeled ranges

Species group	Number species	Number genera	Total number localities	Median number records per species	No. data sources collaborating Institutions	Number Maxent models formed	Median distributional area, (km^2)
Amphibians	177	30	1060	2	9	85	399
Birds	115	69	2437	15	15	99	21,075
Mammals	55	29	618	7	12	47	24,156
Plants	435	66	3040	3	50+	264	3543

For each of the 782 species of endemic plants and animals, we compiled locality records from an exhaustive search of specimen records in 81 local and international natural history collections and herbaria, published records, and for birds and mammals only, observational data. Specimen searches were carried out 2004-2006 with Peruvian, Bolivian and international institutions, individuals, and from published sources (see Additional File 5). The majority of specimens were collected in the 1990's and 2000's, yet dates ranged wider for published sources that we validated with national gazetteers of collecting locations [56]. The oldest localities for example, were collected for mammal species in the early part of this century [57]. Because many specimen labels did not include global positioning system-based coordinates for the collecting locations, we identified the most reliable localities based on their described location and geo-referenced them using standardized methods [58], and additional resources such as consultation with the collector, and geographic gazetteers (e.g.[56]). To further assure the creation of an accurate locality database, we then asked taxonomic specialists familiar with the species and geography to review mapped localities to ensure the creation of an accurate locality database. We buffered the study area by 100 km for the endemic species data gathering and modelling to avoid edge effects.

Predictive Distribution Modelling

We used spatial environmental layers describing climate, topography, and vegetation within our study area at 1-km^2 resolution together with the field locality data to develop species distribution models (Table 2). The WorldClim climate data [59] is currently the best available for this region yet it has its own inaccuracies as will any future downscaled version, because meteorological information is scarce in many areas of the study area. To maintain consistent spatial resolutions, we resampled the most accurate elevation data for the region (NASA's Shuttle Radar Topography Mission, SRTM [60]) to match the 1-km^2 resolution of the climate data. Vegetation characterizations were made with a 3-year seasonal time series of satellite-derived MODIS vegetation indices and per cent tree cover [61] at the same resolution. Mapping of detailed ecological systems, discussed below, was conducted separately as an independent characterization at a higher spatial resolution based on NASA's Landsat Thematic Mapper satellite sensors.

Table 2: Environmental predictors and data sources for species distribution modelling

Variable	*Data Source*
Mean annual temperature, mean temperature diurnal range, isothermality, precipitation of wettest and driest month, precipitation seasonality	Worldclim, (Hijmans et al. 2005. www.worldclim.org), 1-km resolution
Topography: Elevation	Shuttle Radar Topography Mission digital elevation data provided by CGIAR (http://srtm.csi.cgiar.org/) resampled to 1-km resolution
Slope	Degree of slope (maximum rate of change in elevation from each pixel to its 8 neighbors) derived from the SRTM digital elevation data
Topographic exposure	Expresses the relative position of each pixel on a hillslope (e.g. ridge, valley, toe slope). Using methods of Zimmermann (2000) on the SRTM digital elevation data with three neighborhood windows of 3x3, 6x6 and 9x9
Percent tree cover	MODIS global vegetation continuous fields sourced fromhttp://glcf.umiacs.umd.edu/data/modis/vcf/data.shtml (Hansen et al. 2003) 1-km resolution, and summarized within 3- and 5- km moving windows
Enhanced Vegetation Index (EVI) Principal component 1 Principal component 2	MODIS vegetation indices 16-Day data product sourced from the NASA EOS data gateway; Principal component analysis of 3 years of 16-day composites. MODIS EVI data summarized within 5 km moving window

There are drawbacks to predictive distribution modelling-for example, models may overestimate species' geographic ranges [62, 63]-as well as advantages, such as reducing the effect of uneven collecting efforts [64]. Nonetheless, distribution modelling is arguably the best approach at present when reliable locality and environment data are available [65]. We chose Maximum Entropy ("Maxent") [66], a statistical mechanics approach, as our modelling algorithm because of its documented success at modelling species with limited locality data, a common problem when working with endemic species [65, 67, 68, 69]. To ensure that Maxent was best suited to modeling distributions of Andean species, we compared the success of Maxent and two new promising methods: Mahalanobis Typicalities (a method adopted from remote sensing analyses), and Random Forests (a model averaging approach to classification and regression trees). We found that Maxent produced more consistent predictions across varying climatic conditions for 16 species [67]. Two to seven taxonomic specialists reviewed each model output to determine thresholds to convert continuous predictions into presence-absence maps based on known areas of absence, and to remove areas of known over-prediction (i.e., where the species was known not to occur). Specialist review is especially necessary when modelling with small sets of locality data [52, 67, 70]. For species known from a single or very few localities, we ran "rule-based" models (instead of Maxent) consisting of the geographic intersection of known ranges in elevation and other environmental variables such as temperature and precipitation.

Areas of Endemism and Irreplaceability

Traditionally, ecologists have overlain distribution maps of species to identify areas of high endemism or species richness [39]. We followed this approach to identify areas of high endemism for each vertebrate and plant group. To identify discrete areas of high endemism we chose an arbitrary threshold value of two-thirds the maximum number of overlapping species for each group and compared these patterns with previous studies, where they exist. This simple threshold could be changed depending on the desire to be more or less inclusive in identifying areas of high endemism.

To highlight areas harbouring species with very restricted ranges, and therefore of potentially greater conservation significance, we created maps of summed irreplaceability for each group using the C-Plan Software [71]. Summed irreplaceability is the likelihood that a given analysis unit should be protected to achieve a specified conservation target for the study area [54]. We used 10-km^2 analysis pixels and defined 25 of these pixels for each species as a conservation "target". If a given species was found present in < 25 of the 10-km^2 pixels, we set the target as the number of pixels in which the species occurs. For each species, irreplaceability for each pixel ranges from 0 to 1. Low values of irreplaceability indicate that for a species there are many other (replaceable) sites that may be conserved (in other words that a species occurs in many pixels), whereas high values indicate there are very few sites available (irreplaceable) because the species have very narrow ranges. The final irreplaceability number is the result of summing irreplaceability values for all species occurring at each location, thereby emphasizing the locations with the higher number of narrow-range endemics.

Ecological Systems

To complement the endemic species information, we produced a detailed map of natural vegetation types at a scale of 1:250,000 (25 ha minimum mapping unit). We applied a hemisphere-wide vegetation classification system [72] that is the terrestrial classification employed as a standard in North America in U.S. federal mapping projects [73, 74] and an emerging standard in Latin America [75]. The classification relies on the concept of terrestrial ecological systems [73], which are groups of vegetation communities that tend to co-occur in landscapes as a result of the action of common ecological processes, substrates, and/or environmental gradients. The ecological system classification allows for effective integrated vegetation mapping, at desired levels of thematic detail, permitting planners to prioritize across borders and across large regions. The species distribution models did not use this map as a predictor variable, thus the map provides an independent characterization of areas where endemics reside. In addition to analysing protection gaps and representativeness of the systems, we examined the overlap between ecological systems and areas of high endemism. Our goal was to identify if any systems were disproportionally represented in endemic areas compared to their distributions across the study area.

To create the ecological systems map, we incorporated existing vegetation maps where possible, and with in-country mapping teams of local field and botanical experts; we applied one cohesive classification system across the two countries. The mapping relied on field work, visual interpretation of Landsat TM and ETM+ satellite images in the Peruvian lowlands and areas of Bolivia, and spatial modelling and image classification for upland areas in Peru. Though more advanced mapping methods exist (e.g., [76]), we found our methods to be appropriate for these landscapes and the limited data availability, as well as more accessible to the in-country mapping teams. For ecological system characterization as well as accuracy assessment, we developed a rapid field survey protocol for more than 2000 points across the study area using spatial optimization to identify candidate clusters of points. Field observations and aerial transects of high-resolution digital photos of remote and inaccessible areas provided the basis for map validation and accuracy assessment. Details of the mapping methods, classification system and accuracy assessment can be found in Additional File 1.

Gap Analysis

We conducted a gap analysis (*sensu*[77]) by examining the representation of terrestrial ecological systems, species distributions, and areas of high endemism and irreplaceability with respect to existing national-level protected areas. We included all designated nationally administered areas corresponding to World Conservation Union (IUCN) categories I-VI (IUCN 1994), as well as those that have not yet been scored against the IUCN criteria. This covered national parks, communal reserves, protected forests, integrated management areas, and other national sanctuaries. Rather than limiting our analysis to those areas with IUCN categories reflecting the strictest levels of protection, we took an inclusive approach, recognizing that in this region effective protection can vary in any category. We used digital maps of protected area boundaries from 2007 provided by our in-country collaborators as they were more current than the World Database of Protected Areas WDPA [78] at the time. National level protected area boundaries have not changed in the region at the time of publication of this article; however improvements have been made to the WDPA information. Regional protected areas have experienced shifts in jurisdiction, area, and level of protection. While including regional protected areas in this analysis would be advantageous, information on protection levels and boundaries of regional areas is incomplete in some areas and inconsistent across country borders.

RESULTS

The datasets and individual species maps for most of the analyses described here are publically accessible (in both graphic and geospatial format) on the project website (http://www.natureserve.org/andesamazon). The supporting Additional Files 1, 2, 3, 4, 5, 6, contain supplementary results in detail.

Endemic Species

We compiled 7154 unique records of existing specimen localities to create distribution models for all 115 birds, 55 mammals, 177 amphibians, and 435 plants included in our endemic species analysis (Table 1; see Additional File 1). Sample sizes of unique localities for modelling distributions of individual species were highest for birds, followed by mammals, plants, and amphibians. There were 3 mammal and 3 bird species having just one reliable locality, whereas 123 plant and 65 amphibian species were limited to one location, none of which were predicted with distribution modelling. Modelled distribution sizes varied from just 2 km^{2}for the plant *Centropogon bangii*, to 690,992 km^2, or 55% of the study area, for the frog *Colostethus trilineatus*. On average, endemic mammals tended to have the largest geographic distributions, followed by birds, plants and amphibians (Figure 3). Maxent models produced satisfactory distribution maps, according to expert reviewers and model evaluation techniques, for 67% of the species. We produced distributions for the remaining species, which had too few known localities for Maxent models, using rule-based models. Expert review was essential for eliminating areas from the distribution where the species was known not to occur for reasons of competition or geographic isolation.

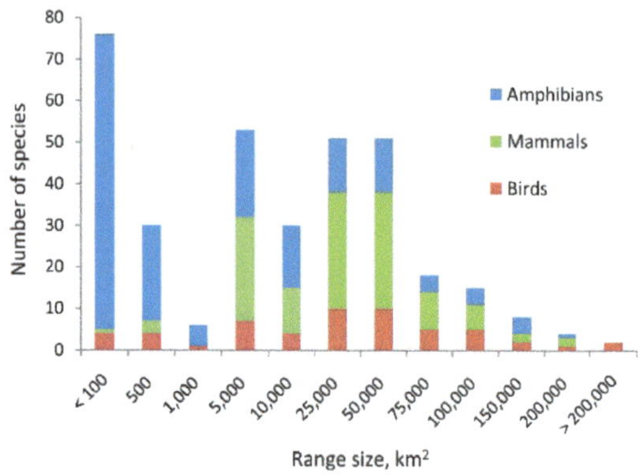

Figure 3: Distribution of predicted range sizes by species groups.

Areas of Endemism and Irreplaceability

Areas with the highest numbers of endemic species lie along mid to upper elevations on the eastern slope of the Andes, yet patterns vary by taxonomic group. Both birds (25-38 species per 1-km^2 grid cell) and mammals (17 - 20 species per cell) followed this trend (Figures 4a, b) with peaks of endemic richness encompassing elevations between 2500 and 3000 m and extending almost the entire length of the study area. Amphibians, by contrast, displayed peaks of endemism (21 - 29 species per 1-km^2 cell) on lower slopes, between 1000 and 1500 m elevation. These areas were concentrated in southern Peru, northern Bolivia, and in an isolated endemic area in the northern Peruvian department of San Martin (Figure 4c). Combining all vertebrate species reveals high concentrations between 2000 and 3000 m elevation (Figure 5) with highest concentrations (75 to 78 overlapping species) in Bolivia's Cochabamba and Tiraque *Cordilleras* (mountain ranges) and extensive areas of high value along Peru's Vilcabamba Cordillera. We found that the different plant groups varied widely in endemic patterns among themselves and with respect to vertebrates. Areas of high *Fuchsia* endemism, for example, were at similar elevations as birds and mammals, but with local concentrations in the departments of Cusco (Peru), and Cochabamba (Bolivia) (Figure 4d). Endemic species of Aquifoliaceae, Chrysobalanaceae,*Inga*, Loasaceae, and Malpighiaceae were concentrated in the northern portion of the study area, whereas endemic Brunelliaceae, Campanulaceae, Ericaceae, Marcgraviaceae, *Mimosa*, and Passifloraceae were concentrated in the south. We found concentrations of endemic Acanthaceae in both the north and south. Endemic species of Anacardiaceae, Chrysobalanaceae, *Inga*, and Malpighiaceae were concentrated in the lowlands, whereas Acanthaceae and Cyatheaceae occurred largely at mid elevations (around 1000 m); endemic species in the remaining nine groups occur mostly above 2000 m (maps of all plant species can be found here: http://www.natureserve.org/aboutUs/latinamerica/maps_plants_intro.jsp).

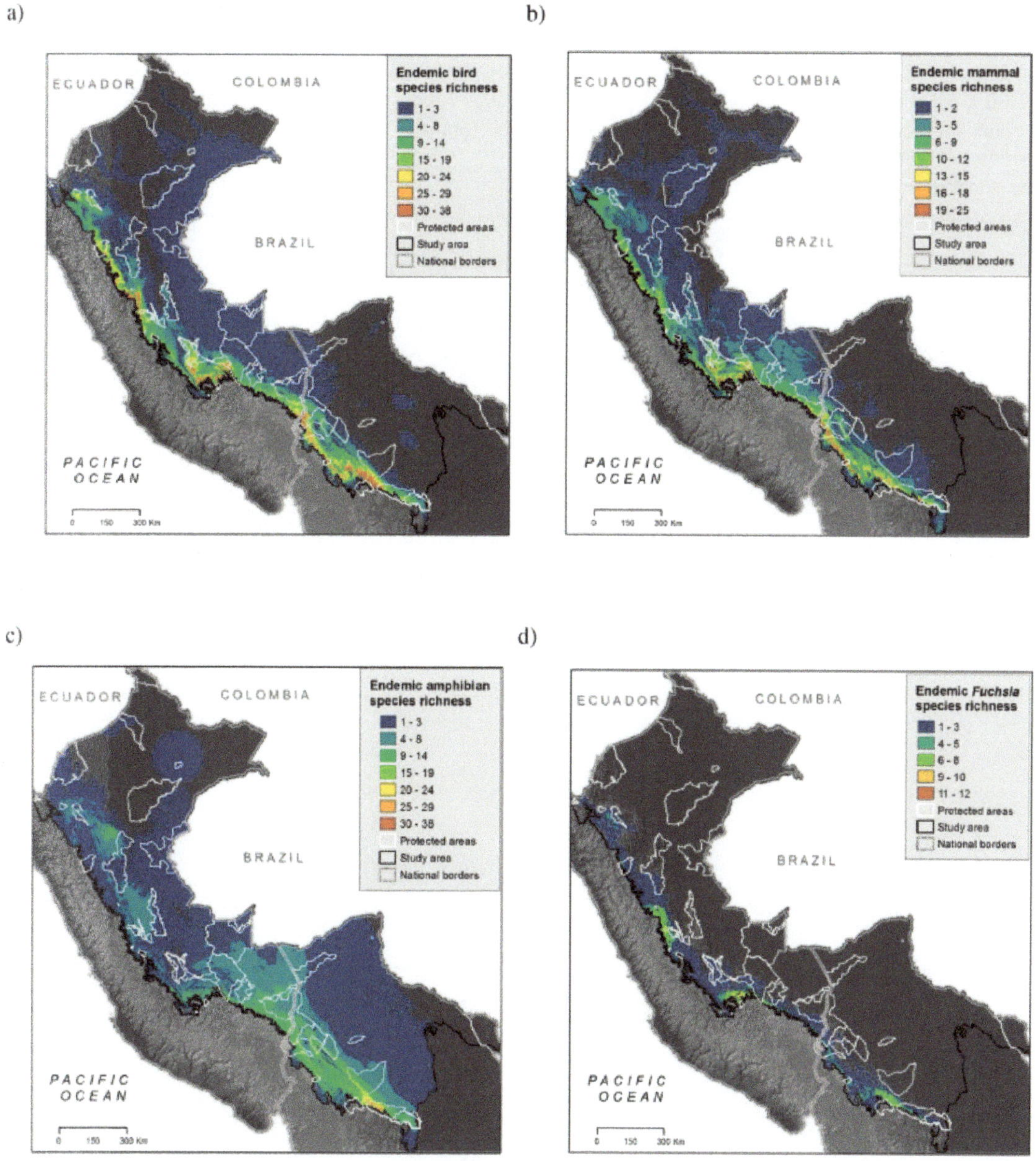

Figure 4: a-d - Endemic species richness. Overlapping distribution maps for different species groups: a) birds, b) mammals c) amphibians d) Fuchsia genus plant species. Fuchsia is shown as an example of one of the 15 groups modelled (Seehttp://www.natureserve.org/andesamazon for maps of individual species and all plant groups).

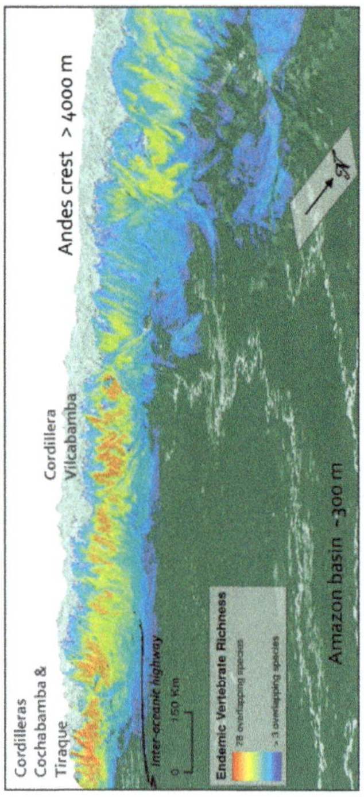

Figure 5: Endemic vertebrate species richness. Combined endemic mammal, bird, and amphibian richness over a three-dimensional oblique perspective. Viewpoint is from northeastern Peru looking south across the Amazon basin towards the southern Peruvian and northern Bolivian Andes.

Summed irreplaceability analysis which highlights areas with the greatest numbers of narrow-ranging species, shows different key areas than the endemic areas analysis. Similar to the endemic areas, many of the peaks of summed irreplaceability occurred in the higher elevation slopes along the Andean cordillera (Figure 6a-d, areas over threshold value shown). Endemic richness of birds and mammals overlapped more than other groups yet summed irreplaceability showed differences between these two taxonomic groups, as well as for amphibians. The northern portion of the study area in the Peruvian department of Amazonas (Cordillera de Colán and Alto Mayo) is highly irreplaceable for plants, amphibians, and birds but was not identified as an endemic area by the simple overlay of species ranges (Figures 4a, c, d); this emphasizes the large number of very restricted range species that occur there. Summed irreplaceability also highlighted some lowland areas for species groups in which most other species occurred at higher elevations. For instance, birds have high irreplaceability in north-eastern Peru, where a number of species are restricted to the lowland white-sand forests near Iquitos. Similarly, there are two restricted range primate species in the Beni savanna of Bolivia, emphasizing the irreplaceability of that region for mammals. Detailed descriptions of locations of the areas of high endemism and irreplaceability for all species groups can be found in Young et al. (2007).

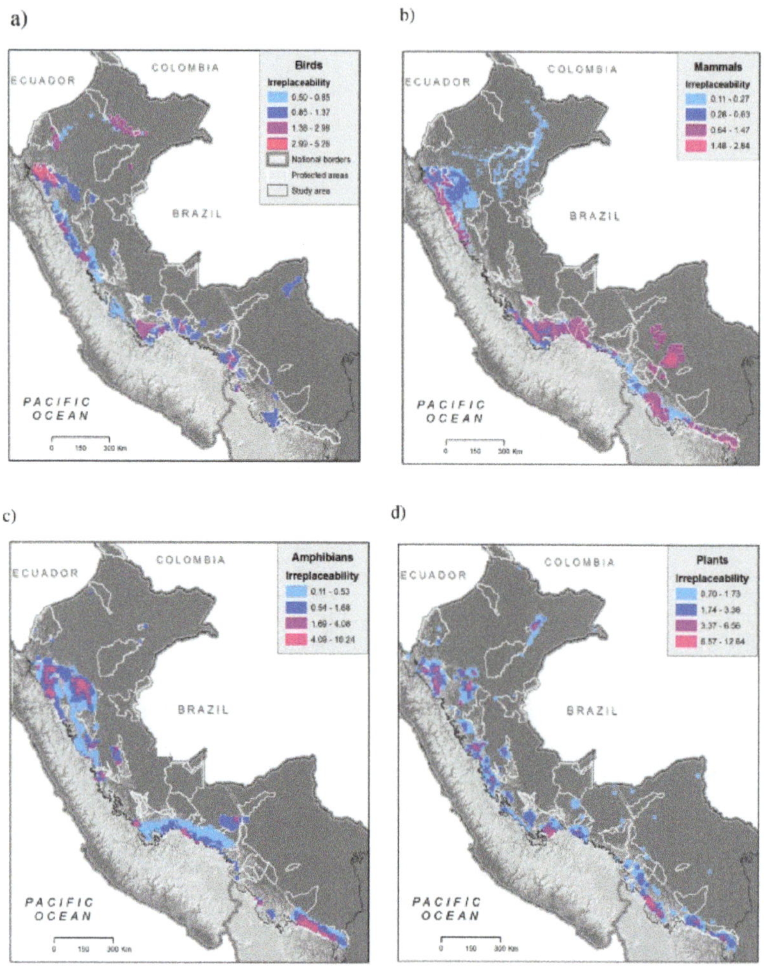

Figures 6: a-d - Summed irreplaceability analysis for different species groups; a) birds, b) mammals c) amphibians d) plants from all 13 groups.

Discrete centres of endemism (Figure 7), covered 23,844 km^2 for birds, 11,655 km^2 for mammals, 2781 km^2 for amphibians, and 67,676 km^2 for plants. (We only included 13 groups for plants as Anacardiaceae and Cyatheaceae did not have more than two co-occurring endemic species anywhere in the study area.) Combining all plant and animal endemic areas results in a region covering 78,790 km^2 or 6.3% of the study area. In contrast, the intersection of endemic areas for the three vertebrate groups covers a mere 140 km^2, highlighting differences among these groups.

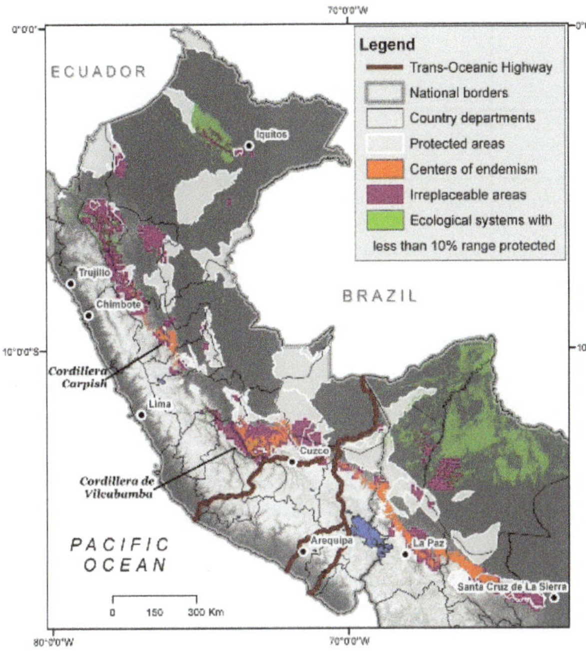

Figure 7: Biodiversity indicators. Discrete centres of vertebrate endemism and high levels of summed irreplaceability (all species groups), and ecological systems with less than 10% of their ranges protected.

Ecological Systems

We distinguished 91 unique ecological systems and complexes across the basin, ranging from flooded savanna systems to xeric shrub types (Figure 8 shows an area in detail for northern Peru; see [79] for a description of each ecological system). The systems represent unique vegetation communities, further distinguished by bioclimate, geomorphology, substrate, flooding regime, river type (black, white, mixed water) and regional compositional differences. Half of the ecological systems consist of different forms of wetlands and cover 30% of the study area and systems with bamboo-dominated forests cover over 71,500 km^2. Forty-two of the ecological systems are unique to the Amazon basin of Peru and Bolivia. The Andean uplands region of Peru, typified by steep elevation gradients and subsequent vegetation zonation [80], represents only 12% of the study area yet harbours 37% of the different ecological systems. Accuracy of the ecological system map varied by region and the type of validation data. For detailed classes of ecological systems (not including areas converted to human uses), accuracy ranged from 62 to 91% by mapping region, while a map legend of 20 coarser groupings of systems defined by ecological similarity had accuracies ranging from 81 - 90% (See Additional File 1 and 4).

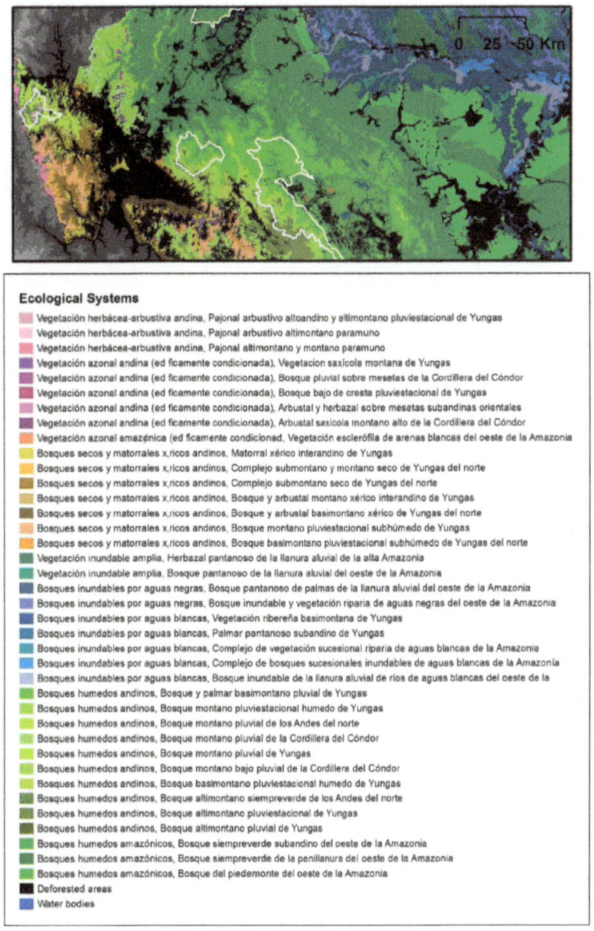

Figure 8: Ecological systems detail of subarea in northern Peru.

Combining the areas of endemism of the three vertebrate groups creates a region covered by 16 ecological systems (Table 3). The montane pluvial forest of the *Yungas* (montane and cloud forests of the Andean-Amazon slope in Peru and Bolivia) covers ~36% of the vertebrate endemic area, yet only makes up 1.7% of the study area. This ecosystem together with the three other Yungas forest types, (lower mountain pluvial forest, montane humid pluviseasonal, upper montane pluvial forests) cover an overwhelming proportion, (77%) of the vertebrate endemic areas but themselves cover just 7% of the study area. Ecological systems that occurred in highly irreplaceable areas, were more evenly distributed in terms of system type (Table 3 second column); western Amazon sub-Andean evergreen forest covered the highest percentage (12.6%) of highly irreplaceably areas followed by the Yungas lower mountain pluvial forest (9.3%).

Table 3: Ecological systems that overlap vertebrate endemic areas and irreplaceable areas

Ecological system	Percent of endemic area covered by system	Percent of irreplaceable areas covered by system	Percent of study area covered by system	Percent of system range that is protected
Montane pluvial forest of the Yungas	35.7	1.3	1.7	34
Lower montane pluvial forest and palm grove of the Yungas	16.1	9.3	3.5	41
Montane humid pluviseasonal forest of the Yungas	14.9	6.3	1.1	13
Upper montane pluvial forest of the Yungas	10.4	1.4	0.6	22
Upper montane pluviseasonal forest of the Yungas	5.2	2.6	0.6	9
Converted lands	4.6	14.7	6.0	4
Low montane subhumid pluviseasonal forest of the southern Yungas	3.3	0.8	0.6	16
Lower montane humid pluviseasonal forest of the Yungas	3.3	2.3	0.8	18
Southwestern Amazon subandean evergreen forest	2.8	5.2	5.9	42
High-Andean and upper montane pluvial grassland and shrubland of the Yungas	2.3	3.4	0.4	30
Western Amazon subandean evergreen forest	0.0	12.6	5.0	38
Southwestern Amazon piedmont forest	0.0	7.4	2.6	39
Southwestern Amazon subandean evergreen seasonal forest	0.0	5.6	1.7	48
Western Amazon semideciduous azonal forest	0.0	5.4	1.0	1
Lower montane humid pluviseasonal forest of the Yungas	0.0	2.3	0.8	18
Lower montane pluvial forest of the Condor Mountain Range	0.0	2.0	0.2	42

Systems shown cover at least 2% of vertebrate endemic (2,7676 km^2) or irreplaceable areas (150,500 km^2); ordered by coverage of endemic areas.

Gap Analysis

National protected areas cover approximately 12% of the study area, resulting in variable levels of protection for endemic species and their ecosystems. Of the endemic species examined, 327 (42%) have less than 10% of their distributions within protected areas (Table 4 see Additional File 3). About a third of all endemic species (226) occur completely outside of protected areas. As for discrete areas of endemism, amphibian areas receive the greatest protection: 67% of the area occurs within existing protected areas. Protection for the other endemic areas was lower (birds, 7%; mammals, 29%; plants, 24%). Only 20% of the combined endemic centres occurred within national-level protected areas (Figure 7). Fewer than 20% of all combined irreplaceable areas are under national protection, with protection varying by species groups (birds, 17%; mammals, 18%; amphibians, 17%; plants, 15%) (Figures 6a-d). Five of the seventeen ecological systems that cover the areas of endemism (Table 3) have less than 5% of their extents protected across the study area. About half of the 91 ecological systems have 10% or less of their extents covered by protected areas, with 26 of these systems having less than 2% under legal protection (Table 5; Figure 9; see Additional File 3).

Table 4: Coverage of endemic species ranges by national-level protected areas; IUCN I - VI (IUCN, 1994)

Percent range in IUCN I-VI protected area	Birds	Mammals	Amphibians	Plants
> 75	3	3	21	15
51 to 75	5	2	11	23
26 to 50	40	23	33	97
10 to 25	44	13	29	92
< 10	23	14	83	207
No protection	5	5	72	144
Total number of species	115	55	177	435

Table 5: Terrestrial ecological systems having less than 2% protection in the study area

Ecological system	Area (ha)	Percent of study Area	Area protected (ha)	Percent protected
Complex of non-alkaline savannas of the Beni transitional to the Cerrado	2,221,743	1.8	459	0
Cerrado complex of the northern Beni	1,766,905	1.4	0	0
Western Amazon semideciduous azonal forest	1,276,552	1.0	14,533	1
Complex of non-alkaline savannas of the Beni	585,143	0.5	0	0
Central-south Amazon Palm dominated forest	578,331	0.5	0	0
Chiquitania and Beni seasonally flooded herbaceous oligotrophic savanna	506,966	0.4	0	0
Beni seasonally flooded palm grove and savanna of the alkaline flatlands	226,672	0.2	6	0
Chiquitania and Beni "Cerradão"	214,452	0.2	0	< 1
Beni seasonally flooded herbaceous mesotrophic savanna	208,539	0.2	217	< 1
Montane interandean xeric forest and shrubland of the Yungas	205,749	0.2	40	< 1
Interandean xeric scrub of the Yungas	152,396	0.1	0	0
Beni and Chiquitania open hydrophytic savanna	145,912	0.1	0	0
Lower montane xeric forest and shrubland of the northern Yungas	137,919	0.1	101	< 1
Beni mixed-water riparian vegetation and forests complex	120,637	0.1	0	0
Northern Yungas dry submontane complex	95,189	0.1	0	0
Cerrado hydrophytic savannah with termite mounds	63,280	0.1	0	0
Chiquitania and Beni semideciduous subhumid forest	48,789	< 0.1	0	0
Beni clear and dark-water riparian forests and vegetation complex	35,684	< 0.1	0	0
Central-south Amazon ridges lithomorphic scrub	21,028	< 0.1	0	0

Northern Yungas dry montane and sub-montane complex	19,602	< 0.1	0	0
Yungas ridge pluviseasonal forest	16,994	< 0.1	325	1
Montane lithomorphic vegetation of the Yungas	10,296	< 0.1	0	0
Western Beni seasonally flooded thorn forest of the alkaline flatlands	10,009	< 0.1	0	0
Upper montane pluvial *Polylepis* forest of the Yungas	8423	< 0.1	73	1

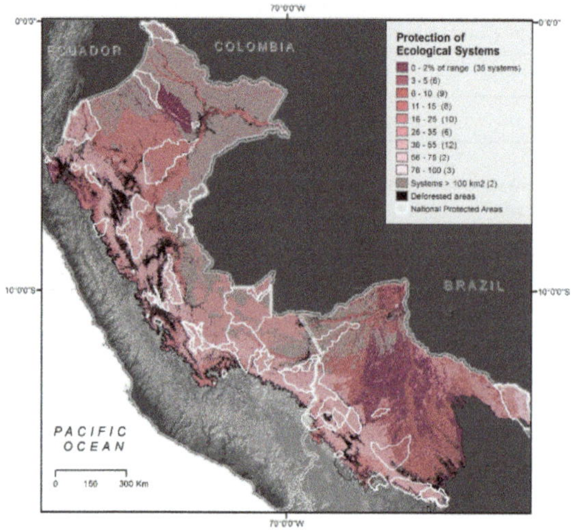

Figure 9: Ecological systems protection. Percentage of each system's range protected in study area.

Several areas of endemism and irreplaceability without current national-level protected status are worth highlighting (Figure 7). In northern Peru, areas near the cities of Iquitos and Tarapoto host unique concentrations of endemic plants. The Tarapoto region also has a large irreplaceable area for amphibians. The Carpish Hills in the Department of Huanuco host many endemic plants (Acanthaceae, Aquifoliaceae and*Fuchsia*) and are highly irreplaceable for endemic birds (up to 32 ranges overlap) but are completely unprotected. The Cordillera de Vilcabamba is a major area of endemism for birds, mammals and plants (*Fuchsia*). It also constitutes the largest cohesive irreplaceable area for birds and mammals in the study area, and is highly irreplaceable for some plants. Currently the Cordillera of Vilcabamba has only one protected area, the Machu Picchu Historical Sanctuary, which covers just 326 km^2, and is highly impacted by tourism. The north-eastern corner of the Department of Puno has numerous endemic birds and mammals and is also unprotected. However, many of the ranges of these species extend into Bolivia where they are protected in Madidi National Park.

In Bolivia, the cordilleras near La Paz have high levels of bird, mammal and plant endemism (8 of the 13 plant groups analysed), and scored as highly irreplaceable for endemic mammals and plants. Most of these cordilleras are not protected, although a small area that is irreplaceable for amphibians coincides with the 608-km^2 Cotapata National Park (Figure 5). In central Bolivia, unprotected endemic areas for birds, mammals, and amphibians occur in the Cordillera de Cocapata-Tiraque and Cochabamba Department, between protected areas.

DISCUSSION

Our results, at a conservation practitioner's scale, identify geographic areas in the eastern slopes of the Peruvian and Bolivian Andes with high concentrations of endemic species, areas with high irreplaceability, gaps in protection for both species and ecosystems, and ecological systems where these endemic species reside. Our focus on a variety of vertebrate and plant groups underlines the variation in spatial distribution patterns among different taxa. The geographical extents and levels of current protection of the ranges of species, endemic areas, irreplaceable areas, and key ecological systems also vary widely.

Mapping species distributions is inherently limited in terms of a true representation of biodiversity. As a one dimensional map of potential habitat based on climate, elevation and vegetation, the distribution modelling omits species interactions such as predation and competition, effect of human edges along habitat, and the effects of climate change [63, 81]. However it is a large step forward for this region where current conservation analyses are obliged to rely upon generalized hand-drawn maps of species ranges, or species lists for very large multi-country geographical units (e.g. Hotspots or Ecoregions) that were not intended nor appropriate for regional or landscape level applications [11]. Our mapping of ecological systems, for example, identified ~90 ecological systems; the same area is covered by parts of 12 ecoregions (*sensu*[82]).

The locations of high endemism (Figure 4) agree with past studies for taxa that have been examined previously, yet earlier studies were conducted with much less data availability and at much coarser spatial resolution. The high levels of endemic bird richness found in the northern part of the study area are consistent with previous work [36, 40, 83]. However, our study revealed previously unrecognized areas of bird endemism in Peru: the southern Huánuco region, the western Cordillera de Vilcabamba, and the region along the Río Mapacho-Yavero east of Cuzco (Figure 4, 7; see [84] for details). This study is the first to reveal detailed patterns of endemic species for mammals and amphibians (see [85] for location descriptions), and therefore few comparisons with past studies can be made. However the areas of high endemic mammal richness in Peru corroborate the one regional study of similar scope [42] and the mid-elevation concentration of endemic amphibians coincides with the less spatially explicit suggestions of [44] and [45]. Centres of plant endemism varied among groups and families, yet the pattern for one group (*Ericaceae*) did correspond to a previous study [49]. Other existing analyses use such coarse resolution (e.g., the 1°×1° Flora Neotropica grid [47]) that comparisons are too general to be meaningful. For most plant groups, this study is the first to assess spatial patterns of endemism in the eastern Andean basin of Peru and Bolivia.

Despite the increased level of detail in spatial scale that our dataset provides, continued work needs to focus on refining these biodiversity data to even finer spatial scales (e.g. 1:100,000) and higher levels of accuracy. The dataset and analyses we have produced are tied to the time of specimen collections and to the quality of available data. As more specimen locations are collected in the future with increasingly accurate locational and elevational information (using a precise global positioning system), distribution models could be re-run and models validated. Geographical collection bias, a problem for presence-only distribution models could be addressed in future modelling efforts by the selection of pseudo-absence data having similar bias as the presence data [86]. More precise geographical climate data could refine the spatial resolution of model predictions; there will be an increasing prevalence of 'downscaled' geographical climate data thanks to higher spatial resolution digital elevation models (SRTM and ASTER). However the overall limitation is the lack of adequate meteorological stations in the region. Other layers that would be useful to incorporate upon their refinement would be a characterization of soils or geology. We successfully modelled all endemic vertebrates yet, additional models of plant species distributions should be realized. Considering there are over 5000 endemic plant species in the country of Peru (of which approximately 3200 fall within the altitudinal range of our study area) [87], our 435 species represents a small fraction of endemics to Peru and/or Bolivia in the Amazon watershed.

Our country wide analysis could be refined to department scale using land tenure information and local to regional protected areas and resource concessions. Current maps of forest deforestation and degradation would aid in calculating the remnant ranges for each species as well as ecological systems. Further analysis could be made in terms of the complementarity of species assemblages and their relationship to ecological systems and levels of protection, whose results could further guide priorities. However, the greater battle for biodiversity conservation lies in managing elements beyond our datasets and analyses, as described below.

The geographical patterns of endemism, irreplaceability, and ecosystems revealed here pose several challenges for conservation planning in the region (Figure 7). The most obvious challenge is the geographic configuration of the locations of endemic or irreplaceable areas. Although we mapped only a small subset of the biodiversity that occurs in the region, we found striking geographic differences in endemic species concentrations across taxonomic groups. The difficulty of using surrogates of one species group for another has been recognized [15, 52, 53], and our findings underscore the need for a large portfolio of protected areas and other protection mechanisms to conserve diverse elements of biodiversity.

Second, the gap analysis demonstrates that many areas where concentrations of endemic species occur remain unprotected today. Considering ongoing threats in the region from infrastructure development [88], oil extraction [89], gold mining [90, 91], illicit crops [36], and the continually advancing agricultural fronts, more carefully situated protected areas and novel land use regulation strategies will be necessary to safeguard substantial amounts of biodiversity.

Third, although we use protected area coverage to evaluate conservation coverage, we acknowledge that protection status does not necessary translate into actual protection on the ground. Indeed, resource extraction and degradation is continuing in many legally protected lands in the study area [92]. Nevertheless, these reserves have the potential to protect important segments of endemic and irreplaceable areas, suggesting that strengthening the capacity of relevant authorities to improve protection is an important and continuing challenge.

Fourth, large reserves will probably be insufficient to maintain all biodiversity. Although large reserves often provide the best means for maintaining well-functioning ecosystems [93], the pattern of endemism we document, in which microendemic species are scattered across the landscape and not always concentrated geographically, will require multi-pronged conservation efforts. Restricted-range species that occur far from the major areas of endemism or irreplaceability, such as the two primates in the Bolivian Beni, would benefit from a wider network of smaller reserves, perhaps established by departmental, provincial, or municipal governments or private entities. Current trends toward the decentralization of responsibility for natural resource management to provincial governments may provide a useful institutional context for the establishment of some of these smaller, but nonetheless critical reserves [94].

Our finding that highly endemic areas disproportionally occupy a handful ecological systems presents yet a fifth challenge. Ecological systems characterize broad, integrated units of biodiversity and can be used as a coarse filter for conservation. While maintaining representation of all systems in landscape-level protection plans [95], planners may need to balance the need to protect endemic species with the need for a representative sample of ecosystem type and function as well as other targets such as endangered species or carbon sequestration. On the other hand, these particular ecological systems could be considered surrogates for areas of high endemism. The systems are advantageously close together in the Yungas region, are relatively limited in extent (totalling 7% of the study area), and have individual ranges that are < 35% protected.

A final challenge is continued climate change. We know that because of climate change, the ranges of many species will shift across the landscape and possibly out of protected areas [96, 97]. Evidence is accumulating that along the Andean slope, species shifts are already occurring [98, 99]. Yet the variation in projections of future South American climate makes assessment of the effects on species' distributions difficult [100]. The steep elevation (and therefore climate) gradients in the Andes, where most endemic species are located, suggest that such displacements may take place over relatively small distances. Extinctions are most likely in species inhabiting the highest-elevation habitats, which occur above our study area [100]. Nevertheless, planners should consider adding upslope buffers to conservation areas designated using current distributions of endemic species, and future research could model these species distributions under future climate scenarios.

To complement the further creation and effective management of protected areas, other alternative approaches, which will result in the maintenance of key ecosystems, should expand and continue. These approaches include, strategic conservation on private lands and brokering conservation agreements with private companies, effective land use planning and possibly carbon accounting at the regional government level for both public and private lands, and payments for ecosystem services (e.g. water provision, ecotourism recreation, carbon storage through forests: Reducing Emissions from Deforestation and forest Degradation, REDD). However priority areas for ecosystem services concessions may not necessarily overlap with priorities for biodiversity conservation (e.g.[101]).

CONCLUSIONS

We believe these spatial datasets provide a substantive base upon which to make decisions and move forward for further protection. The approach to developing these datasets described here, relying on existing environmental data sources, data in natural history collections, and in-country expertise to identify endemic species distributions, concentrations and gaps in protection across national borders is applicable to many regions of the world where survey efforts are incomplete. Our results demonstrate that even under these conditions, conservationists can develop spatial datasets for multiple taxonomic groups at a scale useful to guide planning.

ACKNOWLEDGEMENTS

We are deeply indebted to the Gordon and Betty Moore Foundation for financial support and technical guidance during the project. The project would not be possible without the generous data sharing, opinions, advice and participation of our long list of collaborators, found athttp://www.natureserve.org/aboutUs/latinamerica/col_institutions.jsp. Funds from NatureServe, Duke University's Nicholas School of Environment, and USAID's Emerging Pandemic Threats PREDICT Program supported the publication of this manuscript. The datasets for most of the maps and analyses described here are accessible in both graphic and GIS format at http://www.natureserve.org/andesamazon.

We greatly appreciate the critiques of the anonymous reviewers and Nigel Pitman that helped shape and improve this manuscript.

REFERENCES

1. Myers N, Mittermeier RA, Mittermeie CG, da Fonseca GAB, Kent J: Biodiversity hotspots for conservation priorities. Nature. 2000, 403: 853-858. 10.1038/35002501.
2. Rodrigues ASL, Akcakaya HR, Andelman SJ, et al: Global gap analysis: Priority regions for expanding the global protected-area network. BioScience. 2004, 54: 1092-1100. 10.1641/0006-3568(2004)054[1092:GGAPRF]2.0.CO;2.
3. Orme CDL, Davies RG, Burgess M, et al: Global hotspots of species richness are not congruent with endemism or threat. Nature. 2005, 436: 1016-1019. 10.1038/nature03850.
4. Brooks TM, Mittermeier RA, da Fonseca GAB, Gerlach J, Hoffmann M, Lamoreux JF, Mittermeier CG, Pilgrim JD, Rodrigues ASL: Global biodiversity conservation priorities. Science. 2006, 31: 58-61.
5. Ceballos G, Ehrlich PR: Global mammal distributions, biodiversity hotspots, and conservation. Proceedings of the National Academy of Sciences. 2006, 103: 19374-19379. 10.1073/pnas.0609334103.
6. Olson DM, Dinerstein E: The global 200: priority ecoregions for global conservation. Annals of the Missouri Botanical Garden. 2002, 89: 199-224. 10.2307/3298564.
7. IUCN: Guidelines for protected areas management categories. Cambridge, UK and Gland, Switzerland: International Union for Conservation of Nature. 1994
8. Patterson BD, Ceballos G, Sechrest W, Tognelli MF, Brooks T, Luna L, Ortega P, Salazar I, Young BE: Digital distribution maps of the mammals of the Western Hemisphere. Edited by: NatureServe. 2007, Arlington, V.A.: NatureServe, 2.0
9. Ridgely RS, Allnutt TF, Brooks T, McNicol DK, Mehlman DW, Young BE, Zook JR: Digital distribution maps of the birds of the Western Hemisphere. Edited by: NatureServe. 2007, Arlington, VA, 2.1
10. Schipper J, Chanson JS, False Chiozza: The status of the world's land and marine mammals: diversity, threat, and knowledge. Science. 2008, 322: 225-230. 10.1126/science.1165115.
11. Londoño-Murcia MC, Tellez-Valdés O, Sánchez-Cordero V: Environmental heterogeneity of World Wildlife Fund for Nature ecoregions and implications for conservation in Neotropical biodiversity hotspots. Environmental Conservation. 2010, 37 (2): 116-127. 10.1017/S0376892910000391.
12. Freitag S, Nicholls AO, Van Jaarsveld AS: Nature reserve selection in the Transvaal, South Africa: what data should we be using?. Biodiversity and Conservation. 1996, 5: 685-698. 10.1007/BF00051781.
13. Hurlbert AH, White EP: Disparity between range map- and survey-based analyses of species richness: patterns, processes and implications. Ecology Letters. 2005, 8: 319-327. 10.1111/j.1461-0248.2005.00726.x.
14. Ferrier S: Mapping spatial pattern in biodiversity for regional conservation planning: where to from

here?. Systematic Biology. 2002, 51: 331-363. 10.1080/10635150252899806.

15. Leroux SJ, Schmiegelow FKA: Biodiversity concordance and the importance of endemism. Conservation Biology. 2007, 21: 266-268. 10.1111/j.1523-1739.2006.00628.x.

16. Balmford A, Mace GM, Ginsberg JR: The challenges to conservation in a changing world: putting processes on the map. Conservation in a changing world. Edited by: G.M Mace AB, and J.R. Ginsberg. 1998, Cambridge: Cambridge Univ Press, 1-28.

17. van der Werff H, Consiglio T: Distribution and conservation significance of endemic species of flowering plants in Peru. Biodiversity and Conservation. 2004, 13: 1699-1713.

18. Roy MS, Silva JMCD, Arctander P, Garcia-Moreno J, Fjeldså J: The speciation of South American and African birds in montane regions. Avian Molecular Evolution and Systematics. Edited by: Mindell DP. 1997, New York: Academic Press, 325-343.

19. Fjeldså J: Geographical patterns for relict and young species of birds in Africa and South America and implications for conservation priorities. Biodiversity and Conservation. 1994, 3: 207-226. 10.1007/BF00055939.

20. Roy MS, Torres-Mura JC, Hertel F: Molecular phylogeny and evolutionary history of the tit-tyrants (Aves: Tyrannidae). Molecular Phylogenetic Evolution. 1999, 11: 67-76. 10.1006/mpev.1998.0563.

21. García-Moreno J, Fjeldså J: Chronology and mode of speciation in the Andean avifauna. Bonner Zoological Monographs. 2000, 46: 25-46.

22. Dingle C, Lovette IJ, Canaday C, Smith T: Elevational zonation and the phylogenetic relationships of the Henicorhina wood-wrens. Auk. 2006, 123: 119-134. 10.1642/0004-8038(2006)123[0119:EZATPR]2.0.CO;2.

23. Patton JL, Smith MF: MtDNA phylogeny of Andean mice: a test of diversification across ecological gradients. Evolution. 1992, 46: 174-183. 10.2307/2409812.

24. Young KR: Biogeographical paradigms useful for the study of tropical montane forests and their biota. Biodiversity and conservation of neotropical montane forests. Edited by: Churchill; SP, Balslev; H, Forero; E, Luteyn JL. New York. 1995, The New York Botanical Garden

25. Young KR, Ulloa C, Luteyn JL, Knapp S: Plant evolution and endemism in Andean South America: an introduction. Botanical Review. 2002, 68: 4-21. 10.1663/0006-8101(2002)068[0004:PEAEIA]2.0.CO;2.

26. Hughes C, Eastwood R: Island radiation on a continental scale: Exceptional rates of plant diversification after uplift of the Andes. Proceedings of the National Academy of Sciences. 2006, 103: 10334-10339. 10.1073/pnas.0601928103.

27. Donoghue MJ: A phylogenetic perspective on the distribution of plant diversity. Proceedings of the National Academy of Sciences. 2008, 105: 11549-11555. 10.1073/pnas.0801962105.

28. Lynch JD, Duellman WE: The Eleutherodactylus of the Amazonian slopes of the Ecuadorian Andes (Anura: Leptdactylidae). Miscellaneous Publication of the University of Kansas Natural History Museum. 1980, 69: 1-86.

29. Graham CH, Ron SR, Santos JC, Schneider CJ, Moritz C: Integrating phylogenetics and environmental niche models to explore speciation mechanisms in Dendrobatid frogs. Evolution. 2004, 58: 1781-1793.

30. Lynch JD: Origins of the high Andean herpetological fauna. Págs. High Altitude Tropical Biogeography. Edited by: Vuilleumier F, Monasterio M. 1986, Oxford: Oxford University Press, 478-499.

31. Santos JC, Coloma LA, Summers K, Caldwell JP, Ree R, Cannatella DC: Amazonian amphibian diversity is primarily derived from late Miocene Andean lineages. PLOS Biol. 2009, 7: 0448-0461.

32. Killeen TJ, Douglas M, Consiglio T, Jørgensen PM, Mejia J: Dry spots and wet spots in the Andean hotspot. Journal of Biogeography. 2007, 34 (8): 1357-1373. 10.1111/j.1365-2699.2006.01682.x.

33. Hoorn C, Wesselingh FP, ter Steege H, Bermudez MA, Mora A, et al: Amazonia through time: Andean uplift, climate change, landscape evolution, and biodiversity. Science. 2010, 330: 927-931. 10.1126/science.1194585.

34. Bush M, Lovejoy T: Amazonian conservation: pushing the limits of biogeographical knowledge. Journal of Biogeography. 2007, 34: 1291-1293. 10.1111/j.1365-2699.2007.01758.x.

35. Rodriguez LO, Young KR: Biological diversity of Peru: determining priority areas for conservation. Ambio. 2000, 29: 329-337.

36. Fjeldså J, Alvarez MD, Lazcano JM, Leon B: Illicit crops and armed conflict as constraints on biodiversity conservation in the Andes region. Ambio. 2005, 34: 205-211.

37. Fjeldså J, Lambin E, Mertens B: Correlation between endemism and local ecoclimatic stability documented by comparing Andean bird distributions and remotely sensed land surface data. Ecography. 1999, 22: 63-78. 10.1111/j.1600-0587.1999.tb00455.x.

38. Hellmayr CE: Ueber neue und selteneVögel aus Südperu. Verh Ornithol Ges Bayern. 1912, 11: 159-163.

39. Müller P: The dispersal centers of terrestrial vertebrates in the Neotropical realm. Biogeographica. 1973, 2: 1-244.

40. Stattersfield AJ, Crosby MJ, Long AJ, Wege DC: Endemic bird areas of the world. 1998, Cambridge, UK.: BirdLife International

41. Pacheco V, Quintana HL, Hernandez PA, Paniagua L, Vargas J, Young BE: Mammals. Endemic species distributions on the east slope of the Andes in Peru and Bolivia. Edited by: Young BE. Arlington, Virginia: NatureServe. 2007, 40-45.

42. Pacheco V: Mamíferos del Perú. Diversidad y conservación de los mamíferos neotropicales. Edited by: Ceballos; G, Simonetti J. Mexico City, Mexico. 2002, CONABIO-UNAM, 586-

43. Doan TM, Arizábal W: Microgeographical variation in species composition of the herpetofaunal communities

of Tambopata region, Peru. Biotropica. 2002, 34: 101-117.

44. Duellman WE: Distribution patterns of amphibians in South America. Patterns of distribution of amphibians. Edited by: Duellman WE. Baltimore, MD. 1999, Johns Hopkins Univ Press, 255-328.

45. Reichle S: Distribution and conservation status of Bolivian Amphibians. 2007, Germany: Rheinische Friedrich Wilhelms Universitaet

46. Kessler M: The elevational gradient of Andean plant endemism: varying influences of taxon-specific traits and topography at different taxonomic levels. Journal Biogeography. 2002, 29: 1159-1166. 10.1046/j.1365-2699.2002.00773.x.

47. Knapp S: Assessing patterns of plant endemism in Neotropical uplands. Botanical Review. 2002, 68: 22-37. 10.1663/0006-8101(2002)068[0022:APOPEI]2.0.CO;2.

48. León B, Young KR: Distribution of pteridophyte diversity and endemism in Peru. Pteridology in perspective. Edited by: Camus; JM, Gibby; M, Johns RJ. Kew, U.K. 1996, Royal Botanic Garden, 77-91.

49. Luteyn JL: Diversity, adaptation, and endemism in Neotropical Ericaceae: biogeographical patterns in the Vaccinieae. Botanical Review. 2002, 68: 55-87. 10.1663/0006-8101(2002)068[0055:DAAEIN]2.0.CO;2.

50. Guisan A, Thuiller W: Predicting species distribution: offering more than simple habitat models. Ecology Letters. 2005, 8: 993-1009. 10.1111/j.1461-0248.2005.00792.x.

51. Guisan A, Zimmermann NE: Predictive habitat distribution models in ecology. Ecological Modelling. 2000, 135: 147-186. 10.1016/S0304-3800(00)00354-9.

52. Kremen C, Cameron A, Moilanen A, Phillips S, Beentje H, Dransfeld J, Fisher BL, Glaw F, Hijmans R, Lees D, et al: Aligning conservation priorities across taxa in Madagascar with high-resolution planning tools. Science. 2008, 320: 222-226. 10.1126/science.1155193.

53. Prendergast JR, Quinn RM, Lawton JH, Eversham BC, Gibbons DW: Rare species, the coincidence of diversity hotspots and conservation strategies. Nature. 1993, 365: 335-337. 10.1038/365335a0.

54. Ferrier S, Pressey RL, Barrett TW: A new predictor of the irreplaceability of areas for achieving a conservation goal, its application to real-world planning, and a research agenda for further refinement. Biological Conservation. 2000, 93: 303-325. 10.1016/S0006-3207(99)00149-4.

55. IUCN, Conservation International, NatureServe: Global amphibian assessment. 2006, IUCN (International Union for Conservation of Nature, Conservation International, NatureServe, Version 1.1

56. Stephens L, Traylor ML: Ornithological gazetteer of Peru. Museum of Comparative Zoology. Edited by: University H. Boston, MA. 1983

57. Thomas O: New mammals from Peru and Bolivia, with a list of those recorded from the Inambari River, upper Madre de Dios. Ann Mag Nat Hist. 1901, 7 (5): 148-153.

58. MaNIS/HerpNet/ORNIS: Mammal Networked Information System: Georeferencing Guidelines. 2001

59. Hijmans RJ, Cameron SE, Parra JL, Jones PG, Jarvis A: Very high resolution interpolated climate surfaces for global land areas. International Journal of Climatology. 2005, 25: 1965-1978. 10.1002/joc.1276.

60. Farr TG: The Shuttle Radar Topography Mission. Rev Geophys. 2007, 45 (RG2004):

61. Hansen M, DeFries R, Townshend JR, Carroll M, Dimiceli C, Sohlberg R: Vegetation Continuous Fields, MOD44B, 2001 Percent Tree Cover, Collection 3. Edited by: Univ Maryland CP. College Park, MD 2003.

62. Graham CH, Ferrier S, Huettman F, Moritz C, Peterson AT: New developments in museum-based informatics and applications in biodiversity analysis. Trends in Ecology and Evolution. 2004, 19: 497-503. 10.1016/j.tree.2004.07.006.

63. Loiselle BA, Howell CA, Graham CH, Goerck JM, Brooks T, Smith KG, Williams PH: Avoiding pitfalls of using species distribution models in conservation planning. Conservation Biology. 2003, 17: 1591-1600. 10.1111/j.1523-1739.2003.00233.x.

64. Nelson BW, Ferreira C, da Silva M, Kawasaki ML: Endemism centres, refugia and botanical collection density in Brazilian Amazonia. Nature. 1990, 345: 714-716. 10.1038/345714a0.

65. Elith J, Graham CH, Anderson RP, et al: Novel methods improve prediction of species' distributions from occurrence data. Ecography. 2006, 29: 129-151. 10.1111/j.2006.0906-7590.04596.x.

66. Phillips SJ, Anderson RP, Schapire RE: Maximum entropy modeling of species geographic distributions. Ecological Modelling. 2006, 190: 231-259. 10.1016/j.ecolmodel.2005.03.026.

67. Hernandez PA, Franke I, Herzog SK, Pacheco V, Paniagua L, Quintana HL, Soto HA, Swensen JJ, Tovar C, Valqui TH, et al: Predicting species distributions in poorly-studied landscapes. Biodiversity and Conservation. 2008, 17: 1353-1366. 10.1007/s10531-007-9314-z.

68. Hernandez PA, Graham CH, Master LL, Albert DL: The effect of sample size and species characteristics on performance of different species distribution modeling methods. Ecography. 2006, 29: 773-785. 10.1111/j.0906-7590.2006.04700.x.

69. Wisz MS, Hijmans RJ, Li J, Peterson AT, Graham CH, Guisan A: Predicting Species Distributions Working Group N: Effects of sample size on the performance of species distribution models. Diversity and Distributions. 2008, 14: 763-773. 10.1111/j.1472-4642.2008.00482.x.

70. Loiselle BA, Jørgensen PM, Consiglio T, Jiménez I, Blake JG, Lohmann LG, Montiel OM: Predicting species distributions from herbarium collections: does climate bias in collection sampling influence model outcomes?. Journal of Biogeography. 2008, 35: 105-116.

71. Pressey RL, Watts M, Ridges M, Barrett T: C-Plan conservation planning software. User Manual. 2005, New South Wales, Australia: New South Wales Department of Environment and Conservation

72. Josse C, Navarro G, Comer P, Evans R, Faber-

Langendoen D, Fellows M, Kittel G, Menard S, Pyne M, Reid M, et al: Ecological systems of Latin America and the Caribbean: a working classification of terrestrial systems. 2003, Arlington, VA: NatureServe

73. Comer P, Faber-Langendoen D, Evans R, et al: Ecological systems of the United States: a working classification of U.S. terrestrial systems. 2003, Arlington, VA: NatureServe

74. Comer P, Schulz K: Standardized ecological classification for meso-scale mapping in southwest United States. Rangeland Ecology Management. 2007, 60: 324-335. 10.2111/1551-5028(2007)60[324:SECFMM]2.0.CO;2.

75. Sayre R, Bow J, Josse C, Sotomayor L, Touval J: Terrestrial ecosystems of South America. North America land cover summit: a special issue of the Association of American Geographers. Edited by: Campbell JC, Jones KB, Smith JH. 2008, Washington, DC, 131-152.

76. Lowry J, Ramsey RD, Thomas K, et al: Mapping moderate-scale land-cover over very large geographic areas within a collaborative framework: a case study of the Southwest Regional Gap Analysis Project (SWReGAP). Remote Sensing of Environment. 2007, 108 (108): 59-73.

77. Scott MJ, Davis F, Cusuti B, Noss R, Butterfield B, Groves C, Anderson H, Caicco S, D'Erchia F, Edwards TC, et al: GAP analysis: a geographic approach to protection of biological diversity. Wildlife Monographs. 1993, 123: 1-41.

78. IUCN, UNEP-WCMC: The World Database on Protected Areas (WDPA). 2010, Cambridge, UK: UNEP-WCMC

79. Calle Josse: Ecological Systems of the Amazon Basin of Peru and Bolivia: Classification and Mapping. 2007, Arlington, Virginia: NatureServe

80. Ellenberg H: Vegetationsstufen in perhumiden bis perariden Bereichen der tropischen Anden. Phytocoenologia. 1975, 368-387. 2

81. Sieck M, Ibisch P, Moloney K, Jeltsch F: Current models broadly neglect specific needs of biodiversity conservation in protected areas under climate change. BMC Ecology. 2011, 11 (1): 12-10.1186/1472-6785-11-12.**View ArticlePubMedPubMed Central**

82. Olson DM, Dinerstein E, Wikramanayake ED, Burgess ND, Powell GVN, Underwood EC, D'Amico JA, Itoua I, Strand HE, Morrison JC, et al: Terrestrial Ecoregions of the World: A New Map of Life on Earth. Bioscience. 2001, 51 (11): 933-938. 10.1641/0006-3568(2001)051[0933:TEOTWA]2.0.CO;2.

83. Graves GR: Linearity of geographic range and its possible effect on the population structure of Andean birds. Auk. 1988, 105: 47-52.

84. Young BE, Franke I, Hernandez PA, Herzog SK, Paniagua L, Soto A, Tovar C, Valqui T: Using spatial models to predict areas of endemism and gaps in the protection of Andean slope birds. Auk. 2009, 126: 554-565. 10.1525/auk.2009.08155.

85. Endemic species distributions on the east slope of the Andes in Peru and Bolivia. Edited by: Young BE. 2007, Arlington, VA: NatureServe

86. Phillips SJ, Dudík M, Elith J, Graham CH, Lehmann A, Leathwick J, Ferrier S: Sample selection bias and presence-only distribution models: implications for background and pseudo-absence data. Ecological Applications. 2009, 19 (1): 181-197. 10.1890/07-2153.1.

87. Leon B, Pitman N, Roque J: Introducción a las plantas endémicas del Perú. Rev Peru Biol. 2006, 13 (2): 9-25.

88. Delgado C: Is the Interoceanic Highway exporting deforestation?. Master's thesis. 2008, Durham: Duke University

89. Finer M, Jenkins CN, Pimm SL, Keane B, Ross C: Oil and Gas Projects in the Western Amazon: Threats to Wilderness, Biodiversity, and Indigenous Peoples. PLoS ONE. 2008, 3 (8): e2932-10.1371/journal.pone.0002932.

90. Fraser B: Peruvian gold rush threatens health and the environment. Environmental Science and Technology. 2009, 43: 7162-7164. 10.1021/es902347z.

91. Swenson JJ, Carter CE, Delgado CI, Domec JC: Gold mining in the Peruvian Amazon: global prices, deforestation, and mercury imports. PLoS One. 2010, 6 (4): e18875-

92. Killeen TJ, Calderon V, Soriana L, Quezeda B, Steininger MK, Harper G, Solorzano LA, Tucker TJ: Thirty years of land-cover change in Bolivia: exponential growth and no end in sight. Ambio. 2007, 36: 600-606. 10.1579/0044-7447(2007)36[600:TYOLCI]2.0.CO;2.

93. Margules CR, Pressey RL: Systematic conservation planning. Nature. 2000, 405: 243-253. 10.1038/35012251.

94. Ferroukhi L: Municipal forest management in Latin America. 2003, Bogor, Indonesia.: CIFOR and IDRC

95. Groves CR, Jensen DB, Valutis LL, Redford KH, Shaffer ML, Scott JM, Baumgartner JV, Higgins JV, Beck MW, Anderson MG: Planning for biodiversity conservation: putting conservation science into practice. Bioscience. 2002, 52: 499-512. 10.1641/0006-3568(2002)052[0499:PFBCPC]2.0.CO;2.

96. Parmesan C: Ecological and evolutionary responses to recent climate change. Annual Review Ecology Systematics. 2006, 37: 637-669. 10.1146/annurev.ecolsys.37.091305.110100.

97. Williams P, Hannah L, Andelman S, Midgley G, Araújo M, Hughes G, Manne L, Martinez-Meyer E, Pearson R: Planning for climate change: identifying minimum dispersal corridors for the Cape Proteaceae. Conservation Biology. 2005, 19: 1063-1074. 10.1111/j.1523-1739.2005.00080.x.

98. Feeley KJ, Silman MR, Bush MB, Farfan W, Cabrera KG, Malhi Y, Meir P, Revilla NS, Quisiyupanqui MNR, Saatchi S: Upslope migration of Andean trees. Journal of Biogeography. 2011, 38 (4): 783-791. 10.1111/j.1365-2699.2010.02444.x.

99. Forero-Medina G, Terborgh J, Socolar SJ, Pimm SL: Elevational Ranges of Birds on a Tropical Montane Gradient Lag Behind Warming Temperatures. PLoSOne. 2011, 6 (12): e28535-

100. Magrin G, Gay García C, Cruz Choque D, Giménez JC, Moreno AR, Nagy GJ, Nobre C, Villamizar A: Latin

America. Climate change 2007: impacts, adaptation and vulnerability. Climate change 2007: impacts, adaptation and vulnerability. Edited by: Parry; ML, Canziani; OF, Palutikof; JP, Linden; PJvd, Hanson CE. Cambridge, U.K. 2007, Cambridge Univ Press, 581-615.

101. Chan KMA, Shaw MR, Cameron DR, Underwood EC, Daily GC: Conservation Planning for Ecosystem Services. PLOS Biol. 2006, 4 (11): e379.-

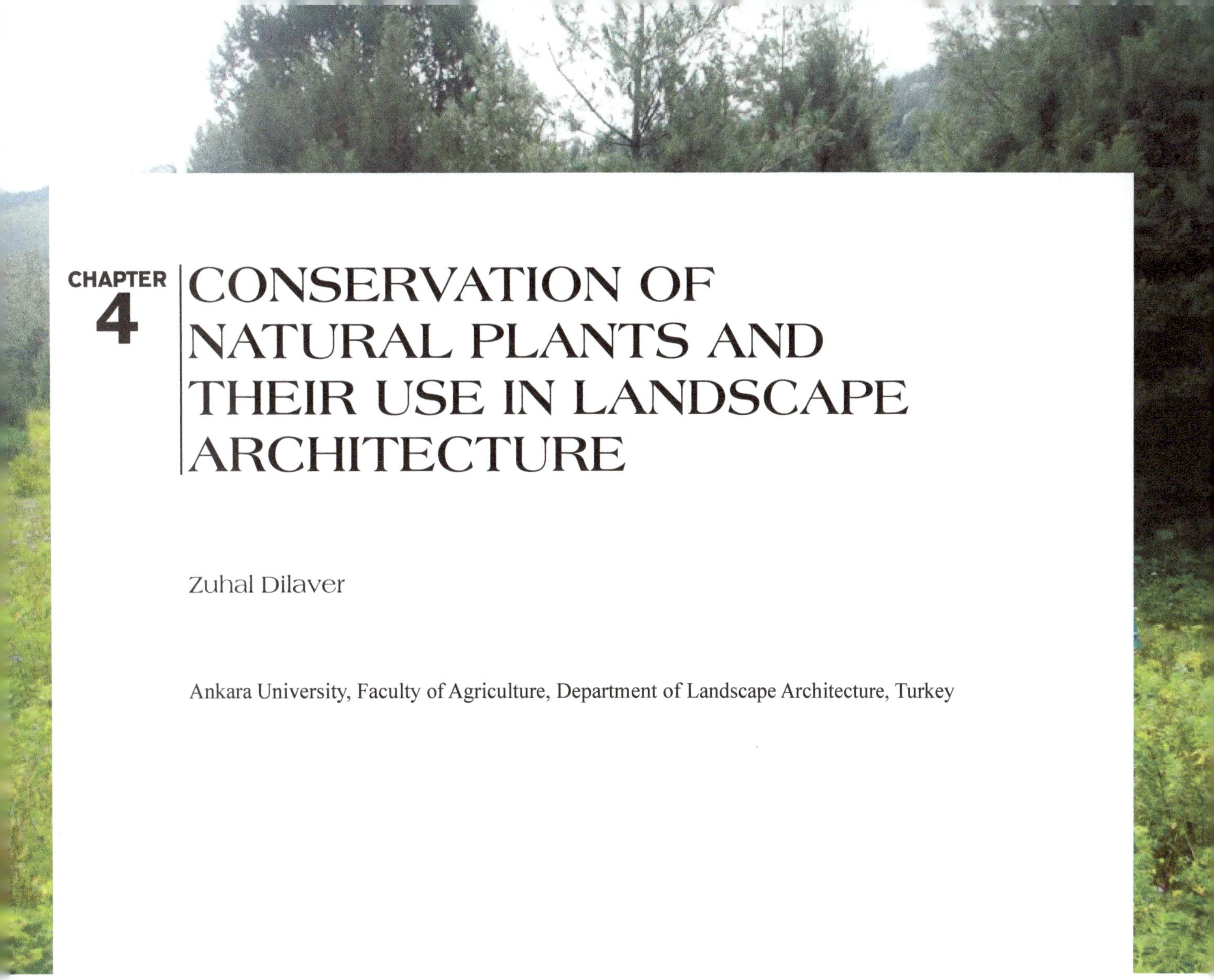

CHAPTER 4

CONSERVATION OF NATURAL PLANTS AND THEIR USE IN LANDSCAPE ARCHITECTURE

Zuhal Dilaver

Ankara University, Faculty of Agriculture, Department of Landscape Architecture, Turkey

INTRODUCTION

Biological diversity means the diversification of living organisms in all sources including the terrestrial, marine and other aquatic ecosystems and the ecological structures that are part of these ecosystems. It refers to the differences exhibited by the living environments (habitats or, more broadly, ecosystems) of species in terms of various biotic and abiotic factors. It also refers to the differences between living things in ecosystems and between living and non-living things, which vary according to place and time and to all of genes, species, ecosystems and functions [1].

Biological diversity can be divided into three hierarchical categories: genetic diversity, species diversity and ecosystem diversity.

Genetic Diversity means the diversity within a given species. It is measured by the genetic difference within a certain species, population, variety, sub-species or race.

Species Diversity means the diversity of the species in a particular region or in the entire world. "Species richness", i.e. the number of species in a region is the most commonly used measure in this regard.

Ecosystem Diversity is about the whole created by the community of organisms that interact with each other as an ecological unit and by their physical environment. The ecosystem consists of factors, such as fire, climate and the food cycle, which are separate from the community level and which are non-living themselves but still they affect the formation, structure and interaction of the community of living things. The conservation of biological diversity at the ecosystem level includes the conservation of the food chain and of the energy flow. At this level, it is necessary to protect not only species or groups of species but also characteristics and processes [1].

As human activity continues to spread throughout the earth, natural areas are changed, resulting in increasing extinctions of plants, animals, and other types of species. Currently, many experts believe such extinctions are occurring at the fastest rate in history and perhaps the fastest rate since the extinction of dinosaurs 65 million years ago. This loss of biological diversity is accelerating as desertification, deforestation (especially in the tropics), degradation of oceans and water resources, atmospheric change, and other environmental changes continue rapidly [2].

Biological diversity is important to human welfare for many reasons. Agricultural crops derive from wild species, and the high-yielding hybrids of modern agriculture depend on continuing revitalization from wild genetic stock. Furthermore, future crop species that could be used directly or modified by biotechnology are lost when entire ecosystems are wiped out. Plants are the basis of prescription drugs; a number of plants

Citation: Zuhal Dilaver, Conservation of Natural Plants and Their Use in Landscape Architecture, DOI: 10.5772/55767.

discovered in tropical rainforests or other wild areas have made significant contributions to treatment of serious diseases. Also, loss of species often means the breakdown of ecosystems that may have important roles in regulating rainfall, controlling floods, producing oxygen and storing carbon, affecting both regional and global climate [2].

At the World Summit on Sustainable Development held in 1992 in Rio de Janeiro, it was agreed that the decrease of biological diversity is a major problem and can only be prevented through international efforts. The Summit ended with the signing of important global conventions including the Convention on Biological Diversity, to which Turkey is a party by signing and ratifying it through the No. Law 4177 of 29 August 1996. The Convention entered into force in our country on 14 May 1997 [1].

The three main goals of the Convention on Biological Diversity (CBD) are:

- Conservation of biological diversity;
- Sustainable use of biological resources; and
- Fair and equitable sharing of the benefits arising from the utilization of genetic resources [1].

The Convention dictates that each country must identify its biological resources for which special protection measures are needed and which have a greater potential for sustainable use and also identify and monitor the categories and processes of actions that may have a negative effect on conservation and sustainable use [1].

BIOLOGICAL DIVERSITY IN TURKEY

Turkey displays the character of a small continent in terms of biological diversity. Among the reasons for this are the following facts:

- the country has three different types of bio climate and three Biogeographical Zones, namely Euro-Siberian, Mediterranean and Irano-Turanian,
- its topographic, geological, geomorphologic and soil diversity,
- the existence of different types of aquatic bodies such as the sea, lakes, rivers, and fresh water, salt water and mineral water lakes,
- the altitude differences that range between 0 and 5,000 meters,
- the country has deep canyons and very different types of ecosystems,
- it was less affected by the glacial period in comparison with European countries,
- the existence of the Anatolian Diagonal which links Northern Anatolia to Southern Anatolia resulting in ecological and floristic differences,
- the country is at the point where three continents intersect [1].

The Euro-Siberian Biogeographical Zone extends throughout Northern Anatolia and through those parts of the Thracian Region which face the Black Sea. This climatic region has the highest rainfall and is largely covered with forests. The Mediterranean Biogeographical Zone, on the other hand, covers all areas on the Mediterranean cost and the western parts of Thrace and includes very different types of ecosystems. The Irano-Turanian zone is the largest of the Biogeographical Zones, starting in Central Anatolia and extending as far as Mongolia. The continental climate and steppe flora are predominant in this zone [1].

ECOSYSTEM DIVERSITY

Turkey has agricultural, forest, mountain, steppe, wetland, coastal and marine ecosystems and different forms and combinations of these ecosystems [1].

Agricultural Ecosystems: In Turkey different agricultural ecosystems are seen in coastal regions, inner regions and transition regions. Cultivated areas, located mostly in steppe zones, constitute about 35 percent of Turkey's total surface area [1].

Steppe Ecosystems: There are about 21 million hectares of steppes and grasslands, defined as areas covered with herbaceous plants in Turkey. The steppe ecosystems are prevalent especially in Central Anatolia, in the high mountain floors of the Aegean and Mediterranean Regions and in a major part of Eastern Anatolia. The most characteristic feature of the steppe ecosystem is the predominance of annual or perennial herbaceous plants. The floristic composition of the steppe vegetation is very rich and includes many endemic plants. The steppe formation can be divided into two as "Plain Steppe" and "Mountain Steppe" in Turkey depending on the topographical structure of the area in which it lies. The plain steppe provides a habitat for species such as halophilic plants, members of the Chenopodiaceae, Juncaceae and Cyperaceae, *Peganum harmala* L., *Veroniva cymbalaria* Bodard, *Thymus L.* and *Salvia*L. The mountain steppe on the other hand, hosts the species of *Astragalus* L., *Onobrychis cornuta* (L.) Desv., *Acantholimon* Boiss., *Asphodelus* L. and *Thymus* L. In the Eastern Anatolia mountain steppe,*Ferula* L. becomes more predominant unlike in other regions. Alpine and sub-Alpine meadows cover large areas in the higher parts of the Eastern Black Sea Mountains and in the northern and northeastern parts of Eastern Anatolia [1].

Forest Ecosystems: In Turkey, forest ecosystems cover a total area of 21,188,747 hectares. Forests consisting of broad-leaved trees are more widespread in Turkey. Coniferous trees occur at all altitudes from sea level to the highest limit where forests exist. In the Aegean and Mediterranean regions, there are humid and semi-humid coniferous and dry forests (*Quercus* L, *Pinus nigra* Arnold and *Pinus brutia* Ten.) as well as shrubs and maquis. These rich forest ecosystems of Turkey provide habitats for a great number of endemic plant species, important bird species and other wildlife species. These ecosystems also include the wild relatives of many cultivated plants which are important for agricultural biological diversity [1].

Mountain Ecosystems: In Turkey, there are mountain systems formed by folding, faulting and volcanism. The types of the mountain ecosystems differ according to biogeographical regions, to patterns of formation and to altitudes [1].

Inland Waters Ecosystems: Turkey has very important inland water resources to maintain biological diversity with its rivers and lakes covering an area of about 10,000 km2. In studies conducted so far, 135 wetlands of international significance have been identified and 12 of them designated as Ramsar sites. In Turkey, there are 7 drainage basins including 26 river basins, and the ground waters are estimated at 94 billion m^3. The average annual rainfall is about 640 mm, roughly one third of which reaches water reserves and thus contributes to the maintenance of wetlands. In Turkey's wetlands, plants such as *Typha* L., *Phragmites* L, *Schoenoplectus* (REICHB.) PALLA and *Juncus* L. form large communities. In addition, there are also plants that cover the water surface such as *Nymphae* L. and underwater plants that grow in shallow lakes such as *Phodophyllum* L., *Wolffia* Horkel ex Schleid.,*Lemna* L. and *Ceratophyllum* L., *Myriophyllum* L. and *Potamogeton* L. [1].

Coastal and Marine Ecosystems: Different characteristics of the seas surrounding Turkey, namely the Black Sea, the Marmara, the Aegean and the Eastern Mediterranean, have resulted in the diversification of the biological resources they contain. The Mediterranean, which has the highest salinity and temperature among the Turkish seas, is the area with the richest biological diversity. Coastal ecosystems are highly special ecosystems as they are important sudden transition zones (ecotones) where marine and terrestrial ecosystems intersect. Coastal ecosystems form 4.1 % of the terrestrial resources that make up the country's surface area. The fact that the patterns in which mountains come down to the sea, and the coastal topography, differ from each other in the coastal areas of our country have resulted in various coastal ecosystems such as dunes, caves, deltas, lagoons, marshes and calcerous terraces. Among all these coasts, particularly the coastal areas in the Eastern Mediterranean region are rich ecosystems with very high flora and fauna diversity [1].

SPECIES DIVERSITY

Turkey has a relatively rich flora (about 12 000 species) and still a great number of species are being described (Figure 1,2,3,4). In this flora, there are a lot of interesting species such as halophytic species, semi-desert plants, carnivorous plants and nickel hyperacumulators. In Turkey, the rate of endemism is relatively high when compared with other European countries. It is about 34 percent in Turkey while it is 14.9 percent in Greece, 2.9 percent in France, 18.6 percent in Spain and 0.1 percent in Poland. The number of endemic species in Turkey is greater than 3000 [3].

Figure 1: *Iris schachtii* Markgraf (Z. Dilaver).

Figure 2: *Crocus danfordiae* Maw.(Z. Dilaver).

Figure 3: *Astragalus anthylloides* Lam. (Z. Dilaver).

Figure 4: *Scutellaria orientalis* L. (Z. Dilaver).

In terms of endemic plant species, the richest family of Turkey is Compositae, which has approximately 425 endemic species. The richest genus in terms of endemic species, on the other hand, is *Astragalus* L. with 250 species. It is followed by *Verbascum L.* with 185 species [4].

Among the phytogeographical regions, the Irano-Turanian is the region having the highest number of endemic. Mediterranean Region is the second and the Euro-Siberian Region is the third in this regard. Turkey›s richest region in terms of endemic plants unique to that region alone is the Mediterranean region with 800 or so species. In Eastern Anatolia 375, in Central Anatolia 275, in Black Sea 210 and in the Aegean Region 150 endemic species grow. Marmara (70) and Southeast Anatolia (35) Regions are the poorest geographical regions of Turkey in terms of endemics. The other endemics are distributed in more than one geographical area [4].

"Red List of Threatened Species" is the most well-known study of the World Conservation Union (IUCN), which was last published in 2004. This red list is accepted to be the most comprehensive source about the latest status of threatened plant and animal species. IUCN cited the danger categories as:

1. Extinct,
2. Extinct in the wild,
3. Critically endangered
4. Endangered
5. Vulnerable
6. Lower risk
7. Data deficient
8. Not evaluated [5].

In our country, after being completed 9 volume of Flora of Turkey as a main source in 1985, a list similar with this was started to be constituted with the help of Turkish Association for Conservation of Nature and Natural Resources. It was published in 1989 with the name of "List of Rare Threatened and Endemic Plants in Turkey". After this publication, more and more floristic studies -especially the project investigating endemic plants of Turkey in detail- showed that some data in the first list should be changed or at least could be updated. Thereupon "Red Data Book of Turkish Plants" which was prepared and published in 2000 became a national list using new threat categories of IUCN. This source is because it shows Turkey's biological diversity and the status of species in terms of the danger categories. In order to protect biological diversity it is necessary to know about it and to create its database. In this study 3504 endemic and 1096 rare plants were assessed and listed in accordance with appropriate danger categories. According to this list, 46.6 percent of 3504 endemic plants and 77.4 percent of 1096 rare plants is under threat. On the other hand, 13 plant taxon in total, of which 12 was endemic, and one was rare, already extinct. While 270 endemic and 244 rare plant taxon could not be placed under a danger category because of lack of data and were listed under DD-Data deficient category [6].

Table 1: Taxon numbers of species and subspecies of various plant groups; endemism, rare and endangered species, extinct species

Plant Groups	*Defined Species/ subspecies*	*Endemic Species*	*Rare and Endangered Species*	*Extinct species*
Algae	*2.150*	----	*unknown*	*unknown*
Lichen (Lichenes)	*1000*	----	*unknown*	*unknown*
Moss (Bryophytes)	*910*	*2*	*2*	*unknown*
Pteridophytes Ferns	*101*	*3*	*1*	*unknown*
Gymnospermae (Gymnosperms)	*35*	*5*	*1*	*unknown*
Monocotyledonous (Monocotyledons)	*1.765*	*420*	*180*	-
Dicotyledonous (Dicotyledons)	*9.100*	*3500*	*1100*	*11*

GENETIC DIVERSITY

Plant genetic diversity is of great importance for both Turkish and world agriculture. Turkey has a very special position in terms of plant genetic resources. Turkey is situated at the intersection of two important Vavilovian gene centers: Mediterranean and Near East. These two regions have a key role in the emergence of cereals and horticultural plants. Some of the cultivated plant species of Anatolian origin are the following: *Linum* L., *Allium* L., *Hordeum* L., *Triticum* L., *Avena* L., Cicer L., *Lens*Mill., *Pisum* L., *Vitis* L., *Amygladus* L, *Prunus* L., *Beta* L., etc. Turkey, situated where two different gene and diversity centres overlap, is the gene and origin centre of the following cultivated plants among others: *Triticum* L, *Hordeum* L, *Secale* L., *Avena* L., *Linum* L., *Allium* L., *Cicer* L., *Lens*Mill., *Pisum* L., *Medicago* L. and *Vicia* L. Turkey is the home of many decorative plants including*Tulipa* L. and *Galanthus* L. [1].

Turkey is also quite rich in gene resources, including the valuable gene resources of *Cedrus libani* A. Rich., *Picea orientalis* (L.) LINK and *Fagus orientalis* LIPSKY together with 5 *Pinus* L., 4 *Abies*Mill., 20 *Quercus* L. and 8 *Juniperus* L. species among the local forest trees of national and global importance. The important forest trees are as follows: *Pinus* L. species (*Pinus brutia* Ten., *P. nigra*Arnold, *P. sylvestris* L., *P. halepensis* Mill. and *P. pinea* L.), *Abies* Mill. species (*Abies nordmanniana*(STEV.) SPACH *subsp. nordmanniana* (STEV.) SPACH, *A. nordmanniana* (STEV.) SPACH *subsp. bornmulleriana* (mattf.) COODE ET CULLEN, *A. nordmanniana* (STEV.) SPACH *subsp. equitrojani*(ASCHERS. ET SINT. EX BOISS.) COOD ET C, *A. cilicica* (ANT. ET KOTSCHY) CARR. *subsp. cilicica* (ANT. ET KOTSCHY) CARR., *A. cilicica* (ANT. ET KOTSCHY) *subsp. isaurica* COODE ET CULLEN), *Cedrus libani* A. Rich., *Fagus orientalis* LIPSKY, *Picea orientalis* (L.) LINK, *Tilia*L., *Alnus* Mill. (2 species, a total of 6 taxons), Juniperus L. (8 species), and Quercus L. (about 20 species) [1].

CONSERVATION OF BIOLOGICAL DIVERSITY

There are two different methods that are widely accepted in the world, used in order to conserve biological diversity. First of these is the "In Situ" conservation that aims at protecting the plants within their own natural growth areas. The other system is the "Ex Situ" conservation that envisages the protection of biological diversity features outside their natural living areas [7].

Despite it is possible to protect a large number of plant species by moving them to the areas outside their ecosystems by means of ex-situ conservation methods, the most important way to protect plant species is to protect them in the habitats they live. The reason for this is that all living things depend on each other in an ecosystem. No species in nature has an isolated environment and the species interact with other in many ways. Another advantage of in situ conservation is that protecting an efficient species population in its own habitat is easier when compared with ex situ conservation. As a result, ex situ conservation methods appears to be only as supporting methods alongside the in situ conservation. However, the botanical gardens allowing you to see many species of plants together and learn about are important, especially in terms of education and scientific research. Likewise, seeds and tissue banks is increasingly gaining importance with the fast growing biotechnology methods. However, in species with the physical environment as a whole to protect ecosystems, the future will be guaranteed [8].

IN SITU CONSERVATION

In the "In Situ" system, the restricted areas where the genetic material is found are protected by minimizing the human and animal effects [9].

The rapid consumption of natural resources, the increasing amount of pollution and environmental problems that have reached a global dimension have brought the obligation for countries to act in cooperation. Today, protected areas are vital component of all global and National nature conservation efforts. Many areas that have rich biodiversity and natural - cultural values have been designated as a national park and other protection statues to support conservation [10].

Protected areas are the cornerstone of biodiversity conservation; they maintain key habitats, provide refugia, allow for species migration and movement, and ensure the maintenance of natural processes across the landscape. Not only do protected areas secure biodiversity conservation, they also secure the well-being of humanity itself. Protected areas provide livelihoods for nearly 1.1 billion people are the primary source of drinking water for over a third of the world's largest cities and are a major factor in ensuring global food security. Well managed protected areas harboring participatory and equitable governance mechanisms yield significant benefits far beyond their boundaries, which can be translated into cumulative advantages across a National economy and contribute to poverty reduction and sustainable development including achievement of the Millennium Development Goals [10].

Definition of Protected Area According to the IUCN (1994): An area of land and/or sea especially dedicated to the protection and maintenance of biological diversity, and of natural and associated cultural resources, and managed through legal or other effective means [11].

Definition of Protected Area According to the IUCN (2008): A clearly defined geographical space, recognized, dedicated and managed, through legal or other effective means, to achieve the long-term conservation of nature with associated ecosystem services and cultural values [12].

IUCN divided the protected areas into the following categories [12]:

- Category I a: Strict nature reserve

Category I a - strictly protected areas set aside to protect biodiversity and also possibly geological/geomorphological features, where human visitation, use and impacts are strictly controlled and limited to ensure protection of the conservation values. Such protected areas can serve as indispensable reference areas for scientific research and monitoring.

- Category I b: Wilderness area

Category I b - protected areas are usually large unmodified or slightly modified areas, retaining their natural character and influence, without permanent or significant human habitation, which are protected and managed so as to preserve their natural condition.

- Category II: National park

Category II - protected areas are large natural or near natural areas set aside to protect large-scale ecological processes, along with the complement of species and ecosystems characteristic of the area, which also provide a foundation for environmentally and culturally compatible spiritual, scientific, educational, recreational and visitor opportunities.

- Category III: Natural monument or feature

Category III - protected areas are set aside to protect a specific natural monument, which can be a landform, sea mount, submarine cavern, geological feature such as a cave or even a living feature such as an ancient grove. They are generally quite small protected areas and often have high visitor value.

- Category IV: Habitat/species management area

Category IV - protected areas aim to protect particular species or habitats and management reflects this priority. Many category IV protected areas will need regular, active interventions to address the requirements of particular species or to maintain habitats, but this is not a requirement of the category.

- Category V: Protected landscape/Seascape

Category V - protected areas where the interaction of people and nature over time has produced an area of distinct character with significant ecological, biological, cultural and scenic value: and where safeguarding the integrity of this interaction is vital to protecting and sustaining the area and its associated nature conservation and other values.

- Category VI: Protected area with sustainable use of natural resources

Category VI - protected areas conserve ecosystems and habitats, together with associated cultural values and traditional natural resource management systems. They are generally large, with most of the area in a natural condition, where a proportion is under sustainable natural resource management and where low-level non-industrial use of natural resources compatible with nature conservation is seen as one of the main aims of the area.

Besides these categories, it is observed that countries develop others based on their regulations and institutional structure. Also, there are other protected areas like UNESCO Biosphere Reserves, World Heritage Areas, and RAMSAR Areas. While there are protected areas managed by national legislations and specific laws that belongs to the areas, at the same time, those are affected by the international conventions and agreements which are relevant with the conservation of biodiversity and natural resources, management and sustainable use of the values and sustainable development [13].

Protected. Areas in Turkey

Protection initiatives in Turkey are mostly in the form of in situ protection. There are various protected areas in Turkey which are under different institutions' authority and supervision and defined under certain laws. Turkey has signed many international conventions and protocols on Nature Protection. These conventions are;

- Convention on Biodiversity Conservation (Rio Convention) (1997)
- Cartagena Protocol (2004)
- CITES (1996)
- Barcelona Convention (1988)
- Bucharest Convention (1994)
- Protection of Cultural and National Heritage (1983)
- Convention on Combating Erosion (1998)
- European Landscape Convention (2000)
- Bern Convention(1984)
- Ramsar Convention (1994)
- Kyoto Protocol (2009) [14].

In accordance with these conventions, by 2011, nearly 1800 sites had been identified by the Ministry of Forest and Water as warranted protection under the 1983 law (Table 2) [14].

Turkey has 41 National Parks and approximately 37 percent of their total area is forest ecosystems (Figure 5,6,7). Besides, Turkey has 41 Nature Parks and approximately 25 percent of them are forest ecosystems (Figure 8) [15].

Table 2: Protected areas which identified by the Ministry of Forest and Water [14]

Conservation Status	Number	Related Law
National park	41	Law on National Parks
Nature conservation area	31	Law on National Parks
Natural monument	106	Law on National Parks
Nature park	41	Law on National Parks
Wild life reserve areas	79	Law on Terrestrial Hunting
Conservation forest	57	Law on Forest
Genetic conservation areas	214	Law on Forest
Seed stands	339	Law on Forest
Specially protected areas (SPAs)	14	Law on Environment
Natural sites	947	Law on Conservation of Cultural And Natural Heritage
Ramsar sites	13	Ramsar Convention By-law on Conservation of Wetlands
Biosphere Reserve	1	Law on National Parks -Law on Forest

Figure 5: National Parks in Turkey [16].

Figure 6: Munzur Valley National Park, Turkey [17].

Figure 7: Soğuksu National Park, Turkey [17].

Figure 8: Nature Parks in Turkey [16].

EX SITU CONSERVATION

This method is composed of;

- observation,
- gathering,
- storage, production and replacement,
- assessment,
- documentation,
- distribution and material exchange,
- preservation,
- education
- cooperation and organization steps [7].

Ex Situ conservation actions are an important complementary measure to In Situ habitat management. Ex situ conservation of wild plants is a central and unique role of botanic gardens. They have the appropriate facilities and staff expertise in botany and horticulture to be an 'insurance policy' against plant extinction. The Botanic Gardens ex situ program primarily employs three strategies: seed banking, tissue culture and genetic analysis of endangered plants. Seed banking is one of the most effective and useful ways of conserving genetic diversity ex situ, as the majority of wild plant species from dry land environments produce desiccation tolerant seeds that can be successfully stored for over 200 years. In most outbreeding species, the majority of the genetic diversity of the species may be captured by a single large seed sample. Conservation of population seed samples from these species provides insurance against loss of the wild population, whilst allowing biologists to develop germination protocols, propagation techniques etc. to support use of the species [22]. Tissue culture is the propagation, under controlled laboratory conditions, of rare and endangered plants that are difficult to propagate from seed or whose seed does not store well. Genetic analysis of rare species through various techniques, including, Microsatellites, Sequencing, AFLP (Amplified Fragment Length Polymorphism and ISSR (inter-simple sequence repeats), is helping guide preservation and restoration activities [18].

Ex. Situ Conservation in Turkey

Botanic gardens, which are the most significant applications of Ex situ protection methods, have been newly developing in Turkey. The first botanic garden to be established in Turkey is Istanbul University, Alfred Heilbronn Botanic Garden. It was established in 1935 under Istanbul University, Faculty of Applied Sciences, Department of Botany and it was given its current name in 2003. In this garden, there are 5 thousand genus and 6 thousand species belonging to 127 families, which are endemic and foreign homed. There are also seed banks and herbarium units in the garden. "Istanbul Botanic Garden Seeds Catalog" was published for the first time in 1935. The garden has been exchanging seeds with 373 botanic gardens from 63 countries. Istanbul University, Center for Research and Application of Natural Resources, established in relation with the Botanic Garden, aims at preventing the loss of rare and epidemic species of Turkish plant possessions by making them live in appropriate areas and conservatories [18].

Another important botanical garden in Turkey is Nezahat Gokyigit Botanical Garden (NGBB), located in Istanbul (Figure 9). NGBB is established on 32 hectares of parkland and it was opened to public in 2002. NGBB forms a space for breath-taking offering Istanbul 12 percent of its green space, and it is also a research, education and training center [19].

Figure 9: Nezahat Gökyigit Botanical Garden, Garden of Medicinal and Aromatic Plants [19].

Besides, "National Botanical Garden" is planned to be established in Ankara. The project covering an area of 2000 decare has started in 2008 and the construction work has begun in 2011.

Gene banks are also ex situ conservation units apart from the botanical gardens. Aegean Agricultural Research Institute Gene Bank and Plant Genetic Resources Unit of Field Crops Central Research Institute have been working for the conservation of genetic resources. In addition, the world's 3rdlargest gene bank (Seed Gene Bank) after the U.S. and China was established in Ankara in 2010 with a storage capacity of 250 thousand samples.

USE OF NATURAL PLANTS IN LANDSCAPE ARCHITECTURE IN THE SCOPE OF PRESERVING BIOLOGICAL DIVERSITY

Years before, when the natural areas had not been this much damaged with the pressure from the settlement, industrial and even agricultural areas, people would find it absurd to use the natural vegetation species in plant design work. Furthermore, until recently, many natural samples were seen as weed. But today, it has been realized that especially in developed countries, cities and areas of cultural activities puts nature away from people. For this reason, people want to see natural plant samples around them in order to feel close to nature.

Alpha

The fact that the natural areas are confronted with an increasing pressure increases concerns over plants of natural vegetation. Today, some species are threatened and are in danger of extinction. For some, on the other hand, it is observed that their expansion areas are getting narrower and that they do not grow in areas where they have been previously observed. In this regard, the use of natural vegetation samples which are in danger of extinction or damaged, in landscape architecture is important. Transfer of natural samples to gardens, are important for providing these plants with protection areas, as well as for increasing the awareness of protection and their recognition by the people.

Native plants naturally occur in the region in which they evolved, while non-native plants might provide some of the above benefits, native plants have many additional advantages. Because native plants are adapted to local soils and climate conditions, they generally require less watering and fertilizing than non-natives. Natives are often more resistant to insects and disease as well, and so are less likely to need pesticides. Wildlife evolved with plants; therefore, they use native plant communities for food, cover and rearing young. Using native plants helps preserve the balance and beauty of natural ecosystems [20].

Non-native plants can be invasive. They have few or no naturally occurring measures to control them, such as insects or competitors. Invasive plants can spread rapidly and smother or out-compete native vegetation. Invasive, non-native plants are not effective in providing quality habitat [20].

Use of natural vegetation species in landscape architecture, creates a landscape harmonious with nature, and it is an application compatible with economic conditions. In this way, selection of plant materials that meet environmental conditions with minimal care should be preferred as an economic approach to planning.

In Turkey, due to lack of work for cultivation of natural plants, obtaining plants from nurseries is not yet possible. In this case, it is not possible to provide the supply of plants for applications, as well as natural vegetation cannot become a sector generating income. Over the years, the greatest economic gain has been derived from the plants with bulb, tuber and corm, which have been pulled out from their natural areas. However, with their extensive use in landscape architecture natural plants will be an important branch of ornamental plant industry.

THE UTILIZATION PROCESS OF NATURAL PLANTS

Anatolia, throughout the history, has been the location where many natural species were cultivated for the first time. For example, it is known that the origin of 30 percent of the field corps is Anatolia. Besides the field crops, in many parts of Turkey, many natural plant species are collected directly from nature and used for nourishment. Plants are also used for different purposes other than nourishment [3].

About 1000 species of Geophytes are grown in Turkey which constitute an important part in the biological richness of the country and majority of which have economic potential as ornamental plants and medicinal plants [21]. Collection of certain types of bulbs of these plants was initially in the form of a personal curiosity but afterwards it gained commercial importance with the collection of flower bulbs in bulk. This trade, which was conducted in Netherlands, Denmark, Switzerland, Germany, Italy and in various other European countries, reached fairly large quantities since the 1960s. While up to 15000 *Galanthus* L. bulbs were exported in the early 1970s after 10 years, the amount has exceeded two times of this amount. Although in some species, small-scale domestic production activities are carried out, *Galanthus* L., *Eranthis hyemalis* (L.) Salisb., *Anemone blanda* Schott et Kotschy,*Leucojum aestivum* L., *Cyclamen* L., *Fritillaria* L., et *Sternbergia* Waldst kit, *Lilium candidum* L.,*Tulipa humilis* Herbert L., *Arum, Dracunculus* Miller, *Geranium tuberosum* L., *Colchicum speciosum*Stev., *Muscari* Miller, *Ornithogalum nutans* L., *Scilla bifolia* L., *Gladiolus* L., *Oxalis* L., *Urginea maritima* (L.) Baker, *Pancratium maritimum* L. and *Narcissus* L. are largely collected from nature and became the species traded most [23]. However, in recent periods, controls over the onion pull-outs are from nature are conducted and cultivation of bulbous plants is carried out. Due to the measures taken after 1990s, the populations of geophytes sold abroad were brought back to their former levels and important progress was made in the production and exportation of *Lilium candidum* L., *Sternbergia lutea* (L.) KER-GAWL. EX SPRENGEL, *Fritillaria imperialis* L., *F. persica* L., *Leucojum aestivum*L. and *Cyclamen hederifolium* AITON abroad. [4].

Besides bulbous plants, some plant species are collected from the nature and consumed in the country or sold abroad, because of their medical and aromatic qualities. Plant species with scientifically identified medical feature and which are used by the people are numerous in the flora of Turkey. Aromatic plants such as *Origanum* L. and the plants used as tea like *Salvia* L. and *Sideritis* L. are used in small residential gardens. In addition, with the spread of theme parks, gardens of medicinal and aromatic plants are organized in parks and botanical gardens. However, due to the difficulties in finding local plant samples foreign plants are more commonly found.

The most important condition for the promotion of the use of natural plants is conducting studies for their production; cultivation, variety creation and adaptation, as well as enabling users obtain them through nurseries. In recent years, studies on this issue have been increasing.

Some research has been being carried out at Atatürk Garden Culture Central Research Institute regarding the cultivation of various ornamental plants. In 2002, within the framework of the project on «Detection, Rehabilitation and Growing Techniques of Existing *Paeonia* L. Species in the Flora of Turkey" which was realized by the cooperation of private sector, university and public sector, species belonging to 11 taxon out of 55 populations were collected and preserved in peony gene resources garden [21].

Another important work was the "Cultivation of Some Natural Plants and the Inclusion of New Species and Varieties in Ornamental Plants Industry» project, carried out in 2006-2009. In the framework of the project which was conducted with the participation of 7 research institutes, 8 universities and 18 private sector organizations; 1166 population have been identified covering 241 species (177 geophytes species, 20 dune, and 44 outdoor plant species). Plant samples were subjected to pre-selection, moved to cultivation areas and preserved at responsible institution. Candidates of new varieties of existing species were determined at end of the study. The work under the project is continuing with variety creation activities [21].

In addition, a project has been initiated in 2010 called "Cultivation of Turkey's Geophytes, Providing New Species and Cultivars to the New Sectors. With this work it is aimed to collect the geophytes to be exhibited in "Turkey Geophyte Garden, to develop of varieties from the collected plants and to determine the biological activities of the plants collected. [21].

CONCLUSION

Increasing human activity results in negative effects for the natural areas, threatening the existence of plants, animals, and other types of species. It is generally admitted that the extinctions of plant species are occurring at an unprecedented pace in recent years. Thus, conservation of biological diversity is an issue gaining greater importance in today's world.

Turkey is an important country in terms of biological diversity as it displays the character of a small continent having a rich flora. However, Turkey has been facing the same threats against its biological diversity and its conservation becomes more and more important. As described above, there are different ways and methods used for preserving plant diversity in Turkey. One method is to protect the endemic species and prevent their extinction by making use of these plants in gardens and other living environments of people. By doing so, there would be an important step taken for increasing people's consciousness. As people spend more time with natural plants, their protection motives and knowledge would certainly increase. However, in order for providing these environments to people and increase their access to natural plants, more works and projects for improving cultivation and production of natural plants are needed.

REFERENCES

1. Ministry of Environment and Forestry, The National Biological Diversity Strategy and Action Plan, General Directorate of Nature Conservation and National Parks, Department of Nature Conservation, 978-6-05393-030-3 Tasarim Ofset, Ankara. 2008.
2. S. R. Fletcher, Biological Diversity: Issues Related to the Convention on Biodiversity,CRS Report for Congress. 1995
3. M. Avci, Çesitlilik ve Endemizm Açisindan Türkiye'nin Bitki Örtüsü. Istanbul Üniversitesi Edebiyat Fakültesi Cografya Bölümü, Cografya Dergisi, Sayi 13, Sayfa 2755Istanbul. 2005
4. Türkiye Çevre Vakfi, Türkiye'nin Biyolojik Zenginlikleri. TÇV Yayin 170 9-75725-079-1 2005.
5. T Ekim, ., M Koyuncu, ., M Vural, ., H Duman, ., Z Aytaç, . and N Adigüzel, ., Red Data Book of Turkish Plants. Türkiye Tabiatini Koruma Dernegi, Van Yüzüncü Yil Üniversitesi, 9-75936-110-8 2000.
6. A. Uzun, S. Palabas, S. Terzioglu, R. Ansin, Uluslararasi Doga Koruma Birligi Tehlike Kategorileri ve Türkiye Florasi. Korunan Dogal Alanlar Sempozyumu Sözlü Bildiriler Kitabi, 810Eylül 2005, Isparta. 2005
7. M. Dokuzoguz, Bitki Genetik Kaynaklari. Türkiye'nin Biyolojik Zenginlikleri. Türkiye Çevre Sorunlari Vakfi Yayini, Önder Matbaa, Ankara. 1990
8. S Kaynak, . Biyolojik Çesitlilik ve Çevre Koruma Rehberi. Türkiye Tabiatini Koruma Dernegi, 9-75936-111-6 2001.
9. M. Yücel, Korunan Alanlarin Siniflandirilmasi ve Uzun Devreli Gelisme Planlari Yapiminda Yasanan Sorunlar. Korunan Dogal Alanlar Sempozyumu Sözlü Bildiriler Kitabi, 810Eylül 2005, Isparta. 2005
10. S. C. Gümüs, G. Sen, D. Toksoy, H. Ayaz, and B. Bahat, Nature Conservation and National Parks in Turkey. First Serbian Forestry Congress, 1113November 2010Belgrade. 2010.
11. IUCN, Guidelines for Protected Area Management Categories. IUCN Commission on National Parks and Protected Areas with the assistance of the World Conservation Monitoring Centre, 2-83170-201-1 UK. 1994.
12. N Dudley, . (Editor) Guidelines for Applying Protected Area Management Categories, IUCN, 86 pp., 978-2-83171-086-0 Gland, Switzerland. 2008.
13. 810Yakinkiliç, M.K., Yenilmez Arpa, N. Türkiye'deki Korunan Alanlar ve Ekoturizm. Korunan Dogal Alanlar Sempozyumu Sözlü Bildiriler Kitabi, 8-10 Eylül 2005, Isparta. 2005
14. M Özyavuz, . Protected Areas, Landscape Planning, Dr. Murat Ozyavuz (Ed.), 978-9-53510-654-8 InTech, Available from: http://www.intechopen.com/books/landscape-planning/protected-areas, 2012.
15. E. Ertürk, Protected Area Management in Turkey. Republic of Turkey, Ministry of Environment and

Forestry, Ankara. 2011

16. M.K Yalinkiliç, . Protected Areas of Turkey. General Directorate of Nature Conservation and National Parks. http://cembit.dmi.gov.tr/FILES/doc/korunan- alanlar/turkiyenin_korunan_alanlari_en-US.pdf, Istanbul. 2010.
17. wwwmilliparklar.gov.tr
18. Z. Dilaver, 2007Conservation of Biological Diversity in the World and in Turkey. International Conference on Environment: Survival and Sustainability, Near East University, 19-24 February 2007, Proceedings, 521528Nicosia- Turkish Republic of Northern Cyprus.
19. http://wwwngbb.org.tr, 2013
20. SlatteryBritt E., Kathryn Reshetiloff, and Susan M. Zwicker. Native Plants for Wildlife Habitat and Conservation Landscaping: Chesapeake Bay Watershed.U.S. Fish & Wildlife Service, Chesapeake Bay Field Of?ce, Annapolis, MD. 82 2003
21. Kaya, E. Kir Çiçeklerimizi Etkin Degerlendiriyoruz. Plant Peyzaj ve Süs Bitkileri Dergisi, Yil:3, Sayi:6, Nisan-Temmuz 2012, Adapazari. 2012
22. Wyse JacksonP.S. and Sutherland, L.A. International Agenda for Botanic Gardens in Conservation.Botanic Gardens Conservation International, U.K. 2000
23. T. Ekim, M. Koyuncu, A. Güner, S. Erik, B. Yildiz, Ve Vural, M.Türkiye'nin Ekonomik Deger Tasiyan Geofitleri Üzerinde Taksonomik ve Ekolojik Arastirmalar. Orman Genel Müdürlügü Yayin Sira 669Seri No: 65, Ankara. 1991

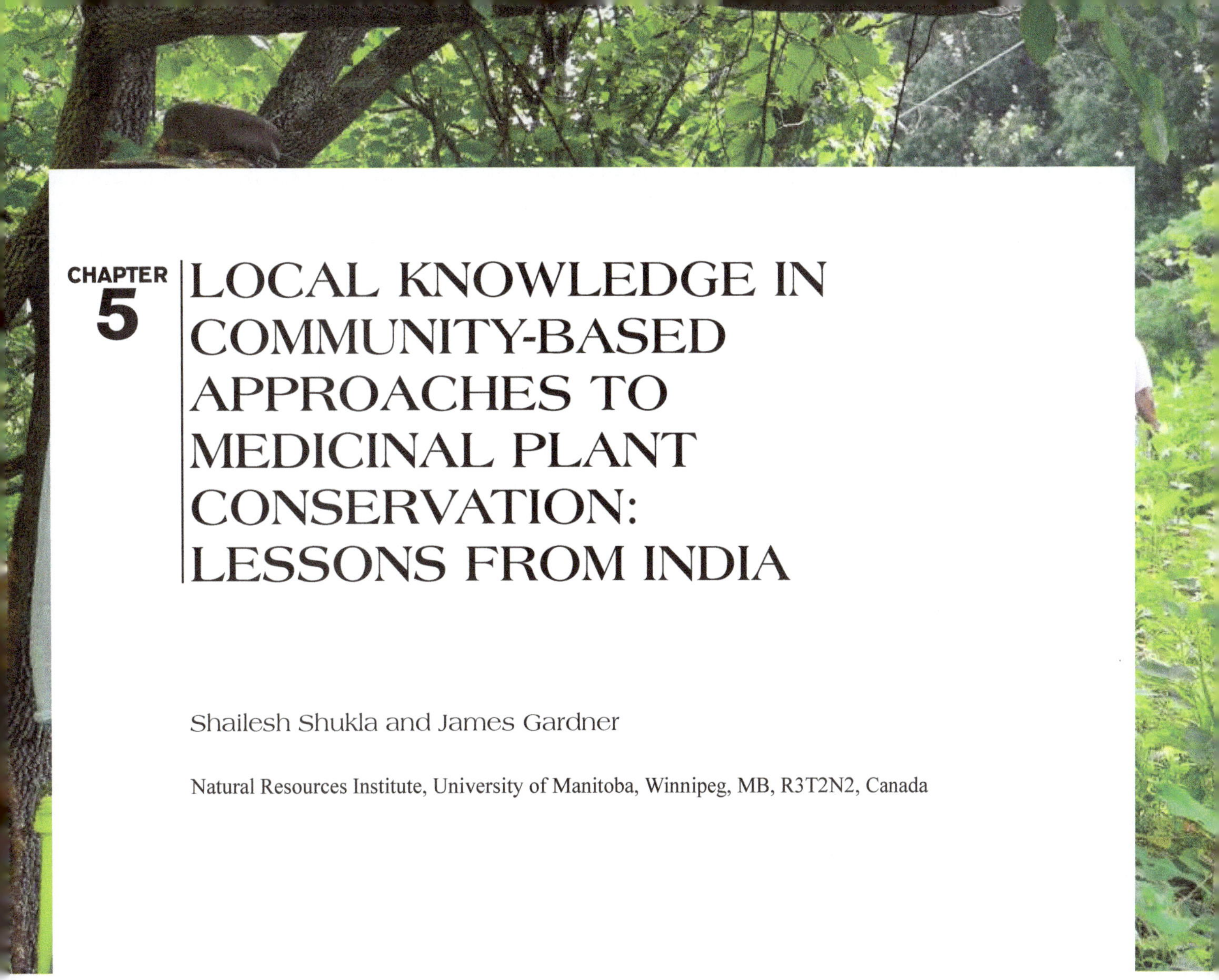

CHAPTER 5

LOCAL KNOWLEDGE IN COMMUNITY-BASED APPROACHES TO MEDICINAL PLANT CONSERVATION: LESSONS FROM INDIA

Shailesh Shukla and James Gardner

Natural Resources Institute, University of Manitoba, Winnipeg, MB, R3T2N2, Canada

ABSTRACT

Background

Community-based approaches to conservation of natural resources, in particular medicinal plants, have attracted attention of governments, non-governmental organizations and international funding agencies. This paper highlights the community-based approaches used by an Indian NGO, the Rural Communes Medicinal Plant Conservation Centre (RCMPCC). The RCMPCC recognized and legitimized the role of local medicinal knowledge along with other knowledge systems to a wider audience, i.e. higher levels of government.

Methods

Besides a review of relevant literature, the research used a variety of qualitative techniques, such as semi-structured, in-depth interviews and participant observations in one of the project sites of RCMPCC.

Results

The review of local medicinal plant knowledge systems reveals that even though medicinal plants and associated knowledge systems (particularly local knowledge) are gaining wider recognition at the global level, the efforts to recognize and promote the un-codified folk systems of medicinal knowledge are still inadequate. In country like India, such neglect is evident through the lack of legal recognition and supporting policies. On the other hand, community-based approaches like local healers' workshops or village biologist programs implemented by RCMPCC are useful in combining both local (folk and codified) and formal systems of medicine.

Citation: Shailesh Shukla and James Gardner, Local knowledge in community-based approaches to medicinal plant conservation: lessons from India, DOI: 10.1186/1746-4269-2-20.

Conclusion

Despite the high reliance on the local medicinal knowledge systems for health needs in India, the formal policies and national support structures are inadequate for traditional systems of medicine and almost absent for folk medicine. On the other hand, NGOs like the RCMPCC have demonstrated that community-based and local approaches such as local healer's workshops and village biologist program can synergistically forge linkages between local knowledge with the formal sciences (in this case botany and ecology) and generate positive impacts at various levels.

BACKGROUND

Purpose and objectives of the research

The contribution of local knowledge systems in conservation and sustainable use of natural resources is gaining wider recognition [1]. Local knowledge associated with the use and conservation of medicinal plants is either codified in ancient scriptures or is folk-based and transmitted through generations in the form of community-based health traditions. The codified knowledge has managed to expand globally through official recognition and some corresponding policy and financial support. The largely undocumented folk-based knowledge and traditions, on the other hand, have survived on their own in the absence of any official recognition and policy and administrative support by governments at the state and national levels [2]. The problems, progress, and prospects of folk and codified streams are therefore distinct, with the former deserving more attention. Despite the lack of official recognition and support, the recent efforts and approaches of grassroots community groups and NGOs in revitalizing and promoting folk knowledge systems of medicine are noteworthy. The purpose of this paper is to describe and analyze such an NGO-led approach from India in order to demonstrate the contributions of local knowledge systems in generating positive outcomes at the local, the regional, the national and the global levels. The specific objectives of the paper are:

i) to describe the community-based approaches taken by the RCMPCC (Rural Commune's Medicinal Plant Conservation Centre) that successfully used folk and codified local medicine systems along with formal medical systems

ii) to understand the outcomes of these community-based approaches at the various levels in order to derive lessons for policy and practice in regard to community-based medicinal plant conservation

The global status of local medicinal plant knowledge systems

Medicinal plants are gaining wider recognition in recent initiatives for conservation and development at the global level. This is evident in the vision and mission statement of World Health Organization (WHO) on health improvement and in community-based conservation initiatives by international organizations, including the World Bank, the International Development Research Centre (IDRC) and UNDP, for example. The effort by the WHO to recognize and promote the use of local medicinal plant knowledge systems in the health sector, particularly in developing countries, is prominent. The terminologies related to a use of plant-based medicine vary in different cultures, countries, and communities. The WHO, in its widely acknowledged report, uses an umbrella term 'traditional medicine' to describe such uses and offers its working definition as 'diverse health practices, approaches, knowledge and beliefs incorporating plant, animal, and/or mineral based medicines, spiritual therapies, manual techniques and exercises applied singularly or in combination to maintain well-being, as well as to treat, diagnose or prevent illness' [6]. The scope of this paper is limited to practices/approaches/knowledge with respect to only plant-based medicines.

The World Bank report [7] indicates that more than eighty percent of the population of South Asia uses plant-based medicines for maintaining and improving their health. The total reported usages of medicinal plants vary. For instance, the WHO [6] study lists 21,000 plants with reported medicinal uses around the world, while Schippmann and co-workers [8] estimated this figure as 52,885. Amidst these conflicting claims on numbers, the use of medicinal plants by local communities or groups has remained high. Such local uses of medicine, are again known by different terms but can be classified broadly into the following three categories[9]: (i) Traditional Systems of Medicine (TSM), with a systematic codified body of knowledge either in the form of pharmacopoeias or ancient scriptures [Ayurvedic, Chinese and Tibetian medicine, Siddha, Unani (Arabic) etc.]; (ii) Traditional Medical Knowledge or Folk Medicine, which is transmitted by oral means and is mostly acquired through learning-by-doing approaches, and (iii) Shamanistic or Spiritual Medicine, with a strong religious/spiritual element and can be practiced only by highly specialized local experts called 'shamans'. In general, except for TSM, the legal and political recognition and support for folk and spiritual medicine at the national and the global level (see, Table 1) has remained weak [10, 11]. The lack of official and governmental support, compounded by the devaluation of the folk knowledge [12] by local communities and societal systems at large, has resulted in the erosion of the local medicinal plant knowledge systems. The issues related to the formal recognition of TSM, and the lack thereof for other medicinal plant knowledge systems (folk and spiritual), have some things in common. However, in most cases, the distinct lack of attention to the latter systems suggests that they deserve further consideration in countries like India, China and South Africa, where a reliance on TSM and folk medicine is more widespread and evident.

Table 1: Comparisons of the features and policy issues of local medicinal plant knowledge: The folk and codified systems of medicine in India

Folk traditions of local medicinal plant knowledge	***TSM of local medicinal plant knowledge***
Originated in communities to meet daily healthcare/survival needs, largely undocumented	Originated by scholars, physicians and seers and documented in manuscripts/Vedic texts(1000–1500 BC), scriptures for human well-being and developed as a classified main branches
Transmission multigenerational and by oral means through learning-by doing and through more than 300 formal educational colleges	Transmission is often institutionalized through written texts and hands-on training
Mainly empirical, adapted	Sophisticated philosophical and theoretical roots with a scope for refinement
No legal status, No budgetary allocation, on the contrary vulnerable to disregard and devaluation	Legal status as 'Indian Systems of Medicine' with five percent of budgetary allocation (health) wider social and official acceptance and recognition
Approximate # practitioners are 600,000 birth attendants, 60,000 bone setters, 100,000 herbal healers, 60,000 healers specialized in treating poisonous snake bites and millions of households/women	Approximately 600,000 registered medicinal practitioners, out of which, 10 percent practice medicine on the basis of TSM.
Uses more than 7,500 medicinal plants	The four streams of Ayurvedic, Unani, Siddha and Tibetan uses approximately 4,500 medicinal plants
POLICY LEVEL ISSUES	
Local state and national incentives for systematic documentation and dissemination needed	Available documentation in Sanskrit at scattered places, interpretation and consolidation in a commonly-understood language will facilitate further use/research
In-depth understanding of and incentives for (local/state/national/global) incentives can facilitate transmission	Formal institutions for transmission are present but are poorly funded
Sustaining interest and apprenticeship of the younger generations is a challenge	Maintaining quality and standards of practitioners is a challenge
Scope of learning from TSM and allopathic medicine system is limited due to access, affordability and literacy issues at the community level	Both TSM and allopathic medicine draw heavily on the folk system for herbal remedies or drugs without giving credit or sharing benefits to local communities
Benefit sharing mechanisms are developing and difficult to implement at community level	Well-established and implemented benefit sharing mechanism in the form of patent/trademarks and other forms of protection
Efficacy, standardization and safety studies using scientific parameters are almost nil due to lack of authentic documentation and neglect by official policies	Efficacy, standardization and safety studies are not encouraged due to high-cost (200,000 US$) and time consuming (8–10 years) scientific validation and language barriers
Collaboration by other stakeholders is difficult and confined to documentation/dissemination efforts	Collaboration is generally encouraged if the epistemological and philosophical foundations are matching

Sources: Compiled based on Shankar (2001)[11], Shankar and Venkatasubramanian (2004)[15] and WHO (2002)[6]

The local medicinal plant knowledge systems in India

In a country like India, where 65 percent of the total population has access to only local medicinal plant knowledge systems [6], and 70 percent of the population lives in villages struggling to access and afford modern allopathic medicines, both TSM and the folk knowledge systems of medicine are of significance[13] TSM exists in the form of well-known classical traditions of Ayruevda, Unani and Siddha, which are characterized by a large number of practitioners trained through formal institutions, a well-codified body of texts either in the form of the scriptures or other written forms and an official recognition as 'Indian Systems of Medicine'[14]. Both Ayurvedic and

Siddha systems of medicine originated more than 3000 years ago and were prevalent in North and South India, respectively. The Unani system of medicine originated in Greece (460–377 BC), and became more popular in India after the establishment of the Central Council for the Research in Indian Medicine and Homeopathy in 1969. Complementary to these codified systems, the folk knowledge systems are largely transmitted through oral means and flourish at the village level [15] with little or negligible support from official channels at the state or the national level. The features and policy issues concerning these systems are compared in Table 1.

Although folk knowledge ranks at the top in terms of the total number of users, the number of medicinal plants used and number of practitioners, the desired policy support for its recognition and development is not evident. There is some policy and legal support for the codified systems of Ayurvedic, Unani and Siddha in India [6] but it is inadequate [15].

International organizations like UNDP, IDRC, OXFAM, WHO, Ford Foundation and the World Bank are the leading sources of funding for broad-based programs dealing with medicinal plant conservation and development in India and Asia [16]. The funding and capacity building support of these agencies to Indian governments and NGOs is predominant in developing state, national, and regional visions and strategies for participatory, comprehensive, and sustainable management of medicinal plants. At a national level, the Department of the Indian Systems of Medicine (Health) and the Medicinal Plant Boards [17] are the most relevant government agencies that deal directly with the medicinal plant sector and associated local knowledge systems. These national level efforts are more recent and more targeted at improvement in codified streams such as TSM. Further, their direction, design and delivery have largely been limited to the state level. At the sub-state level (district, block or taluka and village), the programmatic interventions related to medicinal plant and associated local knowledge systems are organized by local NGOs and community groups.

The most notable and oldest among these NGOs is the Foundation for Revitalization of Local Health Traditions (FRLHT) [2], which has been identified and supported by the Government of India as a 'centre of excellence' in the field of medicinal plants. FRLHT's programs and activities in this sector are largely confined to peninsular and Southern India. Nevertheless, FRLHT has inspired similar initiatives in other parts of India and Asia through networking and capacity building for other NGOs. The RCMPCC, based in the western state of Maharashtra is one such NGO. The RCMPCC has demonstrated that the use of community-based participatory approaches at the local level facilitates learning among various stakeholders and provides a platform for interactions among the formal botanical knowledge, TSM and the folk knowledge systems relevant to medicinal plants at the state level. The learning resulting from these local knowledge-based approaches has transformed the agenda of medicinal plant conservation at the sub-state or state level in Maharashtra and has the potential to enrich and inform stakeholders at the national and global levels.

METHODS

Before the start of field research activities, a series of consultations was held at the Center for Community-based Natural Resources Management at Natural Resources Institute (NRI), University of Manitoba, Winnipeg, Canada between April and October 2003 as a part of the Equator Initiative of the United Nations Development Program (UNDP). This helped in the selection of communities, research objectives and field research methods. The RCMPCC of Pune, India was one of the four Equator Initiative sites selected as part of the study. The purpose, objectives and research design were based on an interactive adaptive approach as suggested by Nelson [3] and were shared with key leaders and field staff of RCMPCC. The RCMPCC had established thirteen project sites in the rural, tribal and forested areas known as medicinal plan conservation areas across the state of Maharashtra. One of their project sites near the village of Amboli was chosen as a case study, on the basis of the following criteria:

a) ***Willingness of the community/project functionaries to participate:***The willingness and oral consent of villagers were addressed and obtained in local dialects during an initial workshop in Amboli that was attended by the village panchayat, i.e., local management committee members. Special efforts (e.g. personal visits or informal conversations while walking in forest) to ensure written consent and active participation of special groups, such as women and local healers, were made.

b) ***Evidence of use of local knowledge in sustainable management of local biodiversity:***Amboli has well-documented evidence for community-based conservation and is rated as one of the best examples of such (267.63 hectares) by the RCMPCC [4].

A review of literature was undertaken on issues related to medicinal plants and associated local knowledge systems to contextualize the problem as described in the background. The field research used a variety of qualitative research tools [5] such as semi-structured interviews with key people (N = 12) from the RCMPCC, local healers (N = 11, seven male and four female), community members (N = 5) and Forest Department officials (N = 3). The selection of the research participants was limited to those who were involved in designing, implementing and/or participating in the two community-based approaches (i.e. *vaidu sammelan* and village biologist program) of RCMPCC. For the purpose of this paper, data relevant to the outcomes (as perceived as 'benefits' or 'visible impact' by the research participants) were collected and analyzed. The senior author stayed in Amboli for more than three months and participated in a village botanist training program organized at Amboli (December 2003) and *vaidu sammelan* (March 2004). The field research was complemented by a review of relevant documents and internal documents (such as proceedings of the village botanists' workshops (December 2000 and November 2001) provided by RCMPCC.

The Joint-Faculty Research Ethics Board at University of Manitoba approved the research protocols (Protocol No. J2003:141) for this study.

RESULTS AND DISCUSSION

This section describes and analyzes two community-based approaches undertaken by RCMPCC, which combine both local (folk and codified) and formal systems of medicine to demonstrate that such creative integration at the local level can generate positive impacts at all levels.

Building on medicinal plant knowledge of local healers: the RCMPCC way

The RCMPCC initiative was envisioned by a Mumbai-based NGO, called Rural Commune for the In-Situ Conservation and Sustainable Utilization of the Medicinal Plant Diversity of Maharashtra, through developing partnerships among the Forest Department, local communities and NGOs. To this end, the RCMPCC, in collaboration with other stakeholders, organized several activities, such as the establishment of a network of 13 Medicinal Plant Conservation Area or MPCAs (each ranging from 250–400 hectares) in Maharashtra. The MPCAs were selected in consultation with the Forest Department, local communities, and available scientific literature, based on the following criteria:

- Relatively undisturbed forest areas representing different bio-climatic zones
- Forest areas with rich biodiversity
- Areas with natural availability of water
- Locally and otherwise known for harboring medicinal plants.

The MPCAs were legally notified by the Maharashtra Forest Department as permanent medicinal plant reserves and this inspired other state governments and the Government of India to include them in their conservation and development agenda. The RCMPCC also completed the documentation of some 150 species in the Medicinal Plant Conservation Areas through participatory approaches like the village biologist program, the Conservation Assessment and Management Program, local healers' conventions and scientific assessments by field botanists. They created village management structures such as Local Management Committees (LMCs) and Self-Help Groups (SHGs) for marketing and local sale of herbal products [18].

A group of individuals known as local knowledge experts or healers and ecological experts called *vaidus* were identified by the RCMPCC. The term *vaidus* is a generic folk term most commonly used in the Marathi and Hindi languages to describe healers and traditional herbal practitioners, used most commonly for, but not limited to, male healers. These *vaidus* possessed extraordinary knowledge and interest in the local plants, fauna and ecosystems. Many of the *vaidus* practiced herbal treatments for diseased humans, livestock, and crops and developed skills in identification and use of locally grown plants. The RCMPCC recognized the creative potential of the *vaidus* as village biologists (VB) or barefoot botanists in the conservation and management of medicinal plant diversity through sustainable uses [19]. To build effective partnerships with the local knowledge experts, the RCMPCC organized two programs: 1) *vaidu sammelan* or local healer's workshops, and 2) the village biologists (previously known as barefoot botanists) training programs. FRLHT initiated the Barefoot Botanist (BFB) program in 1995 with the original aim to enrich the local medicinal plant knowledge of the village *vaidus* with the relevant formal botanical skills. The trained *vaidus* were expected to perform better in their own profession and in providing guidance to eco-tourists and researchers. RCMPCC prefers the use of more inclusive term 'village biologist' instead of 'barefoot botanist'. These programs also helped in mutual learning and the strengthening of the capabilities of formally-trained botanists and local*vaidus* through dialogue and exchange of knowledge. (See, Figure 1)

Figure 1: Local *vaidus*, RCMPCC scientists and school students engaged in knowledge exchange during the village biologist workshop in Amboli (Photo- JPEG format).

The purpose of *vaidu sammelan* was to: 1) document the knowledge of local *vaidus* about plants; 2) encourage value-added activities by promoting local use and sale of herbal products; 3) provide a platform for*vaidus* to demonstrate their products; and, 4) to provide a platform for the *vaidus* and other health practitioners to interact and encourage participation in local biodiversity conservation efforts. The village biologist program was broad in terms of coverage, with 3 or 4 local experts identified at each location. Selection of village biologists was based on: 1) good knowledge of local plants and their uses and cultural significance; 2) interest in local environmental and conservation issues; and, 3) ability to read and write. Most village biologists identified were *vaidus*, forest guards and knowledgeable elders.

Learning from the outcomes of RCMPCC's community-based approaches

The discussion in this section draws mainly on interviews with selected informants (as indicated in methods) and a review of relevant documents.

Table 2 highlights some of the major outcomes of these two methods, along with the policy implications, at various levels of activity. These levels are geographical units or operations at which the various activities of the RCMPCC has generated visible impacts, as perceived by RCMPCC and the communities including the *vaidus*.

Table 2: Major outcomes of RCMPCC's community-based programs

Major Outcome	**Levels of influence**				
Policy implication	Village/Sub-district	District (MPCA)	State (RC-MPCC)	National	Global

Documentation of traditional knowledge of medicinal plants from *vaidus*- both men and women *Systematic documentation created opportunities for collaborative research with TSM*	Role of women *vaidus*recognized Involvement of *vaidus*/VB in village development/conservation activities	Two local language booklets on ethno botanical information about plants and *vaidus*	Systematic data base on 326 medicinal plants, 465 herbal formulations, Illness-specific database on 265 plants, herbarium record of 804 species	Raw drugs formulations of 75 plants available for further scrutiny	VB, Vaidus training institutionalized in GEF programs in nine states
Local assessment of Rare, Endangered and Threatened (RET) species *Potential of Local medicinal knowledge alongside TSM as inputs in official planning*	Local monitoring by panchayats (village councils), LMCs, SHGs. RET List maintained at two MPCA by local communities with the support of VBs/*vaidus*	Training of 36 selected VB and district forest officers on RET monitoring Incorporating list in district working plans of Forest Department	Prioritized species identified and raised in seven forest nurseries of the state Helped FD in species recovery and enrichment programs	Unique species highlighted	Unique specie of Global importance identified at Leghapani MPCA
Recognition and use of *vaidus* *Official recognition of Local knowledge experts (restricted up to sub-state level)*	*Vaidus* Involved in selection of appropriate seedlings in forest nurseries *Vaidus* and VB got 'official' recognition to practice their knowledge	Employing*vaidus* as field guides/eco-tourism Honouring VB/ *vaidus* in two workshops	State level database of*vaidus* and village botanists according to their area of expertise (RCMPCC) CAMP exercises	Display and dialogue with other national level *vaidus*through two national herbal expos	Local *vaidus*guided ITTO (Japan) research scientists as research anchors in vegetation mapping
Opportunities for collaborative research *Collaborative Research mainly initiated at national/below level and often self-inspired/funded*	A herbarium preparation techniques learned by VB and *vaidus* through botanists Validation of *vaidus*/ VB's local knowledge through sharing/exchange with botanists	Selected VB/ *vaidus* as trainers for other MPCA selected Study in sacred groves by RC-MPCC	Market study of 22 medicinal plants Biodiversity assessment of select MPCAs RCMPCC as a resources institution in State medicinal Plant Board	Pilot project on standardization/cultivation processing of prioritized species in six MPCAs supported by Department of Science and Technology	
Transmission of folk knowledge facilitated *Systematic efforts for understanding transmission are lacking at State/ National/Global levels*	People Biodiversity register at one village Herbarium sheets demonstration in three schools	11 demo gardens, 10 interpretation center highlighting contributions of VB/*vaidus*,40 home herbal gardens		Ministry of tribal affairs supported study on people's biodiversity register	

Source: self-compiled

At the village level, the local medicinal plant knowledge of the *vaidus* is widely recognized as 'valid' alternative systems of knowledge. 'This is the most commonly available knowledge in our village and people often seek our help, even during odd times such as night. We treat many villagers who have snake bites during night or are injured by black bear when we work in our farms near forest› (SA, local vaidu, Amboli, March, 2004). Amboli has a Public health centre but the continuous availability of allopathic practitioners is normally not the case. The accessibility of *vaidus* at anytime, therefore, is regarded as one of the major benefits for the local communities. ‹Approaches such as the *vaidu* sammelan give our *vaidus* a community-recognized local licence to practice› (MG, Community leader, Amboli, January, 2004). In addition, the village biologist program provides avenues to facilitate the use of the local medicinal plant knowledge of the *vaidus* by the local formal institutions in two ways: i) *vaidus'* knowledge about the rare and endemic medicinal plants was used in the selection of plant species in the nurseries by the Forest Department ; ii) *vaidus* learned the herbarium preparation techniques from the botanists during the village biologist program and applied their training by conducting demonstrations of local plants through these herbarium sheets in secondary schools. The People's Biodiversity Register program has been initiated at Chavni village in the Amba Valley MPCA through consultation with the *vaidus* and villagers. This 100 page register was prepared in English and Marathi language. It contains useful information about the local biodiversity in and around Chavni village and details about the *vaidus*. The register has been useful in local biodiversity assessment, conservation and monitoring.

The *vaidus* of all thirteen Medicinal Plan Conservation Areas were involved in the identification of Rare, Endangered and Threatened (RET) plant species. At the Amboli and Leghapani MPCAs, *vaidus* generated a list (in the local language) and photographs of these species, for regular monitoring and regeneration in villagers' home gardens. 'We share the photographs and lists of rare plants with tourists and researchers, so they become sensitive to these plants when they move around in forest' (KG, local *vaidu*, Amboli, March 2004).

Participation of women *vaidus* in these programs has helped in their improved recognition. 'In the past, our expertise was perceived to midwifery skills. But now we also treat people as male *vaidus* do. Besides, our participation in programs like *vaidu sammelan* gives us a confidence and authority to take part in village conservation and development activities related to medicinal plans along with male *vaidus*' (JS, women *vaidu*, Amboli, March 2004). The women *vaidus* in particular and *vaidus* in general, are being regarded as equal partners, not only in generating a useful knowledge base about medicinal plants but also in developing an agenda for their conservation and sustainable use.

At the district level, the Forest Department publicized the contributions of the *vaidus* through a special publication or in their working plans. For example, the district Forest Department of Sindhudurg and RCMPCC published a Marathi language booklet with the list and uses of 100 local medicinal plants. The Latin, English and vernacular names of the plants are listed along with their family names. In addition, a list of 113 *vaidus*with their areas of specialization and addresses is provided in the book. The book has been widely circulated among state forest officials, other state government departments, universities and NGOs. 'We have publicly honoured *vaidus* from Amboli. Some *vaidus* that were listed in the booklet are now being invited by NGOs and Other government departments to participate in training and share their stories' (RN, Forester, Sidhudurg Division, December 2003). The Forest Department has also recognized *vaidus* as eco-guides in Koynanagar. RCMPCC complemented these efforts by involving *vaidus* in developing and maintaining demonstration gardens, home herbal gardens and interpretation centres and in documenting sacred groves (the local areas that were conserved by villagers for religious or spiritual purposes).

At the state level, the collective documentation and collaborative research with the *vaidus*, helped in creating a database of medicinal plants and herbal formulations. These databases provided useful information about the Rare, Endangered and Threatened (RET) plant species that were prioritized for conservation and regeneration in the working plans of Forest Department. 'The research, documentation and database development efforts of RCMPCC with *vaidus* and communities, earned us an honour of ‹Resource institution› in the State Medicinal plant board and Forest Department› (RCMPCC management board representative, Pune, March 2004).

At the national level, these approaches mobilized funding support for the two studies from the Department of Science and Technology and the Ministry of Tribal affairs on standardization of selected medicinal plants in six MPCAs and a biodiversity register in one MPCA, respectively. *Vaidus* from all thirteen MPCAs (three from each site) participated in the National Herbal Expo in 2000 and 2001 where they had shown their herbal products and had an opportunity to interact with government officials, NGO representatives and *vaidus* from other states.

The most visible national and global impacts of these approaches were the inclusion and recognition of *vaidus*as important project stakeholders in the new conservation and development funding plans that nine other states of India have submitted to the Global Environment Facility (GEF) for funding. The unique endemic plant species of *Ceropegia* spp. was discovered in Leghapani MPCA. [20]. ‹Our *vaidus* have guided researchers form International Tropical Timber Organisation (ITTO) from Japan and helped them organize transects by identifying suitable pockets› (MA, Key leader, RCMPCC, December, 2003).

These impacts and outcomes constitute, what scholars of community-based conservation describe as ‹benefits›- equity and empowerment being the key considerations [21]. The *vaidus* and their medicinal plant knowledge, have been recognized, legitimized and in some cases built upon to meet the medicinal plant conservation and development goals at various levels.

It is evident from the material in Table 2, that the activities of the RCMPCC generated some useful outcomes at the village, regional, state and national levels, rather than the global level. The national level outcomes (largely driven by the RCMPCC) include: the mobilization of funding commitments through pilot research projects on RET species, a biodiversity register in one MPCA and the recognition of the contributions of the*vaidus* in country-wide publications and national exhibitions. These outcomes have yielded some legitimization to and official use of the local medicinal plant knowledge and its holders (*vaidus*). At the state level, the recognition and support to the *vaidus* reflected through examples of the documentation on botanical and local nomenclature for plants, validation of medicinal uses of plant (as reported by the (*vaidus*) by Ayurvedic practitioners and botanists and collaboration with formal scientists for collective identification of the priority areas/species for conservation, simple resource mapping and herbarium development. At the global level, there has been some acknowledgment of the contributions of *the vaidus* as evident from their involvement in scientific studies.

The availability and acceptability of *vaidus* at the village level and their legitimacy, recognition and partnerships at the state and national levels, created a sense of empowerment (that they can decide and implement local conservation and development agenda) and equity (that their knowledge can inform and interact with knowledge of formally trained botanists or foresters). While these most sought-after benefits of community-based conservation may be achieved at the village and district levels, their fullest potential is yet to be realized at the state, national and global level.

CONCLUSION

In India, although the majority of the population still relies on local knowledge systems to meet their health needs, the official policies and national support structures are inadequate for TSM and almost absent for folk medicine. NGOs like the RCMPCC have demonstrated that community-based approaches such as the *vaidu*sammelan and the village biologist programs can provide a platform on which holders of local medicinal plant knowledge systems (both folk and TSM) can interact with the holders of formal knowledge (e.g. botanists and other scientists). These approaches have also generated positive outcomes at different levels, such as the legitimization and recognition of the folk knowledge of the village *vaidus* in the district, state and (externally-aided) national plans of the Forest Department, the mobilization of collaborative research and funding commitments by the government departments, NGOs and research institutions at the state and the national levels and, most importantly, the pre-testing of community-based educational models for facilitating transmission of folk knowledge associated with uses of medicinal plants at the village or the sub-state levels. These outcomes show a way to achieve the larger goals of equity and empowerment as conceived in community-based conservation. In this case of community-based medicinal plant conservation, however, the achievement of these important goals is limited to local, district and, to a certain extent, state level. In order to make these goals more durable and widespread, such community-based approaches that build on local medicinal plant knowledge systems need to be encouraged with supportive policy and legislative measures at the national and the global levels.

ACKNOWLEDGEMENTS

Dr John Sinclair and Dr Fikret Berkes provided continuous guidance and friendly support throughout this research. Ms Kritin Kent provided very useful and constructive editorial comments. The funding support from Social Science and Humanities Research Council of Canada (SSHRC) and IDRC, Canada is gratefully acknowledged. Author wishes to acknowledge the generous support and cooperation of RCMPCC, Pune particularly Mr Muneer Alavi, Mr Brigadier Kaul, Dr Suresh Jagtap, Mr Vivek Gourbroome, Mr Satish Eklunchwar, Dr Vaishali Gawandi. Dr. Utkarsh Ghate and Dr. Darshan Shankar provided valuable insights about local knowledge systems and access to FRLHT, Bangalore. The officials and staff from the Maharashtra Forest department particularly, Mr. Bhangre, Mr. Majumdar, Mr Gogte, Mr.Khaire, Mr. Rajadne, Mr. Shivram, Mr Kashiram, have been helpful in providing useful data on the study. Finally, without the warm support and cooperation of the local management committee, healers and villagers of Amboli particularly Mr Dilip Sawant, Mr Arun Chawan Mr Anthony, Mr Anirudhha, Ms Merry D'souza, Ms Anita Sawant, Ms Dixita Guruv, Mr Shailesh Mehtar, Mr Prashant and Mr Suresh the local field work in Amboli would not have been possible. The review by two anonymous referees provided valuable direction in improving the draft. Usual disclaimers apply.

REFERENCES

1. Berkes F: Sacred Ecology: Traditional Ecological Knowledge and Resources Management. 1999, Philadelphia: Taylor and Francis
2. Foundation for Revitalization of Local Health Traditions(FRLHT): Bangalore: FRLHT, (Accessed 22 June 2005), [http://www.frlht.org.in/]
3. Nelson JG: Research in human ecology and planning: an interactive, adaptive approach. The Canadian Geographer. 1991, 35 (2): 114-127.
4. Rural Commune's Medicinal Plant Conservation Center: Conservation of Medicinal Plants in Maharashtra State, Status and Progress Paper, Mimeo, Pune: RCMPCC. 2003
5. Silverman D: Doing qualitative research: A Practical handbook. 2005, London; Thousand Oaks: SAGE
6. World Health Organization: Traditional Medicines Strategy 2002–2005. 2002, Geneva: WHO
7. Nickel W, Sennhauser E: Medicinal Plants: Local heritage with Global importance. World Bank Group, South Asia, Accessed 21 August 2005, [http://lnweb18.worldbank.org/sar/sa.nsf/0/fae63d87e2bd14038525687f0057e0d1?OpenDocument]
8. Schippmann U, Leaman DJ, Cunningham AB: Impact of cultivation and gathering of medicinal plants on biodiversity: global trends and issues. 2002, Inter-Department Working Group on Biology Diversity for Food and Agriculture, Rome, Italy: Food and Agriculture Organizations (FAO)
9. Hamilton A: Medicinal plants and conservation: issues and approaches. 2005, Surrey, U.K.: International plants Conservation Unit, WWF-UK
10. Shankar DS: Revitalizing local health traditions. Food for thought: ancient visions and new experiments of rural people. Edited by: Haverkort B, Wim Hiemstra. 1999, London: Zed Books, 43-51.
11. Shankar DS: Agenda for Revitalization of Indian Medical Heritage. 2001, New Delhi, India: Voluntary Health Association of India (VHAI)
12. Gupta AK: Conserving Biodiversity and Rewarding Associated Knowledge and Innovation system: Honey Bee Perspective. 1999, Invited paper for the first Commonwealth Science Forum-Access, Bio prospecting, Intellectual Property Rights and Benefit Sharing and the Commonwealth, Ahmedabad: SRISTI, [http://www.sristi.org/pub.html]
13. Kala CP: Revitalizing Traditional Herbal Therapy by Exploring Medicinal Plants: A Case Study of Uttaranchal State in India. Proceedings of the International Conference on Indigenous knowledge: Transforming the Academy: 27–28 May 2004. Edited

by: Martezki A, Semali L. 2004, Pennsylvania, Penn State University, 15-22. (Accessed 19 January 2005), [http://app.outreach.psu.edu/IndigenousKnowledges/default.asp?WhichPage=Proceedings]

14. Balasubramaniam AV: People's Knowledge in Health care. Reading Material in Local Health Traditions. Edited by: TSM, Foundation for Revitalization of Local Health Traditions. 2000, Bangalore: FRLHT, 17-26.

15. Shankar DS, Venkatasubramanian P: The evolution of global standards for traditional medicines. IK Dossier on Science Development Network. 2004, (Accessed 27 August 2005), [http://www.scidev.net/dossiers/index.cfm?fuseaction=policybrief&dossier=7&policy=58]

16. Bodeker G: Medicinal plants: towards sustainability and security. 2002, Oxford, UK: Green College

17. National Medicinal Plants Board's website. 2004, (Accessed 21 October 2005), [http://www.nmpb.nic.in]

18. Shukla SR, Gardner JS: Lessons from the Equator Initiative: Rural Commune's Medicinal Plant Conservation Centre, Pune, India: A Technical Report. 2004, Winnipeg: Centre for Community-Based Resource Management, Natural Resources Institute University of Manitoba, (Accessed 24 June 2005), [http://www.umanitoba.ca/institutes/natural_resources/nri_cbrm_projects_eiprojects.html]

19. Rural Commune's Medicinal Plant Conservation Center: Village Biologist Concept Note. 1999, Mimeo Pune: RCMPCC

20. Jagtap S, Deokule S, Watve A: Occurrence of threatened fragrant Ceropegia in Toranmal Forest, Maharashtra. Current Science. 2004, 27 (5): 553-554.

21. Berkes F: Re-thinking community-based conservation. Conservation Biology. 2004, 18 (3): 621-630. 10.1111/j.1523-1739.2004.00077.x.

CHAPTER 6

CONSERVATION AND UTILIZATION OF AFRICAN ORYZA GENETIC RESOURCES

Peterson W Wambugu[1,3], Agnelo Furtado[1], Daniel LE Waters[2], Desterio O Nyamongo[3] and Robert J Henry[1]

[1]Queensland Alliance for Agriculture and Food Innovation, The University of Queensland, Brisbane, St Lucia, Qld, Australia

[2] Southern Cross Plant Science, Southern Cross University, Lismore, NSW, Australia

[3] Kenya Agricultural Research Institute, Nairobi, Kenya

ABSTRACT

Africa contains a huge diversity of both cultivated and wild rice species. The region has eight species representing six of the ten known genome types. Genetic resources of these species are conserved in various global germplasm repositories but they remain under collected and hence underrepresented in germplasm collections. Moreover, they are under characterized and therefore grossly underutilized. The lack of in situ conservation programs further exposes them to possible genetic erosion or extinction. In order to obtain maximum benefits from these resources, it is imperative that they are collected, efficiently conserved and optimally utilized. High throughput molecular approaches such as genome sequencing could be employed to more precisely study their genetic diversity and value and thereby enhance their use in rice improvement. Oryza sativa was the first crop plant to have its reference genome sequence released marking a major milestone that opened numerous opportunities for functional characterization of the entire rice genome. Studies have however demonstrated that one reference genome sequence is not enough to fully explore the genetic variation in the Oryza genus, hence the need to have reference sequences for other species in the genus. An overview of the state of conservation and utilization of African Oryzais hereby presented. Progress in the release of reference genome sequences for these species is also highlighted.

INTRODUCTION

The *Oryza* genus has two cultivated species, *Oryza sativa* and *Oryza glaberrima*, and about 24 wild species (Lu and Jackson 2009; USDA-ARS 2013) representing ten rice genome types (Ge et al. 2001). With a total of eight species of both cultivated and wild rice species, representing six out of the ten known genome types (Table 1 and Figure 1), the African region is, arguably, one of the greatest sources of diversity in the rice gene pool. It is the only region where the two cultivated species co-exist by growing sympatrically. Unlike the Australian *Oryza* which has been genetically isolated from domesticated rice (Henry et al. 2009; Waters et al. 2012), African wild rice is found growing in contact with the cultivated taxa allowing extensive hybridization between cultivated and some of the wild rice species especially those in the AA genome (Lu and Snow 2005). Despite this possible hybridization, African wild rice species remain a largely untapped source of useful genetic

Citation: Peterson W Wambugu, Agnelo Furtado, Daniel LE Waters, Desterio O Nyamongo and Robert J Henry, Conservation and utilization of African Oryza genetic resources, DOI: 10.1186/1939-8433-6-29.

and allelic diversity although they have made an important contribution in rice improvement (Brar and Khush 2002; Jena 2010; Khush et al. 1990).

Table 1: African *Oryza* species, their chromosome numbers and genome types

Species	2n	Genome type
Oryza sativa L.	24	AA
Oryza longistaminata A. Chev. et Roehr.	24	AA
Oryza glaberrima Steud.	24	AA
Oryza barthii A. Chev.	24	AA
Oryza punctata Kotschy ex Steud.	24	BB
Oryza schweinfurthiana Prod.[1]	48	BBCC
Oryza eichingeri A. Peter.	24	CC
Oryza brachyantha A. Chev. et Roehr.	24	FF

[1]*Oryza schweinfurthiana* Prod is also considered the tetraploid form of *Oryza punctata* Kotschy ex Steud. This paper makes no further reference to *Oryza schweinfurthiana* Prod. as there is negligible amount of data available on this species.

Figure 1: Undehusked seeds of African *Oryza* species. (a) *Oryza longistaminata* (b) *Oryza glaberrima* 1 (c) *Oryza glaberrima*2 (d) *Oryza brachyantha* (e) *Oryza eichingeri* (f) *Oryza punctata* (g) *Oryza barthii.*

Despite the long history of rice research covering diverse areas and the demonstrated value of the African *Oryza*species, there is still little information on these species compared to others in the genus (Goicoechea 2009). World agriculture is facing unprecedented challenges largely attributed to climate change and the need to ensure proper conservation and sustainable utilization of genetic resources of crop wild relatives has never been more important (Brar 2004; FAO 2010). In order to ensure that maximum benefits accrue to humankind from these resources, it is imperative that they are collected, efficiently conserved and optimally utilized. The continued dramatic decrease in the costs of sequencing and genotyping has revolutionized genetic resources conservation and utilization and application of high throughput genomics is increasingly becoming common in genebank activities (McCouch et al.2012). While information on the utilization and conservation of other wild species have been reviewed (e.g. Henry et al. 2009; Song et al. 2005), no such efforts have been made for the African *Oryza* species. This paper provides an overview of the status of conservation and utilization of African *Oryza* species and identifies gaps in knowledge and opportunities for further research.

REVIEW

Phylogeny of African *Oryza*

Efficient conservation and utilization of the *Oryza* gene pool will require a clear understanding of the phylogeny and evolutionary relationships of the various species in addition to a reliable taxonomic and biosystematics framework. In the last two decades, efforts have been made in studying the phylogenetic relationships in the *Oryza* genus using a variety of approaches. These have mainly entailed the use of molecular markers such as Random Amplified Polymorphic DNA (RAPD; Bautista et al. 2001; Ishii et al. 1996; Ren et al. 2003), Simple Sequence Repeats (SSR; Ishii et al. 1996; Ren et al. 2003), Restriction Fragment Length Polymorphisms (RFLP; Bautista et al. 2001; Lu et al.2002; Wang et al. 1992), Amplified Fragment Length Polymorphism (AFLP; Park et al. 2003a), Inter Sequence Simple Sequence Repeats (ISSR; Joshi et al. 2000), Short Interspersed Elements (SINEs) and Miniature Inverted-Repeat Transposable Element (MITE) insertions (Cheng et al. 2002; Iwamoto et al. 1999; Mochizuki et al. 1993). In addition, morphological and cytological studies have also been conducted (Lu et al. 2000). These studies have significantly increased our understanding of the evolutionary relationships in the *Oryza* but many questions remain unanswered (Tang et al. 2010; Zou et al. 2008). Resolving apparent phylogenetic incongruences in the *Oryza* genus must therefore remain a major focus of research for some African species.

Among the African species, *Oryza brachyantha* has been the most affected by the conflicts resulting in several inconsistent phylogenetic placements depending on the phylogenetic approach used. For example, a phylogenetic study based on *Adh1*, *Adh2* and *matK* genes conducted by Ge et al. (1999) resulted in 3 different phylogenies all with inconsistent positioning of *Oryza brachyantha*. The *Adh2* phylogeny suggested a strong relationship between*O. brachyantha* and the diploid HH genome which was not supported by the *Adh1* phylogeny. The 3 phylogenies constructed by these authors depicted *O. meyeriana* and *O. granulata* (GG) as the most basal taxa, a finding that was inconsistent with the phylogenetic analysis by Mullins and Hilu (2002) where *O. brachyantha* appeared as the most basal lineage in the whole genus. Due to this discordance, Ge et al. (1999) recommended the use of additional genes in future phylogenetic studies. The most tenuous placement of *O. brachyantha* was by Wang et al. (1992) and Nishikawa et al. (2005), who, despite *O. brachyantha* being one of the most divergent species, placed it closest to the *Oryza sativa* complex. According to Goicoechea (2009), this seemingly controversial placement of*Oryza brachyantha* could be attributed to sequence conservation between AA and FF genomes in some of the loci under investigation. Just like *O. brachyantha* (FF), the AA genome species have also suffered from phylogenetic incongruence. For example, based on MITE-AFLP and RFLP data, Wang et al. (1992) and Park et al. (2003b) found*O. meridionalis* to be the most distinct species. This was contradicted by Ren et al. (2003), who, by conducting an SSR and RAPD analysis, found *Oryza longistaminata* to have undergone significant differentiation from the other AA genome species, a finding that was consistent with other previous studies (Cheng et al. 2002; Iwamoto et al.1999). Figure 2 shows a consensus tree of the AA genome based on *trn* L-*trn* F sequences. Understanding the causes of these different conclusions may help to resolve them.

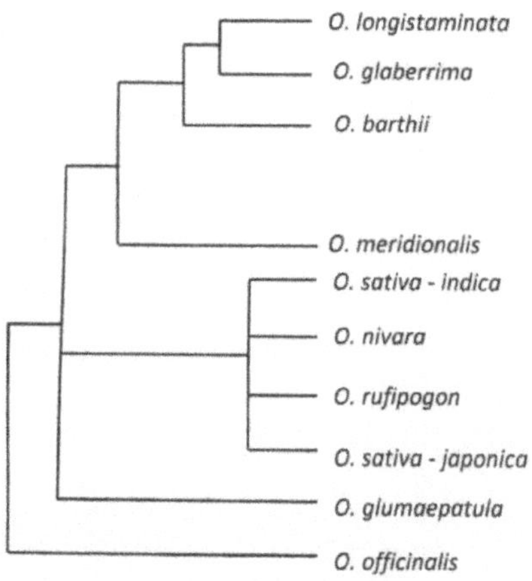

Figure 2: Consensus tree of the *A* A genome rice species based on *trn* L- *trn* F sequences (Adapted from Duan et al.2007).

The differential results in the various phylogenetic analysis can be attributed to misidentification of accessions, gene choice, use of insufficient data, introgression and hybridization, rapid speciation as well as lineage sorting (Wang et al. 1992; Zhu and Ge 2005; Zou et al. 2008). Results from many studies provide evidence which shows that some of the previously unresolved phylogenetic relationships could be resolved by the use of more data, better selection of genes as well as better phylogenetic methods (e.g. Cranston et al. 2009; Zou et al. 2008). The importance of proper choice and use of a sufficient number of unlinked genes in resolving phylogenetic studies was demonstrated by a recent study that used a total of 142 single copy genes to resolve the phylogeny of all the diploid genomes of*Oryza* (Zou et al. 2008). Moreover, the use of chloroplast DNA as opposed to nuclear data has been suggested as one way to reduce incongruence in phylogenetic relationships (Takahashi et al. 2008; Waters et al. 2012). Chloroplast DNA is haploid, non-recombinant and is generally maternally inherited thus making it a useful tool for molecular systematics (Small et al. 2004; Waters et al. 2012). A recent study of the *Oryza* phylogeny generated a better resolved phylogeny based on 20 chloroplast fragments (Tang et al. 2010). However, the potential of chloroplast DNA in resolving phylogenetic relationships as highlighted by these studies is contradicted by other studies where the use of chloroplast fragments has resulted in inconsistent phylogenies or lack of sufficient resolution (Guo and Ge 2005). It would appear that this apparent phylogenetic inconsistency based on chloroplast data may be due to the number of genes or DNA fragments studied. As with nuclear data, increasing the number of chloroplast loci included may increase resolution. Indeed, Waters et al. (2012) reported that the use of whole chloroplast genome sequences has more power in phylogenetic resolution. However, this molecular tool has not been fully exploited in reconstructing phylogeny of the *Oryza*. On the other hand, with the continued advances in bioinformatics and increased computation capacity, it is expected that the use of whole genome sequences in resolving phylogenetic relationships will soon become feasible, thus increasing resolution. Better information on the phylogenetic relationships in the *Oryza* genus will facilitate more efficient conservation and utilization in rice crop improvement programs.

Germplasm conservation of African *Oryza*

Plant genetic resources underpin world agriculture and their conservation is therefore imperative if food security and sustainable development is to be assured. To define the current status of African *Oryza* germplasm conservation, we review approaches that are currently being used in the conservation of wild rice germplasm ex situ and in situ.Details of available germplasm collections and a brief analysis of collection gaps in ex situ conservation, is presented.

Ex situ conservation

Collection and conservation

Ex situ conservation is the most common germplasm conservation method and can take the form of seed banks, field genebanks or botanic gardens. DNA storage in DNA banks represents another option for the conservation of genetic materials (Henry et al. 2009). Collection of wild rice relatives started in the 1950s reaching a peak in the 1980s and 1990s when the International Board for Plant Genetic Resources (IBPGR, formerly IPGRI and now Bioversity International) was formed. IBPGR, in collaboration with both international and national partners funded and spearheaded most of the African wild rice germplasm collection efforts. This was due to the realization that these valuable resources faced eminent threat of genetic erosion due to both environmental and socio-economic factors. The collected materials were deposited in national genebanks with duplicates being conserved in international genebanks especially the International Rice Research Institute (IRRI;IBPGR 1984). Currently, IRRI holds the largest collection of African *Oryza* totaling to about 3601 accessions (Table 2). Other collections of African*Oryza* can be found in Australian Tropical Grains Germplasm Centre; International Network for the Genetic Evaluation of Rice in Africa (INGER-Africa); Bangladesh Rice Research Institute, Dhaka; SADC Plant Genetic Resources Centre (SPGRC),

Zambia and China National Rice Research Institute (CNRRI) (Agnoun et al. 2012; Hay et al. 2013). Duplicate samples of African *Oryza* collections are shared between IRRI and the Africa Rice Center (previously known as WARDA) in Benin, Africa (Brar and Singh 2011). As in other species, the level of unintended duplication is high in rice thus escalating the costs of conservation. Many strategies and methods have been used to identify duplicates but most of these have been unsuccessful. However, high throughput genotyping and sequencing has the capacity to more precisely identify duplicate samples in the future as well as enable a more accurate analysis of available collection gaps.

Table 2: Regional, International and genebanks outside Africa holding African *Oryza* germplasm

Species	IRRI, Philippines[1]	Africa Rice Centre[1]	Australian Plant DNA Bank[2]	Millennium Seed Bank Project[1]	USDA[3]	ILRI, Ethiopia[1]	National Institute of Genetics[4]	SPGRC
O. glaberrima	2828	3796	7	1	476	1		
O. barthii	331	114	5	5	15	1	386	2
O. longistaminata	285		2	3	9	5	149	54
O. eichingeri	37		6	0	0	0	11	
O. punctata	89		0	1	14	2	18	
O. brachyantha	31		0	1	6	1	16	2

[1]http://www.genesys-pgr.org/.

[2]http://collections.ala.org.au/public/show/co133.

[3] USDA-ARS (2013).

[4] Nonomura et al. (2010).

Collection gap analysis

The size of collections conserved in the genebanks (Table 2) suggests that maximum genetic diversity has been captured. However, effective management of these genebanks and conservation of their resources would benefit from detailed gap analysis which would help guide future conservation priorities. Such an analysis would help to determine if all the genetic diversity found in a taxa is represented in in situ as well as ex situ conservation facilities (Maxted et al. 2008). While the combined use of molecular data, eco-geographic surveys and more accurate geographic referencing is vital in identifying gaps and redundancies in existing collections (FAO 2010), very little of this has been done for African wild *Oryza* held in many national collections. With the rapidly changing nature of African agriculture and in the face of increasing documented threats to biodiversity in the region (Khumalo et al.2012; Wambugu and Muthamia 2009), there is need to use these tools to more precisely and systematically understand and document this genetic diversity as well as the status of its conservation. A comparative study of the diversity conserved ex situ with that found in situ would be particularly useful as it would help identify some possible collection gaps. There however seems to be no published work on such comparative studies for African *Oryza*. The success of a collection gap analysis is partially dependent on the quality and integrity of the available data (Ramirez-Villegas et al. 2010) and currently, paucity in the necessary data remains a major constraint in undertaking a comprehensive collection gap analysis. Available data indicates that globally, there are no collection deficiencies at the taxa level as all the taxa are represented in ex situ facilities. However, an analysis of germplasm collection data (Additional file 1: Tables S1 and S2) and herbarium specimens (Additional file 1: Tables S3 and S4) for individual in-country collections reveals some significant gaps in taxa coverage. For example, in Kenya, where the authors had access to comprehensive and reliable data, there is a clear taxonomic gap in the collection of *Oryza eichingeri*. A herbarium specimen was collected and deposited at Missouri Botanic Garden (http://www.tropicos.org/Specimen/3179202) but no germplasm samples of the same have ever been collected.

In contrast to other species such as *Phaseolus sp*. (Ramirez-Villegas et al. 2010), the number of herbarium specimens in *Oryza* is higher than conserved germplasm accessions, a possible indicator of underrepresentation in ex situ conservation facilities. This observation is reinforced by a further numerical assessment of individual in-country germplasm collections (Additional file 1: Tables S1 and S2) which indicates that some of the wild species especially *Oryza eichingeri* and *Oryza brachyantha* have very few or no accessions in both national and international genebanks. A similar observation of under representation of *Oryza* wild species in ex situ conservation facilities was made by Maxted and Kell (2009). These two species had the lowest sampling representative score (SRS), of 5.9 and 5.6 respectively, with *O. glaberrima* having the highest followed by *Oryza longistaminata*. Sampling representative score (SRS) is an indicator of the adequacy of germplasm holdings for a particular taxon, based on available herbarium specimens and germplasm collections (Ramirez-Villegas et al. 2010). While acknowledging that the number of accessions may not always give an accurate reflection of the available diversity (FAO 2010), this data indicates some in-country collections with clearly evident collection gaps that may need to be filled. This finding aligns well with expert opinions on the existence of gaps in ex situ collection. For example, Hay et al. (2013) noted that there were indications of collection gaps of wild rice species in areas outside Asia such as Africa. Similarly, Ngwediagi et al. (2009) reported wild rice collection gaps in Tanzania. Using molecular approaches, efforts should be made to ensure that any targeted collection efforts arising from these collection gap analysis and recommendations, should only be undertaken if it has been established that it will result in new genetic or allelic diversity. The planned release of reference genome sequences of African *Oryza* is expected to provide a basis on which to assess available genetic diversity using high throughput genomic approaches. As observed by McCouch et al. (2012), due to the continued dramatic decrease in the costs of sequencing and concomitant increase in efficiency, it is currently cheaper to undertake low coverage sequencing of an accession than it is to add an accession of cultivated rice to a germplasm collection. Consequently, it is expected that in future, genotyping using molecular markers or by sequencing will become a routine activity before an accession is banked to ensure that only those with novel alleles or allele combinations are added to the collection. In addition to these ex situ collections, the importance of putting in place extra safeguards to genetic resources, by establishing and maintaining in situ collections is now universally accepted.

In situ conservation

In situ conservation of wild species has for decades now been undertaken to complement ex situ conservation and is known to have particular advantages such as allowing the natural process of evolution to continue. However, despite the demonstrated importance of in situ conservation and several warnings on the alarming rates at which genetic diversity of rice wild relatives is being lost (Song et al. 2005; Vaughan and Chang 1992), there is no documented evidence of any targeted in situ conservation programs for African wild rice species (Brink and Belay2006; Maxted and Kell 2009). This is in contrast to other rice species where several studies have indicated the presence of such programs (Song et al. 2005; Xie et al. 2010). The need to put in place a robust, complementary in situ conservation program for the African rice gene pool cannot be overemphasized and there is great agreement in the literature on the importance of such a program. Vaughan and Chang (1992) noted that ex situ conservation of wild species that exhibit great heterogeneity in their genetic structure is not only expensive but also time consuming, lending support to the need for in situ conservation. Maxted and Kell (2009) noted that in order to establish in situ genetic reserves, detailed genetic studies of wild rice species are vital as they help in identifying priority locations for in situ conservation. Such knowledge, supplemented with information on herbarium specimens (Additional file: 1: Table S3 and S4) and species distribution (Figure 3) will be important in supporting conservation and sustainable utilization management decisions. Detailed genetic diversity studies on the naturally occurring variation of African*Oryza* are however limited, thus constituting a major information gap that greatly hampers establishment of these reserves. Currently, while populations of African wild rice may be found occurring in nationally designated protected areas, these are just an indirect effect of the establishment of these areas and such populations benefit from no active management. In Tanzania, for example, Vaughan and Chang (1992) reported the occurrence of *O. barthii* and*O. punctata* in Ruaha National Park which is a protected habitat. Outside the protected areas, wild rice species may also be found in cultivated farmers' fields, field edges, pasturelands, orchards, recreation parks and roads.

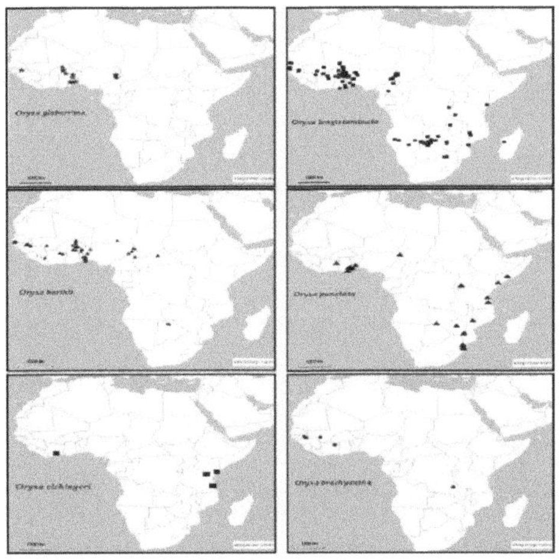

Figure 3: Distribution of *Oryza* species in Africa. Species distribution has been mapped based on records of herbarium specimens which have been preserved in various herbaria globally.

Whether in protected or in non-protected areas, *Oryza* populations are at great risk of extinction due to threats of climate change, overgrazing, flooding, habitat change, invasive alien species and pollution. Climate change probably constitutes the greatest threat, with reports indicating that Africa, especially sub-Saharan Africa which holds considerable diversity of these species, will be adversely affected by climate variability. The challenge of climate change is expected to be serious particularly for wild species and urgent efforts are therefore needed to secure their genetic resources as the likelihood for extinction of narrowly adapted and endemic species is high (Jarvis et al. 2009). However, even in the face of threats posed by climate change, wild rice species are still expected to provide the basis for adapting agriculture to climate change in future. Those naturally occurring in situ populations that have the capacity to withstand the harsh effects of climate variability will have the potential to contribute valuable new traits for rice improvement (FAO 2010; Pettersson et al. 2009). In addition to these threats, an often neglected challenge facing in situ conservation is that of gene flow between cultivated and wild rice species and its impact on genetic diversity and integrity (Figure 4).

While it is generally acknowledged that the transfer of genes from wild to cultivated taxa may have important and beneficial consequences, gene flow in the reverse direction (i.e. from the crop to the wild) may lead to deleterious changes in genetic diversity or even result in extinction of small populations (Ellstrand et al. 1999). It would appear therefore that establishment of genetically isolated reserves is the most viable and effective strategy to conserve and at the same time reduce genetic admixture and its associated consequences. However, such a recommendation needs to be based on comprehensive gene flow studies that allow assessment of the genetic integrity of the populations involved. Such studies have lately been undertaken using high resolution genotyping tools such as Single Nucleotide Polymorphisms (SNPs) and SNP haplotypes as well as sequencing technologies.

Figure 4: Rice field in coastal Kenya planted with *Oryza sativa* landrace with patches of *Oryza punctata* . The sympatric occurrence of cultivated and wild species may lead to gene flow thus affecting the genetic integrity of these species. It could also lead to extinction of less adapted genotypes.

On farm conservation

During the course of farming, farmers have been known to maintain important genetic diversity of *Oryza sativa* and*Oryza glaberrima* within the farming system (Figure 5). This system of conservation, often referred to as on-farm conservation, is usually characterized by local/farmer seed production with little or no acquisition of certified seeds from the formal seed sector. Additionally, the system relies heavily on the use of diverse traditional varieties which usually have high levels of genetic diversity as compared to improved varieties. A study conducted in West Africa shows that about 70% of the farmers grow their traditional rice varieties (Mohapatra 2007). According to a National Research Council report, most of the farmers in this region have reported deliberate mixing of both the cultivated Asian and African rice in their farms so as to foster genetic diversity which occurs as a result of introgression. This has resulted in new types of landraces which, judging by ligule form, grain shape, and panicle type, are intermediate between the cultivated Asian and African rice (National Research Council 1996). These resultant landraces may be low yielding but are able to cope up with many biotic and abiotic stresses due to their heterogeneity and are therefore likely to lead to yield stability. Studies have shown that farmers prefer yield stability to maximum obtainable yields in line with their strategy of risk avoidance (Almekinders and Louwaars 1999; Wambugu et al.2012). This observation is confirmed by reports that farmers in some parts of West Africa such as the *Banfora* area in Burkina Faso are abandoning improved rice varieties in favour of the low yielding but highly adaptable *Oryza glaberrima* varieties (Futakuchi et al. 2012). Barry et al. (2007) observed that the replacement of rice landraces by improved varieties in Africa is less advanced than in Asia. It is therefore important that these landraces are conserved on-farm as well as ex situ and genetic studies undertaken on them as they might harbour useful traits that may find utility in rice improvement.

Figure 5: Phenotypic diversity of both cultivated and wild *Oryza* genetic resources: Upper row, *Oryza sativa* landraces; Lower row, (a) *Oryza glaberrima,* (b, c, d, e, f) African wild *Oryza* species.

Utility of African *Oryza*: achievements and challenges

As already highlighted, a lot of efforts have been made in the collection and conservation of huge collections of African *Oryza* which are currently conserved in various facilities (Table 2), for the benefit of present and future generations. In order to derive maximum benefits from these resources, it is imperative that efforts are made to ensure their optimal utilization in research, crop improvement as well as direct use. Promoting the sustainable use of these plant resources requires an understanding of their value. This section therefore reviews some of the potentially desirable traits they possess as well as some of the progress achieved in incorporating them into commercial varieties.

Genetic potential of African *Oryza*

African cultivated and wild rice species are known to possess enormous genetic diversity (Figure 5) which is of immense genetic value especially on resistance to biotic and abiotic stresses and therefore remain a vital raw material in rice improvement programs (Table 3). Efforts to use these resources have led to significant success in the transfer of useful traits into cultivated rice (Brar and Khush 2002). Hajjar and Hodgkin (2007) indicated that to date, a total of 12 traits in cultivated rice have been improved through the use of wild rice gene pool. One of the most significant and successful uses of wild genes in rice improvement is the transfer of the *Xa-21* gene conferring resistance against bacterial blight resistance which was successfully introgressed from *O. longistaminata* into *Oryza sativa* (Khush et al. 1990). Several improved varieties carrying the *Xa-21* gene have subsequently been released in different countries.

Table 3: Useful traits found in African *Oryza* species

Species	Trait	Reference
O. longistaminata	Resistance to bacterial blight, nematodes, drought avoidance; rhizomatousness	(Brar and Khush 2002; Hu et al. 2011; Jena 2010; Khush et al. 1990; Yang et al.2010)
O. brachyantha	Resistance to bacterial blight, yellow stem borer, leaf-folder, whorl maggot, tolerance to laterite soil	(Brar and Khush 2002; Ram et al. 2010; Yamakawa et al. 2008)
O. glaberrima	Resistance to drought , iron toxicity, nematodes; weed competitiveness; high adaptability to acidic soils showing low levels of phosphorus availability; cultigen; tolerance to waterlogging; crude protein content; cultigen; African gall Midge; stem borers; Rice yellow mottle virus; resistant to nematodes	(Brar 2004; Brar and Khush 19972002; Dingkuhn et al. 1998; Futakuchi et al.2001; Jones et al. 1997; Li et al. 2004; Ndjiondjop et al. 1999; Nwilene et al.2002; Plowright et al. 1999; Sauphanor1985)
O. barthii	Resistance to green leaf hopper, bacterial blight, drought avoidance	(Brar and Khush 2002)
O. punctata	Resistance to brown plant hopper, zigzag leafhopper	(Brar and Khush 2002; Jena 2010)
O. eichingeri	Resistance to brown plant hopper, white-backed plant hopper, green leaf hopper	(Brar and Khush 2002; Yan et al. 1997)

Breeding for higher yields and yield stability remains the major objective of most rice breeding programs worldwide. Though wild rice species are phenotypically inferior and have predominantly been used as a source of genes for resistance to pests and diseases (Hodgkin et al. 2007), they also possess the genetic value necessary to improve the yield potential of cultivated rice (Xiao et al. 1998). However, as noted by Hajjar and Hodgkin (2007), the contribution of wild species in improving yields has been limited or almost non-existent, not only in rice but also in all other economically important crops. In a review of the use of wild relatives in crop improvement in 16 major economically important crops, these authors reported only one case of a released variety bred by incorporating yield enhancing genes from a wild rice species. The rice cultivar, NSICRc112, developed from a cross between *O. longistaminata* and *O. sativa* was released in the Philippines in 2002 and is known to be high yielding (Brar 2004). Similar to the case of yield improving genes, wild species have not contributed any genes to enhance drought tolerance in rice. This apparent lack of contribution of genes to improve quantitatively inherited traits such as yield can be attributed to the fact that it is difficult to phenotypically identify these superior traits in wild species (Bimpong et al. 2011). There is therefore need for more focus to be put on the use of molecular approaches in the identification of yield and drought related Quantitative Trait Loci (QTL) among other quantitative traits. In addition to breeding for stresses and yields, considerable attention has also been given to breeding for rice quality.

Rice quality is an important trait that largely determines its demand both at a household level as well as in international markets. Breeding for rice grain quality is therefore increasingly becoming a priority in breeding programs especially in developed countries. However, as noted by Henry et al. (2009), rice wild relatives have not yet contributed genes to improve the quality of rice. Both wild and cultivated African species however seem to have potential to contribute genes that will improve rice grain quality as well as acting as alternative food sources. In the case of African wild species, some species namely *O. longistaminata*, *O. punctata* and *O. barthii,* are consumed as whole grains in times of food scarcity thereby acting as a safety net for poor communities (Brink and Belay 2006). While it would therefore appear reasonable to suggest that these wild species have capacity for acting as new food crops, their nutritional properties are yet to be studied. Rice is mainly composed of starch and hence an analysis of the starch characteristics of these species and the diversity of genes encoding for these starches would be particularly important but has not been explored in African *Oryza* species. Starch biosynthesis genes have been studied extensively in cultivated rice resulting in identification of huge numbers of Single Nucleotide Polymorphisms (SNPs) in these genes, that have the potential to be used as molecular markers (e.g. Chen et al. 2008; Kharabian-Masouleh et al. 2011; Kharabian-Masouleh et al. 2012; Larkin and Park 2003; Umemoto et al. 2004; Umemoto et al.2008). Kasem et al. (2012) identified a total of 251 SNPs in the exons of three starch genes namely *Grain Bound Starch Synthase I (GBSSI), Starch Branching Enzyme IIb (SBEIIb)* and *Starch Synthase IIa (SSIIa).* This study however included only Australian and Asian wild rice species leaving out African species. Grain morphology is another important quality trait that has an impact on the utilization of rice grains. In a study of grain morphology of a number of Australian wild rice species, Kasem et al. (2010) found a wide variation in grain morphological characteristics, specifically size and shape, but all of which were within standards acceptable to breeders. This study highlighted the potential utility of these wild species as whole grain foods as well as novel sources of important diversity for use in improving both *O. sativa* and *O. glaberrima*. Despite being a cultivated species, *O. glaberrima* is arguably the most genetically promising of all African

Oryza, possessing a rich repertoire of favourable genes and alleles that have the potential of improving a diverse range of agronomically important traits in rice (Table 3). Recent studies of a QTL analysis of an *O. sativa* and *O. glaberrima*cross indicated that *O. glaberrima* contains useful QTL alleles that are likely to significantly enhance traits including yield and other yield components particularly under conditions of drought stress. Some of these alleles are potentially novel and have been shown to be stable across genetic backgrounds (Bimpong et al. 2011). This cultivated African species also seems to be a new source of valuable genes for improvement of specific rice qualities such as eating, cooking and milling properties of rice grain, qualities that are valued in some rice export markets (Aluko et al. 2004). Furthermore, QTL analysis of progenies derived from interspecific crosses between*Oryza sativa* and *Oryza glaberrima* has revealed some loci in *O. glaberrima* that could enhance crude protein content and improve grain shape and appearance (Aluko et al. 2004; Li et al. 2004). Increased protein content is particularly desirable for poor regions such as Asia and Africa where malnutrition is widespread. Interspecific crosses between *O. sativa* and *O. glaberrima* resulted in interspecific hybrids trademarked as New Rice for Africa (NERICA) (Futakuchi and Sié 2009; Jones et al. 1997). NERICA varieties combine some of the stress tolerance traits of *O. glaberrima* with the high yielding potential of *O. sativa* (Jones et al. 1997).

Despite its great potential for utility in rice improvement programs, *O. glaberrima* has some negative traits namely poor yield, seeds that split and shatter easily, a notorious difficulty of milling and plants that lodge easily (Linares2002; National Research Council 1996). Moreover, the brown or red pericarp is also not appealing to most consumers and in most cases the rice has to be polished to remove it (Teeken et al. 2012). It is perhaps due to these reasons that it is rapidly being displaced in West Africa in favour of Asian rice (Jones et al. 1997; Linares2002) and is currently almost becoming extinct (Mohapatra 2010). While this calls for urgent breeding efforts, there is also need to give more attention to the conservation of this germplasm before it gets lost. Moreover, the conserved germplasm needs to be properly characterized in order to unravel its genetic architecture.

Characterization of African *Oryza*

Successful utilization of genetic resources especially of wild species in breeding programs as well as in other research primarily requires some understanding of their phenotypic and genotypic characteristics as this knowledge increases the potential value of these resources. Moreover, this knowledge is important in making important germplasm conservation management decisions. Using AFLPs, Kiambi et al. (2005) studied the genetic diversity and population structure of 48 *O. longistaminata* populations obtained from 8 Eastern and Southern Africa countries. The study revealed higher levels of genetic diversity as compared to some other species in the *Oryza*genus such as *O. glumaepatula*. This diversity was found to be more within than between populations and also to be more in populations within countries than among countries. Similar results were found in a recent study involving 320 accessions of *O. longistaminata* obtained from 8 populations in Ethiopia (Melaku et al. 2013). This study found more within than between population diversity but the level of diversity was higher than the one detected in the study by Kiambi et al. (2005). In another study on genetic diversity and domestication history, Li et al. (2011) found 70% less diversity in *O. glaberrima* as compared to *O. barthii*, its wild progenitor, an indication of severe domestication bottleneck. These and other similar studies have been valuable not only in ensuring effective germplasm utilization but also in setting conservation priorities and defining sampling strategies. It is on the basis of outputs from such genetic diversity and evaluation studies that most of the *Oryza* genomic resources have been developed.

Currently, the *Oryza* research community has access to numerous genomic resources among them a reference sequence, advanced mapping populations, transcriptome data as well as physical and genetic maps. *Oryza sativa*was the first crop plant to have its reference genome sequence released (International Rice Genome Sequencing Project 2005) marking a major milestone that opened numerous opportunities for functional characterization of the entire rice genome. Studies have however demonstrated that one reference genome sequence is not enough to fully explore the genetic variation in the *Oryza* genus (Goicoechea et al. 2010). Consequently, efforts to develop reference genome sequences for some other selected species in the *Oryza* genus have been on-going under the auspices of the International *Oryza* Map Alignment Project (IOMAP). Through this initiative, physical maps have been developed for some selected species among them some African rices namely *O. glaberrima*, *O. punctata* and*O. brachyantha* (Kim et al. 2008). Genome sequences of *O. glaberrima*, *O. barthii*, *O. punctata and O. longistaminata* are available in Genbank though they are yet to be published (Table 4) (Jacquemin et al. 2013). Recently, *O. brachyantha* became the first African species to have its reference genome sequence published (Chen et al. 2013). With the expected release of reference genome sequences for the other wild species and with the continued decline in next generation sequences, it is anticipated that it will become increasingly feasible to study intraspecific genetic diversity by comparing individual genome sequences with reference sequence. Moreover, the release of these reference genome sequences will usher in a new platform that will give impetus to re-sequencing efforts which will in turn yield valuable information on SNPs, insertions, deletions, and other mutations as well as other structural and functional variations.

Table 4: Sequencing status of African *Oryza* genomes

Species	Genome size	Sequencing method	Status
O. sativa ssp. *indica*	≈400 Mb	WGS	2002 (Draft)
O. sativa ssp. *japonica*	≈400 Mb	CBC/PM	2004 (RefSeq)
O. glaberrima	≈354 Mb	BP	2010 (RefSeq)

O. barthii	≈411 Mb	WGS/PM	2012 (RefSeq)
O. brachyantha	≈260 Mb	WGS/PM	2011 (RefSeq)
O. longistaminata	≈352 Mb	WGS	2011 (Draft)
O. punctata	≈423 Mb	BP/WGS/PM	2012 (RefSeq)
O. eichingeri	≈650 Mb	WGS	Sequencing in progress

Key: *WGS* whole genome shotgun, *CBC* clone by clone, *BP* BAC Pool, *PM* physical map integration.
Modified from Jacquemin et al. (2013).

Numerous studies have been undertaken aimed at genome wide-discovery of DNA polymorphisms such as SNPs and InDels in the *Oryza* genus. These studies have led to discovery of huge numbers of DNA polymorphisms (Feltus et al. 2004; Shen et al. 2004; Subbaiyan et al. 2012; Xu et al. 2012). However, despite discovery of these DNA polymorphisms, a significant gap exists in linking them to phenotypic traits so that they can be of more value in crop improvement as well as in genetic diversity analysis. Such information, according to Tung et al. (2010), will lead to more insights on the value of naturally occurring variation and hence lead to better management and utilization of the biodiversity that is conserved in various global germplasm repositories. The integration of this information with genebank accession level information and data is expected to increase the value of genebank collections and thereby boost their utilization in rice improvement as well as other areas of plant science. Among the greatest challenges that genebanks currently face in this endeavor is the inadequacy in bioinformatics skills and computing capacity to handle the huge amount of genomic data that is being generated through sequencing and genotyping. Attempts towards proper integration of genomic and phenotypic data, that allow meaningful downstream analysis first calls for concerted efforts in compiling and publicly sharing genebank's accession level information on characterization and evaluation. The lack of this information arguably presents the greatest obstacle to the effective use of conserved genetic resources (FAO 2010; Khoury et al. 2010; Wambugu et al. 2010). Without evaluation data, for example, it is not possible to link sequence polymorphisms with phenotypic performance. African *Oryza* thus remains greatly under characterized and hence grossly underutilized in rice breeding programs and its imperative that more focus is put in this important area. With the rapidly decreasing costs of next generation sequencing, there is need to start positioning genebanks for the ultra-high throughput genomic era which promises to revolutionize germplasm conservation and utilization as well as all associated activities.

CONCLUSION

Undoubtedly, African *Oryza* genetic resources have numerous traits of potential value in the improvement of cultivated rice. The under representation of these resources in global germplasm repositories and the threats they face especially in the wild, call for concerted efforts, nationally, regionally and internationally, to ensure they are collected and properly conserved. The current advances and cost reduction in next generation sequencing, promises to revolutionize the conservation and utilization of these resources. These sequencing technologies should therefore be fully deployed in the characterization and utilization of these resources for the benefit of human kind.

ACKNOWLEDGEMENTS

The first author thanks the Australian Agency for International Development (AusAID) for financial support through an Australian Development Scholarship (ADS). We also thank Dr. Ruaraidh Sackville-Hamilton from the International Rice Genebank at the International Rice Research Institute for providing information about germplasm collections held at IRRI.

REFERENCES

1. Agnoun Y, Biaou SSH, Sié M, Vodouhè RS, Ahanchédé A: The african rice *Oryza glaberrima* steud: knowledge distribution and prospects. *Int J Biol* 2012, 4: 158–180.
2. Almekinders CJM, Louwaars NP: *Farmers' seed production: new approaches and practices*. London: Intermediate Technology; 1999.
3. Aluko G, Martinez C, Tohme J, Castano C, Bergman C, Oard JH: QTL mapping of grain quality traits from the interspecific cross *Oryza sativa* x *O. glaberrima*. *Theor Appl Genet* 2004, 109: 630–639.
4. Barry MB, Pham JL, Noyer JL, Courtois B, Billot C, Ahmadi N: Implications for in situ genetic resource conservation from the ecogeographical distribution of rice genetic diversity in maritime Guinea. *Plant Genet Resour* 2007, 5: 45–54.
5. Bautista NS, Solis R, Kamijima O, Ishii T: RAPD, RFLP and SSLP analyses of phylogenetic relationships between cultivated and wild species of rice. *Genes genet syst* 2001, 76: 71–79.
6. Bimpong IK, Serraj R, Chin JH, Ramos J, Mendoza EMT, Hernandez JE, Mendioro MS, Brar DS: Identification of QTLs for drought-related traits in alien introgression lines derived from crosses of rice (*oryza sativa* cv. IR64)× *O. Glaberrima* under lowland moisture stress. *J Plant Biol* 2011, 54: 237–250.
7. Brar DS: Broadening the gene pool of rice through introgression from wild species. In *Rice is life: scientific perspectives for the 21st century, Proceedings of the world rice research conference*. Edited by: Toriyama K, Heong KL, Hardy B. Tsukuba, Japan: International Rice Research Institute (IRRI); 2004:157–159.
8. Brar DS, Khush GS: Alien introgression in rice. *Plant Mol Biol* 1997, 35: 35–47.
9. Brar DS, Khush GS: Transferring genes from wild species into rice. In *Quantitatie Genetics, Genomics and Plant Breeding*. Edited by: Kang MS. Wallingford, UK: CABI Publishing; 2002.
10. Brar DS, Singh K: Oryza. In *Wild Crop Relatives:*

Genomic and Breeding Resources, Cereals. Edited by: Kole C. Berlin Heidelberg: Springer; 2011:321–365.

11. Brink M, Belay G: *Plant Resources of Tropical Africa 1. Cereal and Pulses*. Wagenigen, Netherlands: PROTA Foundation, Wagenigen, Netherlands, Backhuys Publishers, Leiden, Netherlands/CTA; 2006:298.
12. Chen M-H, Bergman C, Pinson S, Fjellstrom R: Waxy gene haplotypes: associations with apparent amylose content and the effect by the environment in an international rice germplasm collection. *J Cereal Sci* 2008, 47: 536–545.
13. Chen J, Liang C, Chen C, Zhang W, Sun S, Liao Y, Zhang X, Yang L, Song C, Wang M, Shi J, Huang Q, Liu G, Liu J, Zhou H, Zhou W, Yu Q, An N, Chen Y, Cai Q, Wang B, Liu B, Gao D, Min J, Huang Y, Wu H, Li Z, Zhang Y, Yin Y, Song W, *et al.*: Whole-genome sequencing of *Oryza brachyantha* reveals mechanisms underlying *Oryza* genome evolution. *Nat comm*2013, 4: 1595.
14. Cheng C, Tsuchimoto S, Ohtsubo H, Ohtsubo E: Evolutionary relationships among rice species with AA genome based on SINE insertion analysis. *Genes genet syst* 2002, 77: 323–334.
15. Cranston KA, Hurwitz B, Ware D, Stein L, Wing RA: Species trees from highly incongruent gene trees in rice. *Syst biol* 2009, 58: 489–500.
16. Dingkuhn M, Johnson DE, Sow A, Audebert AY: Relationships between upland rice canopy characteristics and weed competitiveness. *Field Crops Res.* 1998, 61: 79–95.
17. Duan S, Lu B, Li Z, Tong J, Kong J, Yao W, Li S, Zhu Y: Phylogenetic analysis of AA-genome *Oryza* species (poaceae) based on chloroplast, mitochondrial, and nuclear DNA sequences. *Biochem Genet* 2007, 45: 113–129.
18. Ellstrand NC, Prentice HC, Hancock JF: Gene flow and introgression from domesticated plants into their wild relatives.*Annu Rev Ecol Syst* 1999, 30: 539–563.
19. FAO: *Second Report on the World's Plant Genetic Resources for Food and Agriculture*. 2010. (Accessed 18th March 2013)http://www.fao.org/docrep/013/i1500e/i1500e.pdf (Accessed 18th March 2013)
20. Feltus FA, Wan J, Schulze SR, Estill JC, Jiang N, Paterson AH: An SNP resource for rice genetics and breeding based on subspecies indica and japonica genome alignments. *Genome res* 2004, 14: 1812–1819.
21. Futakuchi K, Sié M: Better exploitation of African rice in rice variety development. *Agric Journal* 2009, 4: 96–102.
22. Futakuchi K, Jones MP, Ishii R: Physiological and morphological mechanisms of submergence resistance in African rice (*Oryza glaberrima* Steud.). *Jpn J Trop Agric* 2001, 45: 8–14.
23. Futakuchi K, Sie M, Saito K: Yield potential and physiological and morphological characteristics related to yield performance in *Oryza glaberrima* steud. *Plant Prod Sci* 2012, 15: 151–163.
24. Ge S, Sang T, Lu B-R, Hong D-Y: Phylogeny of rice genomes with emphasis on origins of allotetraploid species. *Proc Natl Acad Sci USA* 1999, 96: 14400–14405.
25. Ge S, Sang T, Lu BR, Hong DY: Phylogeny of the genus *Oryza* as revealed by molecular approaches. In *Rice genetics IV Proceedings of the fourth international rice genetics symposium*. Edited by: Khush GS, Brar DS, Hardy B. Los Banos, The Philippines: IRRI; 2001:89–105.
26. Goicoechea JL: *Structural comparative genomics of four African species of Oryza*. University of Arizona: Dissertation; 2009.
27. Goicoechea JL, Ammiraju JSS, Marri PR, Chen M, Jackson S, Yu Y, Rounsley S, Wing RA: The future of rice genomics: sequencing the collective *Oryza genome*. *Rice* 2010, 3: 89–97.
28. Guo Y-L, Ge S: Molecular phylogeny of Oryzeae (Poaceae) based on DNA sequences from chloroplast, mitochondrial, and nuclear genomes. *Am J Bot* 2005, 92: 1548–1558.
29. Hajjar R, Hodgkin T: The use of wild relatives in crop improvement: a survey of developments over the last 20 years.*Euphytica* 2007, 156: 1–13.
30. Hay FR, Hamilton NRS, Furman BJ, Upadhyaya HD, Reddy KN, Singh SK: Cereals. In *Conservation of Tropical Plant Species*. Edited by: Normah MN, Chin HF, Reed BM. New York: Springer Science; 2013:293–315.
31. Henry RJ, Rice N, Waters DLE, Kasem S, Ishikawa R, Hao Y, Dillon S, Crayn D, Wing R, Vaughan D: Australian *Oryza* : utility and conservation. *Rice* 2009, 3: 235–241.
32. Hodgkin T, Hajjar R, Maxted N, *et al.*: Using crop wild relatives for crop improvement: trends and perspectives. In *Crop Wild Relative Conservation and Use*. Wallingford, UK: CABI; 2007:535–548.
33. Hu F, Wang D, Zhao X, Zhang T, Sun H, Zhu L, Zhang F, Li L, Li Q, Tao D, Fu B, Li Z: Identification of rhizome-specific genes by genome-wide differential expression analysis in *Oryza longistaminata*. *BMC plant biol* 2011, 11: 18.
34. IBPGR: *Collection of crop germplasm: The First Ten Years; 1974–1984*. Rome, Italy: International Board for Plant Genetic Resources; 1984:119.
35. International Rice Genome Sequencing Project: The map based sequence of the rice genome. *Nature* 2005, 436: 793–800.
36. Ishii T, Nakano T, Maeda H, Kamijima O: Phylogenetic relationships in A-genome species of rice as revealed by RAPD analysis. *Genes genet syst* 1996, 71: 195–210.
37. Iwamoto M, Nagashima H, Nagamine T, Higo H, Higo K: p-SINE1-like intron of the CatA catalase homologs and phylogenetic relationships among AA-genome *Oryza* and related species. *Theor Appl Genet* 1999, 98: 853–861.
38. Jacquemin J, Bhatia D, Singh K, Wing RA: The international oryza Map alignment project: development of a genus-wide comparative genomics platform to help solve the 9 billion-people question. *Curr Opin Plant Biol* 2013, 16: 147–156.
39. Jarvis A, Upadhyaya HD, Gowda CLL, Aggarwal PK,

Fujisaka S, Anderson B: *Climate Change and its Effect on Conservation and Use of Plant Genetic Resources for Food and Agriculture and Associated Biodiversity for Food Security*. 2009. . (Accessed on 18th March 2013) http://www.fao.org/docrep/013/i1500e/i1500e16.pdf . (Accessed on 18th March 2013)

40. Jena KK: The species of the genus *Oryza* and transfer of useful genes from wild species into cultivated rice, *O. sativa.Breeding Sci* 2010, 60: 518–523.

41. Jones MP, Dingkuhn M, Aluko/snm GK, Semon M: Interspecific *Oryza sativa* L. X *O. glaberrima* Steud. progenies in upland rice improvement. *Euphytica* 1997, 94: 237–246.

42. Joshi SP, Gupta VS, Aggarwal RK, Ranjekar PK, Brar DS: Genetic diversity and phylogenetic relationship as revealed by inter simple sequence repeat (ISSR) polymorphism in the genus *Oryza*. *Theor Appl Genet* 2000, 100: 1311–1320.

43. Kasem S, Waters DLE, Rice N, Shapter FM, Henry RJ: Whole grain morphology of Australian rice species. *Plant Genet Resour* 2010, 8: 74–81.

44. Kasem S, Waters DLE, Henry RJ: Analysis of starch gene diversity in the wild relatives of *Oryza sativa*. *Trop Plant Biol*2012, 5: 286–308.

45. Kharabian-Masouleh A, Waters DL, Reinke RF, Henry RJ: Discovery of polymorphisms in starch-related genes in rice germplasm by amplification of pooled DNA and deeply parallel sequencing. *Plant Biotechnol J* 2011, 9: 1074–1085.

46. Kharabian-Masouleh A, Waters DL, Reinke RF, Ward R, Henry RJ: SNP in starch biosynthesis genes associated with nutritional and functional properties of rice. *Sci rep* 2012, 2: 557.

47. Khoury C, Laliberté B, Guarino L: Trends in ex situ conservation of plant genetic resources: a review of global crop and regional conservation strategies. *Genet Resour Crop Evol* 2010, 57: 625–639.

48. Khumalo S, Chirwa PW, Moyo BH, Syampungani S: The status of agrobiodiversity management and conservation in major agroecosystems of Southern Africa. *Agric Ecosyst Environ* 2012, 157: 17–23.

49. Khush GS, Bacalangco E, Ogawa T: A New gene for resistance to bacterial blight from *O. Longistaminata*. *Rice Genetic Newsletter* 1990, 7: 121–122.

50. Kiambi DK, Newbury HJ, Ford-Lloyd BV, Dawson I: Contrasting genetic diversity among *Oryza longistaminata* (A. Chev et Roehr) populations from different geographic origins using AFLP. *Afr J Biotechnol* 2005, 4: 308–317.

51. Kim H, Hurwitz B, Yu Y, Collura K, Gill N, SanMiguel P, Mullikin JC, Maher C, Nelson W, Wissotski M, Braidotti M, Kudrna D, Goicoechea JL, Stein L, Ware D, Jackson SA, Soderlund C, Wing RA: Construction, alignment and analysis of twelve framework physical maps that represent the ten genome types of the genus *Oryza*. *Genome biol* 2008, 9: R45.

52. Larkin PD, Park WD: Association of waxy gene single nucleotide polymorphisms with starch characteristics in rice (*Oryza sativa* L.). *Mol Breed* 2003, 12: 335–339.

53. Li J, Xiao J, Grandillo S, Jiang L, Wan Y, Deng Q, Yuan L, McCouch SR: QTL detection for rice grain quality traits using an interspecific backcross population derived from cultivated Asian (*O. sativa* L.) and African (*O. glaberrima* S.) rice. *Genome*2004, 47: 697–704.

54. Li Z-M, Zheng X-M, Ge S: Genetic diversity and domestication history of African rice (*Oryza glaberrima*) as inferred from multiple gene sequences. *Theor Appl Genet* 2011, 123: 21–31.

55. Linares OF: African rice (*Oryza glaberrima*): history and future potential. *Proc Natl Acad Sci U S A* 2002, 99: 16360–16365.

56. Lu B-R, Jackson MT: *Wild rice taxonomy*. 2009. (Accessed 17th April 2013)http://www.knowledgebank.irri.org/extension/wild-rice-taxonomy.html (Accessed 17th April 2013)

57. Lu B, Snow AA: Gene flow from genetically modified rice and its environmental consequences. *Biosci* 2005, 55: 669–678.

58. Lu B-R, Naredo MEB, Juliano AB, Jackson MT: Preliminary studies on taxonomy and biosystematics of the AA genome*Oryza* species (Poaceae). In *Grasses, systematics and evolution*. Edited by: Jacobs SWL, Everett J. Melbourne, Australia: CSIRO; 2000:51–58.

59. Lu BR, Zheng KL, Qian HR, Zhuang JY: Genetic differentiation of wild relatives of rice as assessed by RFLP analysis.*Theor Appl Genet* 2002, 106: 101–106.

60. Maxted N, Kell S: *Establishment of a Global Network for the In Situ Conservation of Crop Wild Relatives: Status and Needs*. 2009. . (Accessed on 20th April 2013) http://www.fao.org/docrep/013/i1500e/i1500e18a.pdf . (Accessed on 20th April 2013)

61. Maxted N, Dulloo E, Ford-Lloyd VB, Iriondo JM, Jarvis A: Gap analysis: a tool for complementary genetic conservation assessment. *Divers Distrib* 2008, 14: 1018–1030.

62. McCouch SR, McNally KL, Wang W, Sackville Hamilton R: Genomics of gene banks: a case study in rice. *Am J Bot* 2012, 99: 407–423.

63. Melaku G, Haileselassie T, Feyissa T, Kiboi S: Genetic diversity of the African wild rice (*Oryza longistaminata* Chev. et Roehr) from Ethiopia as revealed by SSR markers. *Genet Resour Crop Ev* 2013, 60: 1047–1056.

64. Mochizuki K, Ohtsubo H, Hirano H-Y, Sano Y, Ohtsubo E: Classification and relationships of rice strains with AA genome by identification of transposable elements at nine loci. *Jpn J Genet* 1993, 68: 205–217.

65. Mohapatra S: In search of new seeds. In *Rice Today*. International Rice Research Institute (IRRI); 2007. (Accessed on 30th April 2013) http://irri.org/index.php?option=com_k2%26view=item%26id=10360:insearch-of-new-seeds%26lang=en(Accessed on 30th April 2013)

66. Mohapatra S: Pockets-of-gold. In *Rice Today*. International Rice Research Institute (IRRI); 2010. (Accessed on 30th April 2013) http://irri.org/index.php?option=com_k2&view=item&id=8412:pockets-of-

gold&lang=en (Accessed on 30th April 2013)

67. Mullins IM, Hilu KW: Sequence variation in the gene encoding the 10-kDa prolamin in *Oryza* (Poaceae). 1. Phylogenetic Implications. *Theor Appl Genet* 2002, 105: 841–846.

68. National Research Council: *Lost Crops of Africa. Volume 1: Grains*. Washington: National Academy Press; 1996:380.

69. Ndjiondjop MN, Albar L, Fargette D, Fauquet C, Ghesquiere A: The genetic basis of high resistance to rice yellow mottle virus (RYMV) in cultivars of two cultivated rice species. *Plant Dis* 1999, 83: 931–935.

70. Ngwediagi P, Maeda E, Kimomwe H, Kamara R, Massawe S, Akonaay HB, Mapunda LND: *Tanzania: Second Country Report on the State of Plant Genetic Resources for Plant and Agriculture*. 2009. (Accessed on 21st April 2013)http://www.fao.org/docrep/013/i1500e/United%20Republic%20Tanzania.pdf (Accessed on 21st April 2013)

71. Nishikawa T, Vaughan DA, Kadowaki K-i: Phylogenetic analysis of *Oryza* species, based on simple sequence repeats and their flanking nucleotide sequences from the mitochondrial and chloroplast genomes. *Theor Appl Genet* 2005, 110: 696–705.

72. Nonomura K, Morishima H, Miyabayashi T, Yamaki S, Eiguchi M, Kubo T, Kurata N: The wild *Oryza* collection in National BioResource Project (NBRP) of Japan: history, biodiversity and utility. *Breeding Sci* 2010, 60: 502–508.

73. Nwilene FE, Adam A, Williams CT, Ukwungwu MN, Dakouo D, Nacro S, Hamadoun A, Kamara SI, Okhidievbie O, Abamu FJ: Reactions of differential rice genotypes to African rice gall midge in West Africa. *Int J pest manage* 2002, 48: 195–201.

74. Park K-C, Lee JK, Kim N-H, Shin Y-B, Lee J-H, Kim N-S: Genetic variation in *Oryza* species detected by MITE-AFLP. *Genes genet syst* 2003a, 78: 235–243.

75. Park KC, Kim NH, Cho YS, Kang KH, Lee JK, Kim NS: Genetic variations of AA genome *Oryza* species measured by MITE-AFLP. *Theor Appl Genet* 2003b, 107: 203–209.

76. Pettersson E, Lundeberg J, Ahmadian A: Generations of sequencing technologies. *Genomics* 2009, 93: 105–111.

77. Plowright RA, Coyne DL, Nash P, Jones MP: Resistance to the rice nematodes *Heterodera sacchari,* Meloidogyne graminicola and M. incognita in Oryza glaberrima and O. glaberrima x O. sativa interspecific hybrids. *Nematology* 1999, 1: 745.

78. Ram T, Laha GS, Gautam SK, Deen R, Madhav MS, Brar DS, Viraktamath BC: Identification of a new gene introgressed from *Oryza* brachyantha with broad-spectrum resistance to bacterial blight of rice in India. *Rice Genet Newsl* 2010, 25: 57–58.

79. Ramirez-Villegas J, Khoury C, Jarvis A, Debouck DG, Guarino L: A Gap analysis methodology for collecting crop genepools: a case study with phaseolus beans. *PloS one* 2010, 5: e13497.

80. Ren F, Lu B-R, Li S, Huang J, Zhu Y: A comparative study of genetic relationships among the AA-genome *Oryza* species using RAPD and SSR markers. *Theor Appl Genet* 2003, 108: 113–120.

81. Sauphanor B: Some factors of upland rice tolerance to stem-borers in West Africa. *Int J Trop Insect Sci* 1985, 6: 429–434.

82. Shen Y-J, Li X, Yu Q-B, Liu H-J, Chen D-H, Gao J-H, Huang H, Shi T-L, Yang Z-N, Jiang H, Jin J-P, Zhang Z-B, Xi B, He Y-Y, Wang G, Wang C, Qian L: Development of genome-wide DNA polymorphism database for map-based cloning of rice genes. *Plant Physiol* 2004, 135: 1198–1205.

83. Small RL, Cronn RC, Wendel JF: Use of nuclear genes for phylogeny reconstruction in plants. *Aust Syst Bot* 2004, 17: 145–170.

84. Song Z, Li BO, Chen J, Lu BR: Genetic diversity and conservation of common wild rice (*Oryza rufipogon*) in China. *Plant Species Biol* 2005, 20: 83–92.

85. Subbaiyan GK, Waters DL, Katiyar SK, Sadananda AR, Vaddadi S, Henry RJ: Genome-wide DNA polymorphisms in elite indica rice inbreds discovered by whole-genome sequencing. *Plant Biotechnol* 2012, J10: 623–634.

86. Takahashi H, Sato Y-i, Nakamura I: Evolutionary analysis of two plastid DNA sequences in cultivated and wild species of*Oryza*. *Breeding Sci* 2008, 58: 225–233.

87. Tang L, Zou X-h, Achoundong G, Potgieter C, Second G, Zhang D-y, Ge S: Phylogeny and biogeography of the rice tribe (Oryzeae): evidence from combined analysis of 20 chloroplast fragments. *Mol Phylogenet Evol* 2010, 54: 266–277.

88. Teeken B, Nuijten E, Temudo MP, Okry F, Mokuwa A, Struik PC, Richards P: Maintaining or abandoning african rice: lessons for understanding processes of seed innovation. *Hum Ecol* 2012, 40: 879–892.

89. Tung C-W, Zhao K, Wright MH, Ali ML, Jung J, Kimball J, Tyagi W, Thomson MJ, McNally K, Leung H, Kim H, Ahn S-N, Reynolds A, Scheffler B, Eizenga G, McClung A, Bustamante C, McCouch SR: Development of a research platform for dissecting phenotype–genotype associations in rice (*Oryza* spp.). *Rice* 2010, 3: 205–217.

90. Umemoto T, Aoki N, Lin HX, Nakamura Y, Inouchi N, Sato Y, Yano M, Hirabayashi H, Maruyama S: Natural variation in rice starch synthase IIa affects enzyme and starch properties. *Funct Plant Biol* 2004, 31: 671–684.

91. Umemoto T, Horibata T, Aoki N, Hiratsuka M, Yano M, Inouchi N: Effects of variations in starch synthase on starch properties and eating quality of rice. *Plant Prod Sci* 2008, 11: 472–480.

92. USDA-ARS: *National Genetic Resources Program. Germplasm Resources Information Network - (GRIN) [Online Database]*. Beltsville, Maryland: National Germplasm Resources Laboratory; 2013. (Accessed 18 April 2013)http://www.ars-grin.gov/cgi-bin/npgs/html/tax_search.pl (Accessed 18 April 2013)

93. Vaughan D, Chang TT: In Situ conservation of rice genetic resource. *Econ Bot* 1992, 46: 368–383.

94. Wambugu PW, Muthamia ZK: *Second Report on*

the State of Plant Genetic Resources for Food and Agriculture in Kenya. Kenya: FAO; 2009. (Accessed on 18th April 2013) http://www.fao.org/docrep/013/i1500e/kenya.pdf (Accessed on 18th April 2013)

95. Wambugu PW, Mathenge PW, Auma EO, vanRheenen HA: Constraints to on-farm maize (*Zea mays* L.) seed production in Western Kenya: plant growth and yield. *ISRN Agron* 2012, 2012: 1–7.

96. Wang ZY, Second G, Tanksley SD: Polymorphism and phylogenetic relationships among species in the genus *Oryza* as determined by analysis of nuclear RFLPS. *Theor Appl Genet* 1992, 83: 565–581.

97. Waters DLE, Nock CJ, Ishikawa R, Rice N, Henry RJ: Chloroplast genome sequence confirms distinctness of Australian and Asian wild rice. *Ecol evol* 2012, 2: 211–217.

98. Xiao J, Li J, Grandillo S, Ahn S-N, Yuan L, Tanksley SD, McCouch SR: Identification of trait-improving quantitative trait loci alleles from a wild rice relative, *Oryza rufipogon*. *Genetics* 1998, 150: 899–909.

99. Xie J, Agrama HA, Kong D, Zhuang J, Hu B, Wan Y, Yan W: Genetic diversity associated with conservation of endangered Dongxiang wild rice (*Oryza rufipogon*). *Genet Resour Crop Evol* 2010, 57: 597–609.

100. Xu X, Liu X, Ge S, Jensen JD, Hu F, Li X, Dong Y, Gutenkunst RN, Fang L, Huang L, Li J, He W, Zhang G, Zheng X, Zhang F, Li Y, Yu C, Kristiansen K, Zhang X, Wang J, Wright M, McCouch S, Nielsen R, Wang J, Wang W: Resequencing 50 accessions of cultivated and wild rice yields markers for identifying agronomically important genes. *Nat biotechnol* 2012, 30: 105–111.

101. Yamakawa H, Ebitani T, Terao T: Comparison between locations of QTLs for grain chalkiness and genes responsive to high temperature during grain filling on the rice chromosome map. *Breeding Sci* 2008, 58: 337–343.

102. Yan H, Xiong Z, Min S, Hu H, Zhang Z, Tian S, Tang S: The Transfer of Brown Planthopper Resistace from *Oryza eichingeri*to *O sativa*. *J Genet Genomics* 1997, 24: 424–431.

103. Yang H, Hu L, Hurek T, Reinhold-Hurek B: Global characterization of the root transcriptome of a wild species of rice, *Oryza longistaminata*, by deep sequencing. *BMC genomics* 2010, 11: 705.

104. Zhu Q, Ge S: Phylogenetic relationships among A-genome species of the genus *Oryza* revealed by intron sequences of four nuclear genes. *New phytol* 2005, 167: 249–265.

105. Zou XH, Zhang FM, Zhang JG, Zang LL, Tang L, Wang J, Sang T, Ge S: Analysis of 142 genes resolves the rapid diversification of the rice genus. *Genome biol* 2008, 9: R49.

CHAPTER 7

GIS AND EX SITU PLANT CONSERVATION

Nikos Krigas[1, 3], Kimon Papadimitriou[2] and Antonios D. Mazaris[3]

[1]Laboratory of Systematic Botany & Phytogeography, Department of Botany, School of Biology, Aristotle University of Thessaloniki, Greece

[2]Office for Sustainability, Aristotle University of Thessaloniki, Greece

[3]Department of Ecology, School of Biology, Aristotle University of Thessaloniki, Greece

INTRODUCTION

In the frame of the global efforts to halt the biodiversity loss by 2010 and with the aim to develop effective conservation strategies extending beyond 2010, stakeholders have recognized as a priority the *in situ* conservation (on site conservation) of target plant species.

Still, the rapid environmental changes including climate change, habitat loss and alteration, could pose some limitations on our ability to conserve target species effectively *in situ* (Sharrock & Jones, 2009). As a result, conservation biologists, policy makers and managers acknowledge the importance of *ex situ* conservation of target plants in botanic gardens and seed banks as an essential back-up solution (Convention on Biological Diversity [CBD], 1992; Glawka et al 1994; Global Strategy for Plant Conservation [GSPC], 2002; European Strategy for Plant Conservation [ESPC], 2009; Sharrock & Jones, 2009).

For the *ex situ* plant conservation, target species mainly refer to plant taxa (species and subspecies) presenting a narrow distribution in the wild (see Krigas & Maloupa, 2008). This category of plants usually includes:

- Local endemics (endemics of a single mountaintop e.g. *Viola cephalonica* (Katsouni et al., 2009), or endemics of a single island e.g. *Allium samothracicum* (Krigas, 2009), or endemics of a group of nearby areas or islands e.g. *Thymus holosericeus* (Krigas et al. 2010),
- Regional endemics (endemics to small parts of a single country, e.g. endemics of southern Spain, endemics of Peloponnese, southern Greece),
- National endemics or single-country endemics (e.g. Greek endemics, Italian endemics, Spanish endemics),
- Endemics to specified small geographical areas e.g. local Balkan endemics transcending the borders of neighbouring Balkan countries such as *Ranunculus cacuminis* (Krigas & Karamplianis, 2009).

Other target species significant for the *ex situ* plant conservation may include plants which are rare in a certain area (e.g. Europe, Sharrock & Jones, 2009) or plants that are currently threatened with extinction at local, regional or global level i.e. taxa characterized as "Near Threatened", "Vulnerable", "Endangered" or "Critically Endangered" according to the IUCN (2001) criteria. However, given that almost 90% of the Europe's threatened plants are single-country endemics, it should be noted that most of the endemic plants are also threatened with extinction (Sharrock & Jones, 2009). Last but not least, other groups of socioeconomically valuable plants (e.g. saffron) and their progenitors (e.g. plants

Citation: Nikos Krigas, Kimon Papadimitriou and Antonios D. Mazaris, GIS and ex situ Plant Conservation, DOI: 10.5772/50525.

in the genus *Crocus*) may also be considered as target plants for the *ex situ* conservation (Fernández et al., 2011).

GIS has be given a role in analyzing potential and current spatial distribution of target species, locating and assessing the populations of target plant species and assemblages, measuring biodiversity, monitoring biodiversity patterns and identifying priorities for conservation and management (Iverson & Prasad, 1998; Salem, 2003; Powel et al., 2005; Pedersen et al., 2004).

Habitat evaluation or habitat modeling with the use of GIS has the potential to make a substantial contribution to conservation management of target species within an integrated approach and is suitable for setting conservation priorities at multiple spatial scales (Store & Jokimäki, 2003; Powel et al., 2005). GIS have also been used as a tool for specific conservation programmes including comparisons of ecological patterns between local and regional scales, selection of protected areas according to habitat suitability, analysis of the impact of alien species on endemic plants and selection of sites for representative seed collections of target species (Draper et al., 2003). Flexible GIS-based tools have also been developed to exploit static information of botanical collections in an attempt to evaluate species distributional ranges (Schulman et al., 2007; Loiselle et al., 2008), accounting for potential effect of climate change to predicted models (Loiselle et al., 2008). Other GIS-tools and distribution modeling methods have been applied to examine the hypothesis that wild and cultivated plants of certain species may occur in the same types of habitats (Allison et al., 2006).

Currently, the GIS technology has increasingly been used for predictive purposes in species re-introductions. Combining historical information from specimen labels with up-to-date environmental data, GIS can be used to identify the range of environmental conditions in which plants grow, offering an understanding of the ecological requirements of different species. The relevant environmental conditions can then be used to delimit areas of high, moderate or low survival probability (Sawkins, 1999 as cited in Moat & Smith, 2003; Powel et al., 2005). Recent GIS applications further include studies on the assessment of the importance of landscape connectivity, structure and configuration for native and non-native plant communities, dispersal and invasiveness abilities (Minor et al., 2009). Nevertheless, the most frequently encountered obstacle to the use of GIS technology in conservation planning is lack of data, especially distribution data of target species and digital vegetation maps of areas with conservation interest (Sawkins, 1999 as cited in Moat & Smith, 2003).

The GIS technology has also an enormous potential in seed conservation, particularly in targeting collecting needs for botanic gardens and conservation institutes and in identifying what, where and when to collect (Moat & Smith, 2003). Although concerns are raised regarding limitations and biases (Store & Kangas, 2001), some of the GIS applications have been used for their power to pinpoint by spatial niche modeling, new probable locations where rare endemic species may be found in the wild, thus permitting the search for new populations (Jarvis et al., 2005). The GIS could also be used for assessing the sensitivity of target species to climate change identifying potential distributions as well as vulnerable habitats and species (Vanderpoorten et al., 2006). In this way GIS may deliver important information that can drive seed banking priorities and design (Godefroid & Vanderborght, 2010).

Lately, a new application of GIS has been launched for the *ex situ* plant conservation; GIS was used to describe quantitatively and qualitatively the natural habitats of target plants in order to facilitate their propagation and transfer from the wild habitats to man-made habitats like botanic gardens (Krigas et al., 2010).

GIS USED FOR THE DESCRIPTION OF THE NATURAL HABITATS OF TARGET SPECIES

The geographical data associated with plant collections may be considered as a set of facts about the places in which the plants thrive. In the wild, target plants for *ex situ* conservation could thrive literally anywhere. Given the spatial and temporal heterogeneity of the environmental factors and the dissected topography of the landscape, the target plants may well originate in a variety of quite different habitats in which they are adapted to grow naturally. Hence, it seems quite difficult to be able to emulate their preferable conditions when trying to cultivate them in restricted man-made habitats such as botanic gardens (Krigas et al., 2010). A deeper understanding of the ecology and life cycle of such target species has been considered as a key issue towards a species-specific successful propagation and cultivation method (Baskin & Baskin, 1988).

In this framework, the GIS can be used to offer a reliable, quantitative and qualitative description of the *in situ* habitat conditions preferred and/or tolerated by different target plant species in the wild. This novel GIS application is both ecologically meaningful and useful in horticulture and *ex situ* plant conservation (Schulman & Lehvävirta, 2010).

To demonstrate this application a dataset was chosen including target plants originating from sites of varied landscapes that are also subjected to different climatic conditions (Fig. 1): the Aegean Archipelago, Crete, Ionian Islands, and Peloponnese (Greece, Southern Europe). The target plants included in this dataset fall into the conservation priorities of the Balkan Botanic Garden of Kroussia (BBGK, Krigas & Maloupa, 2008, Krigas et al., 2010). To date, the BBGK has organized several botanic expeditions all over Greece in order to collect appropriate plant material for propagation and*ex situ* conservation. All this plant material is currently maintained under *ex situ* conservation (ca. 25% of the known Greek flora which includes at least 6.300 species and subspecies, Krigas et al., 2010).

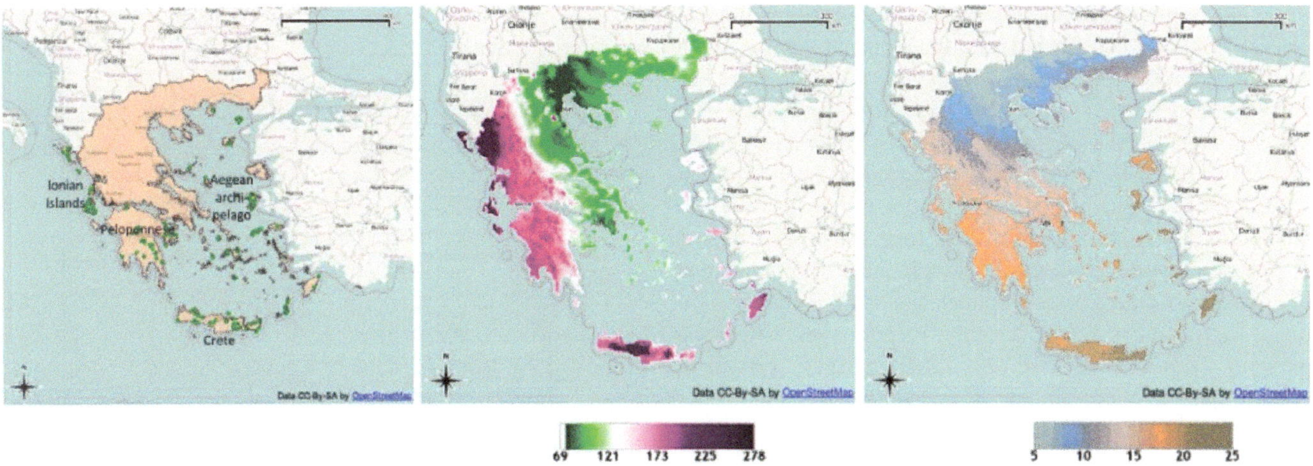

Figure 1: Different target plants (n=256, each dot represents at least one botanic expedition and at least one target plant collected) from various sites originating from the selected area of Greece including the Aegean Archipelago, the Ionian Islands, Crete and Peloponnese (left). All plant material is currently maintained under *ex situ* cultivation at the Balkan Botanic Garden of Kroussia (BBGK). Raster maps displaying the variations of climatic conditions in Greece as expressed by the Emberger Pluviothermic Quotient (middle) and Precipitation Seasonality (right). All values have resulted from map algebra calculations after point sampling for the Greek territory regarding the collection sites of the selected target plants (n=256).

The positions of the collection sites of target plant species were captured in the wild using handheld GPS trackers. The obtained geographic coordinates were consequently imported as point layer into a Geographical Information System (GIS) and point sampling was performed for each one on multiple raster and polygon layers. In order to probe values for topography, terrain, soil, climatic, land cover and habitat attributes at the captured sites, the following datasets were selected (Fig. 2):

a. Raster elevation data from SRTM (USGS 2004), with resolution of 1 km^2 (30 arc second) which were used for terrain and topography attributes.

b. Soil data from the European Soil Database (ESDB) v2.0 (EC, 2004), which is composed of the Soil Geographical Database of Eurasia at a scale of 1: 1,000,000 (version 4 beta) and the Pedotransfer Rules Database (v2.0), with a raster resolution of 1 km^2, presenting 72 soil parameters.

c. Temperature and precipitation data from the WorldClim database (Guarino et al., 2002; Hijmans et al., 2005), with a raster resolution of 1km^2, which has been used for the description of climatic conditions (average values of 50 years).

d. Land cover data from CORINE (Coordination of Information on the Environment) comprehensive hierarchical vector geodatabase at a scale of 1: 100,000 (with a minimum mapping unit of 25ha) which present the spatial distribution of different landcover/land-use types (CORINE land cover classification, EC & ETC/LC, 1999) and soil types (CORINE soil classification EC, 1985).

Additionally, part of the initial data was processed to produce more specific attributes for the collection sites of the target plants, as follows:

a. The elevation data (DEM) was used for raster based terraiJas further used as the base layer for the digitization of the vegetation zones of Greece according to Mavromatis (1980).

b. The rasters describing the spatial variations of precipitation, maximum and minimum temperatures from the WorldClim geodatabase (Guarino et al., 2002), were used in map algebra calculations for the production of a raster map which displays the spatial variation of the Emberger Pluviothermic (Ombrothermic) Quotient as: 2000*P/(Tmax+Tmin)*(Tmax-Tmin), where P: the layer of annual precipitation (mm), T: the layers of mean maximum (Tmax) and minimum (Tmin) monthly temperature of the warmest and coldest month in Kelvin degrees. Precipitation (P) Seasonality was calculated also, using the following map algebra expression: 100* (Pw -Pd) / Py, where Pw and Pd are the raster layers for the spatial variations of precipitation for the wettest and driest quarter of the year respectively, and Py the total annual precipitation (Fig. 1).

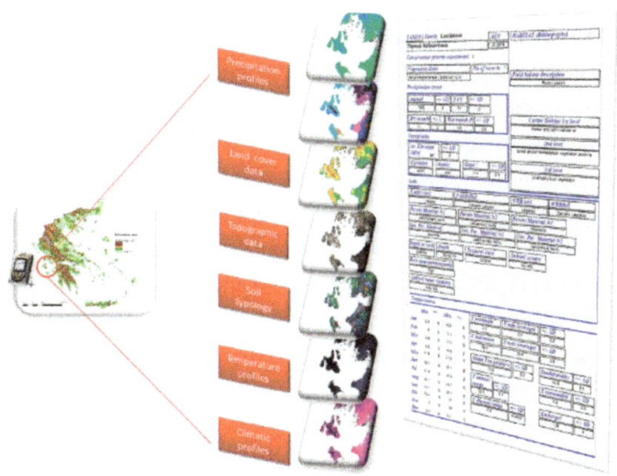

Figure 2: Methodology concept: The GPS data can be used for point sampling in a GIS environment regarding the collection sites of a target species over multiple polygon and raster layers extracted from selected geo-databases. This approach leads to the attribution of selected values regarding precipitation, land cover, terrain, topography, soil typology, temperature and climate to the wild habitats of a target species, resulting in a summarized fact sheet which reflects the ecological preferences of a target species (e.g. *Thymus holosericeus* originating from the Ionian Islands, SW Greece).

Based on this GIS application, simple or advanced ecological fact sheets may be constructed for different target species which actually reveal their preferences in the wild (Fig. 2, 3, 4). An ecological fact sheet illustrates the ecological profile of a wild growing target plant and can be designed in a way to include different kind of information such as: Vegetation zone (different types), Climatic datawhich is a combination of Precipitation and Temperatures (for the different sites: total annual and annual range of precipitation, driest and wettest month, minimum, mean and maximum temperatures, annual and diurnal temperature range, isothermality and seasonality, Emberger pluviothermic quotient), Topographic (for the different sites: elevation, aspect, slope), Habitat (bibliographic, field notes), Land Cover Types (for the different sites: CORINE classification in three levels) andGeological-Pedological data (for the different sites: Food and Agriculture Organization's [FAO] soil classification, World Reference Base Soil classification, dominant parent material in three levels, depth to rock, textural class, topsoil base saturation, subsoil water capacity) (see Fig. 2, 3, 4). \ Additionally, other information may also be associated with the ecological description of the wild habitat of target species such as Taxonomic data i.e. taxon's name, family, accession number (ASN) given, Collection data (e.g. geographical coordinates, description of state, province, exact locality etc), Conservation assessment (e.g. ranking according to priorities assigned), and Mother Plantations (number of total plant accessions or number of records). It should be mentioned that when the number of records introduced in GIS for a specific information field are >1, then any quantitative information may be described by the mean value and its standard deviation (± SD) (see Fig. 2 and 3).

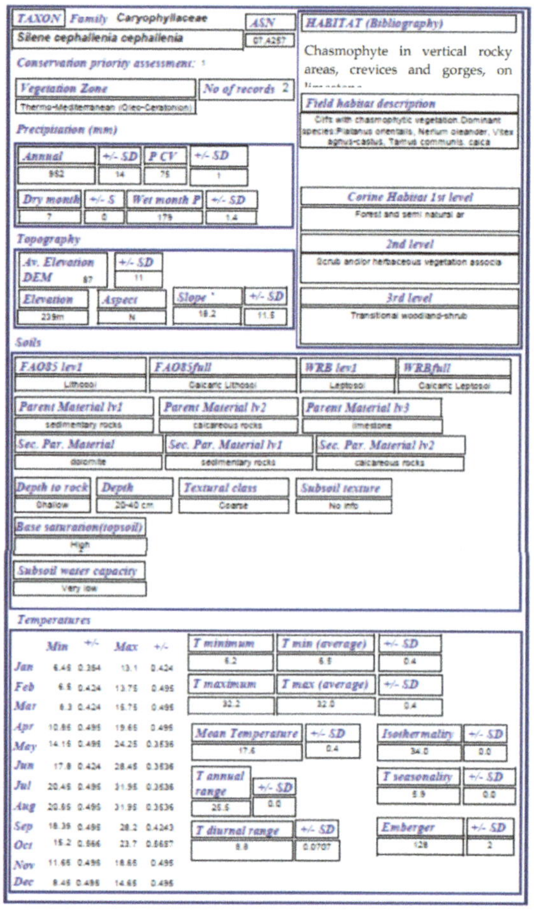

Figure 3: Simple ecological fact sheet for a single-area endemic (*Silene cephallenia* subsp. *cephallenia)*produced after linking its collection data with those of geodatabases. *S. cephallenia* subsp. *cephallenia* is found exclusively at Poros gorge in Cephalonia (Ionian Islands, SW Greece), is protected by the Greek Presidential Decree 67/1981 and recently it has been included in the Red Data Book of Rare and Threatened Plants of Greece as "Critically Endangered" (Karagianni et al., 2009) and in Annex 2 of the European Threatened Species (Sharrock & Jones, 2009).

GIS-FACILITATED DEVELOPMENT OF PROTOCOLS FOR *EX SITU*CONSERVATION OF TARGET SPECIES

Considering the recent challenges in regional conservation planning, it becomes apparent that there is an urgent need for increased applied research in order to develop propagation and cultivation protocols for target plants threatened with extinction, towards species recovery and populations' reinforcements (Bunn et al. 2011, Maunder et al. 2001 Sarasan et al. 2006). Moreover, the GSPC (2002) and the ESPC (2009) have included this urgent need under their conservation targets at European and global level (Target 8). The GIS may serve such a need and may help at the development of species-specific propagation protocols and *ex situ* cultivation guidelines regarding target species. Such methodologies could also provide basic information and criteria for prioritizing collections of threaten or rare species based on their distributional patterns, population status and the genetic and/or geographic representation of current seed bank collections (Draper et al., 2003; Farnsworth et al., 2006) or even prioritizing sites for seed collections (Jarvis et al., 2005; Ramírez-Villegas et al., 2010).

GIS and Seed Germination

The GIS may be used to facilitate the germination of seeds collected from the wild in plant propagation studies as shown by Krigas et al. (2010). This study included plants originating at the Ionian Islands, south-western Greece and provided some example-cases.

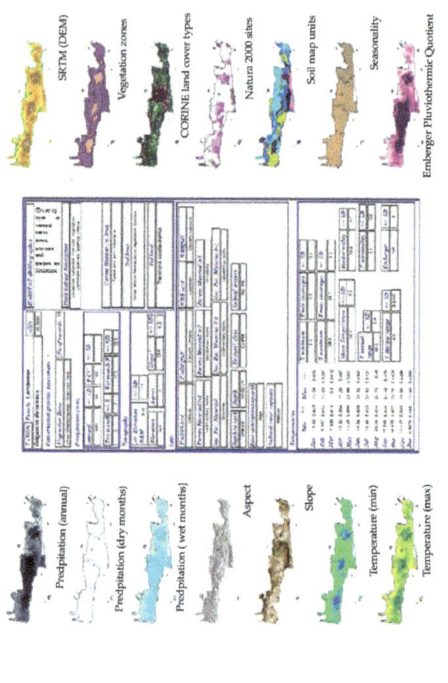

Figure 4: Advanced ecological fact sheet of *Origanum dictamnus* produced after linking data for 18 collection sites with those of

geodatabases, enriched with thematic maps displaying the spatial variations of multiple ecological attributes over the area of Crete (the 18 original collection sites are distributed in three main areas shown with yellow circles). *O. dictamnus* is restricted to Crete (endemic), is protected by the Greek Presidential Decree 67/1981, Bern Convention and the European Directive 92/43/EEC, it has been included in the Red Data Book of Rare and Threatened Plants of Greece as "Vulnerable" (Turland, 1995) and in Annex 2 of the European Threatened Species (Sharrock & Jones, 2009).

Generally, the appropriate season for germination trials inside the greenhouse was selected by comparing the seasonal temperatures profiles of the local greenhouse used with those of the wild habitat of the target species (Ionian Islands). The season deemed as more appropriate was characterized by temperature ranges that could easily mimic those prevailing in the natural environment of the target species.

To illustrate an example, the GIS-derived temperature profiles for the wild habitat of *Silene cephallenia* subsp. *cephallenia* (Figs. 3, 5) have dictated the selection of temperatures to test for seed germination, leading to a germination success of 64% (Krigas et al., 2010). Additionally, its ecological profile has explain the fact that seed germination was inhibited at 10°C, whereas seed germination was considerably increased in only 7 days when higher temperature (21°C±1) was applied, in an attempt to emulate natural conditions.

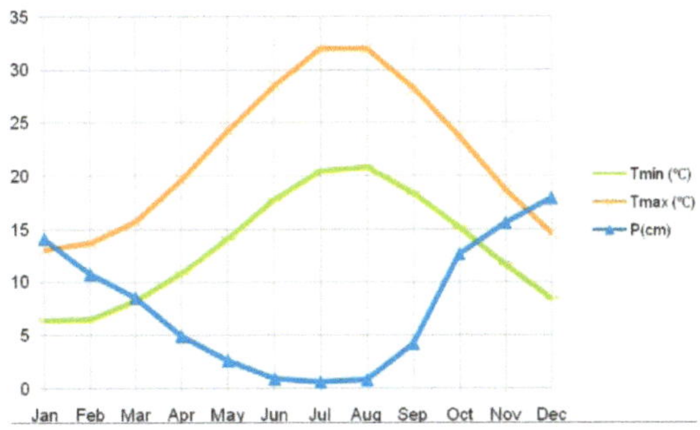

Figure 5: Monthly variations of the climatic conditions (average values of 50 years for precipitation, minimum and maximum temperature) at the original collection sites of *Silene cephallenia* subsp. *cephalleniaderived* from its ecological fact sheet.

GIS and Asexual Plant Propagation by Cuttings

Basically, the asexual propagation and the rooting trials of softwood cuttings from plants are performed in greenhouse conditions using a few individuals to produce large amounts of genetically identical plants. Lately, it has been shown that the GIS may also be used to facilitate the asexual plant propagation by cuttings (Krigas et al., 2010). For example, the GIS-derived data for the wild habitat of *Thymus holosericeus* demonstrated that early spring was the most appropriate season for the rooting trials of this target species (Krigas et al., 2010). During early spring, the temperatures of the selected greenhouse in their case have emulated in the best way the temperatures prevailing in the natural habitat of the target plant. In this study, during propagation the rooting substrate's temperature was kept at 18-21°C in order to accelerate rooting as indicated from previous experience (Maloupa et al., 2008). Nevertheless, the air temperature was kept between 18-25°C, in an attempt to emulate as best as possible the original conditions of the natural habitat which were indicated by the GIS. Furthermore, the relative humidity was kept at 80% for the first 7 days and was reduced gradually to 50% during the second week of rooting. This was followed due to the fact that *Th. holosericeus* is a xerophytic species that grows in areas of very low subsoil water capacity, especially during dry months (Fig. 2). In the case of *Th. holosericeus*, the combination of GIS data with previous experience on rooting of other species of the same genus raised propagation success nearly by 90% (from 45% to 80% rooting).

GIS and *In Vitro* Plant Propagation

The attempts of *in vitro* propagation of rare and threatened plant species are often associated with two inherent problems which may hold back conservation efforts. First of all, there is frequently a lack of published methods regarding the propagation of a certain plant species and secondly, there is often a limited amount of experimental plant material which can be initially available due to scarce populations of limited size found in the wild (Bunn et al., 2011; Krogstrup et al., 2005). Taken these limitations into account, almost every attempt to propagate rare and threatened plants is of great importance but seems difficult to achieve.

Although sophisticated technologies and modern methods have been used to date for the *in vitro* plant propagation (Benson, 1999), the GIS has not been exploited as a tool. Lately, Grigoriadou et al. (2011)have shown how the GIS may also be used to serve the needs and procedures of *in vitro* plant propagation. Using a case-study plant i.e. *Achillea occulta* which is a local endemic of southern Peloponnese (Southern Greece) recently characterized as "Vulnerable" (Constantinidis & Kalpoutzakis 2009), the authors have used GIS in this study for the:

- Selection of temperatures for both greenhouse cultivation and *in vitro* cultures; the GIS-derived ecological profile for *A. occulta* dictated the selection of 22±2 °C day temperature and 15±2 °C night temperature at 16-h photoperiod to be used as most suitable for the development of the *in vitro* cultures.
- Selection of appropriate period for acclimatization and transplanting of plantlets produced *in vitro;* after balancing out the temperature profiles of the wild habitat with those prevailing in the man-made site of the botanic garden (BBGK), the GIS revealed that spring temperatures were more suitable for the active growth and acclimatization of plantlets.
- Selection of growing media and substrates for plantlets produced *in vitro*; the type of commercial peat used with the addition of vermiculite provided a very good imitation of the natural soil conditions as indicated by the GIS-derived ecological profile.
- Selection of appropriate *ex situ* conservation sites for plants raised *in vitro*; after balancing out the temperature profiles of the wild habitat with those prevailing in the available man-made *ex situ*conservation sites of BBGK, the GIS revealed that specific sites seemed more favorable than others for the accommodation of *A. occulta* plants (a site with temperature range as close as possible to the natural temperature range of the wild habitat was chosen).

GIS-DERIVED ECOLOGICAL PROFILES AND GUIDELINES FOR THE *EX SITU* CULTIVATION OF TARGET PLANTS

From the above mentioned and when taking into account the associated technical information (seeKrigas et al., 2010, Grigoriadou et al., 2011) it becomes apparent that GIS may be used for the development of effective protocols concerning the propagation and initial cultivation of plant material derived from target species.

Regardless the method used for plant propagation (cuttings, seeds or *in vitro* techniques), the young individuals produced (plantlets or seedlings) require hardening and acclimatization before their *ex situ*conservation. In this sense, the GIS-derived ecological profiles of the target plants could be used to provide specific guidelines for their effective *ex situ* cultivation (Table 1).

Cultivation Guidelines Regarding Soil Media, Potting Volume, Texture, PH and Drainage

In general, it is known that for the successful *ex situ* cultivation of target plants, the plant medium to be used must be similar to that of the substrate in the plant's natural habitat, providing similar root aeration and drainage conditions. This is quite important since improved drainage conditions during cultivation can actually inhibit fungal disease risks, induce greater rooting depth and enhance general plant health and vigor (Brady & Weil, 2002).

The GIS-derived ecological profile for the wild habitat of *Silene cephallenia* subsp. *cephallenia* (Fig. 3) showed that the plants originally grow on "Calcaric Lithosol" according to the FAO (1985) classification or according to the European Soil Database, on "Calcaric Leptosol (mountainous), shallow (<25 cm) or extremely gravelly soils directly over continuous rock or soils having <20% (by volume) fine earth material" (Krigas et al., 2010). In an attempt to emulate the natural habitat during hardening and acclimatization of plantlets raised from seeds (seedlings), the type of commercial peat used in BBGK for transplantation contained lime and by adding vermiculite, a good approximation of its originally drained soil type was achieved. This fact has actually allowed the plantlets of *S. cephallenia* subsp. *cephallenia* to continue growing without problems observed.

For the further *ex situ* cultivation of *Silene cephallenia* subsp. *cephallenia* and *Thymus holosericeous*, the GIS-derived ecological profiles (Fig. 2, 3) indicated that the basic soil group in which their wild habitats originate from is classified as "Calcaric Lithosol" (stony calcareous with a high concentration of MgCO3 and >15% free CaCO3, developed on dolomitic limestone). Such a substratum is characterized by a good drainage capacity, neutral to alkaline soil pH, low depth and a high ratio of stones to fine earth. The above indicated that for the *ex situ* cultivation of these target plants, a potting medium of relative high pH would be required and a fair degree of drainage was deemed essential. To accomplish this, supplementary perlite, sand and/or fragmented stone were added in their growing media. For these plants, medium sized pots (4.5 l) were suggested as suitable and additional Mg dressing was added in order to match the original chemical composition of their natural habitats.

By adopting these guidelines (Table 1), the natural habitat of the target plants was technically emulated with common commercial materials and the plants have adapted well and continued to grow without apparent problems (Krigas et al., 2010).

Cultivation Guidelines Regarding Temperature Range

When cultivating plants from the wild to man-made habitats, the temperature conditions at the plant's growing site (nursery or outdoors) may often be quite different from those prevailing in the plant's original wild habitat. Although it is known that plants usually have a degree of tolerance for a wide range of temperatures, it is important to know the temperature range that a plant species may tolerate, in order to facilitate its growth by maintaining temperatures within these limits during cultivation.

The GIS-derived ecological profile indicated that *Silene cephallenia* subsp. *cephallenia* (Fig. 3, 5) may experience and/or tolerate in its wild habitat mean temperatures ranging from 6.7 °C to 32.2 °C throughout the year, while for *Thymus holosericeus* the mean yearly temperatures may range from 3.9°C to 29.5 °C (Krigas et al., 2010 and Fig. 2).

For the acclimatization procedure of the seedlings of *S. cephallenia* subsp. *cephallenia* and of the plantlets of *Th. holosericeous* produced in BBGK, the selection of appropriate sites was achieved after balancing out the temperature profiles of the available areas for *ex situ* conservation in northern Greece with the natural range of temperatures of their wild habitats, as revealed with the use of GIS (Table 1). It was suggested that *Th. holosericeus* may be marginally stressed during hot summer days and short term measures could possibly be taken to avoid heat shock (e.g. periodical shading or translocation of mother plants to cooler conditions). However, high summer temperatures did not seem to be a constraint for *S. cephallenia* subsp. *cephallenia*. Moreover, it was suggested that accommodation in a protected greenhouse would be crucial for the winter survival of both plants in northern Greece (Krigas et al., 2010). Indeed, the propagated plants that were transferred indoors (greenhouse at sea level, with controlled winter indoor temperatures ranging from 5 to 25 °C), have actually shown increased height, leaf area and number of flowers during cultivation in comparison to those transplanted outdoors (Krigas et al., 2010).

Cultivation Guidelines Regarding Watering Regimes

When transferring wild plants to man-made sites for their *ex situ* cultivation, the amount of water to be offered to them remains largely unknown. The watering regime followed is based merely on observations regarding the natural habitat of the species in question or exploits previous general horticultural experience associated with the cultivation of other species with presumably similar needs. However, it is generally accepted that the design of an irrigation system for the *ex situ* cultivation of target plants should take into account the natural preferences of the species, the medium type and the irrigation frequency requirements. The latter not only depends on the specific plants in question, but it is also influenced by the growing season and the indoor or outdoor temperatures prevailing in the cultivation sites.

The precipitation profiles derived from GIS for the target plants may indicate differences regarding the water requirements preferred and/or tolerated by each one of them in the wild. For example, the GIS-derived ecological profile of *A. occulta* published by Grigoriadou et al. (2011) for its *ex situ*conservation, indicated a watering regime equal to a mean total annual precipitation of 744 mm (7–136 mm per month) which was followed for the proper cultivation of the *in vitro* propagated plants.

Seasonalvariationofthewateringregimeisequallyimportant as the total amount of water offered to plants. To illustrate an example, the GIS-derived ecological profiles of *Thymus holosericeous* and*Silene cephallenia* subsp. *cephallenia* may be taken into account (Fig. 2, 3, 5, 6). It becomes evident that during the wettest and driest quarters of the year, *Thymus holosericeus*'s habitat receives from 40 to 504 mm, while that of *Silene cephallenia* subsp. *cephallenia* from 26 to 476 mm. During autumn, no noticeable differences exist regarding mean monthly precipitation in the natural habitats of these plants. However, from January to August mean monthly precipitation is consistently higher in the intermediate altitude habitats of *Th. holosericeus*, while until autumn *S. cephallenia* subsp. *cephallenia*(which is naturally found close to sea level) receive comparatively lower precipitation (Fig. 5, 6).

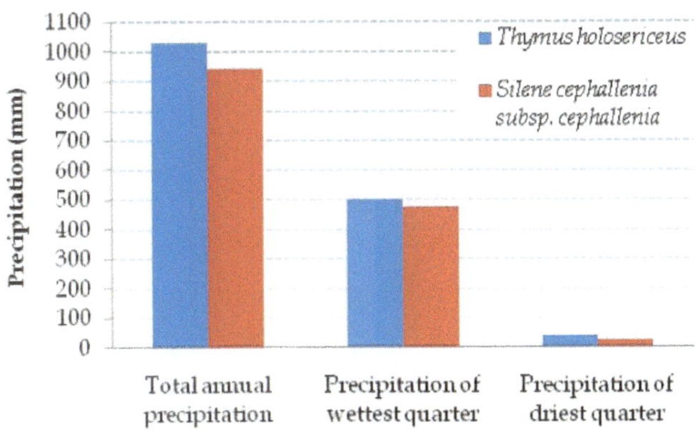

Figure 6: Precipitation profiles for the original collection sites of *Thymus holosericeus* and *Silene cephallenia*subsp. *cephallenia* derived from their ecological fact sheets.

Hence, the GIS-derived precipitation profiles suggest that the watering need (both in terms of water amount received and frequency of watering) is not the same for these plants (Table 1), although both should be under a rather low water regime (regardless of whether it is winter or summer). *S. cephallenia* subsp. *cephallenia* has comparatively lower water demands both annually and seasonally which should result in a more restricted watering schedule, while *Th. holosericeus* has somewhat intermediate watering needs, although a relatively low water regime seems to be equally suitable since it is considered as a xerophytic species (Krigas et al., 2010).

Guidelines Regarding Positioning Of Target Plants in Displays

The qualitative and quantitative description of the natural habitat of a target species may offer information for appropriate spatial and temporal positioning in plant displays of botanic gardens. Different vegetation types where a target plant is naturally found (deciduous or evergreen i.e. depending on the canopy cover of the vegetation) or land cover types (e.g. rock formations, woodland, pastures etc) from which the target species originate from, are attributes that can be exploited as important ecological information. Such attributes (a) may indicate appropriate or unsuitable sites for the *ex situ* cultivation of target plants, (b) could be useful for the selection of the amount of light or shading needed for the plants or to be avoided by them, and (c) could define the species assemblages with common requirements for specific indoors or outdoors plant displays (Table 1).

To illustrate an example, the GIS-derived ecological profile of *Achillea occulta* published byGrigoriadou et al. (2011) for its *ex situ* conservation indicated that (semi-) shady limestone cracks and rock bases with calcaric lithosols should be used as habitats in order to host the propagated plants and a natural positioning at south, south-eastern and south-western exposures of rock formations was chosen as more favorable for plant growth.

POTENTIAL AND IMPLICATIONS FOR THE MANAGEMENT OF LIVING PLANT COLLECTIONS

The novel GIS-facilitated application presented here is a powerful tool able to extract ecologically meaningful environmental information from geodatabases regarding the collection sites of different target plants which are useful in applied research and horticulture. This application is able to identify important ecological differences that can contribute to the development of species-specific baseline plant propagation and cultivation protocols. Given the impracticality and lack of on-the-spot field temperature and precipitation measurements and proper soil sampling followed by laboratory analysis, the nothing (just delete the words) geodatabases with the use of GIS can be used to extract at a fraction of time, information crucial for the success of *ex situ* conservation of target species. This application can be used to (Table 1):

- Understand the amplitude of the *in situ* ecological conditions of different target plants both quantitatively and qualitatively (Fig. 1, 2, 3, 4),

Table 1: Guidelines for the *ex situ* conservation of target plants based on links of their collection data with geodatabases (from Krigas et al., 2010, with modifications)

Variable / Source used	**Selected attributes used**	**Conservation guidelines produced**	Potential to rank or filter plants in terms of different quantitative or qualitative variables or to group target plants sharing common requirements
Soil moisture / ESDB v.2 (EC, 2004)	Available topsoil and subsoil water capacity	Calculation of different watering regimes for different groups or target species	
Soil classes and types / ESDB v.2 (EC, 2004), CORINE Soil Classification (EC, 1985)	Textural class Different soil classes Dominant parent material	Selection of different growing media (regarding texture, pH, drainage) for different groups or target species	
Soil nutrient/ ESDB v.2 (EC, 2004)	Topsoil and subsoil base saturation Cation exchange capacity	Selection of different growing media and development of fertilization regimes for different groups	
Soil limitations / ESDB v.2 (EC, 2004)	Depth to a gleyed horizon Depth to rock Depth of an obstacle to roots Volume of stones	Selection of different growing media (regarding texture, drainage) and potting volume for different groups or target species	
Climate / WorldClim Database (Guarino et al., 2002, Hijmans et al., 2005)	Temperatures Mean minimum or maximum temperatures of the coldest or the warmest month Annual mean temperature range Temperature seasonality Mean diurnal temperature range Precipitation Precipitation of the driest month or the wettest month Mean monthly precipitation Annual precipitation Mean precipitation of driest, wettest, coldest or warmest quarter	Selection of appropriate sites and conditions both in greenhouse and in *ex situ* cultivation sites for different groups or target species Selection of temperatures for seed germination and asexual propagation for different groups or target species Shading and ventilation for temperature regulation Calculation or watering regimes, calibration of different watering regimes in different seasons and/or in different months for different groups or target species	
Topography/ Digital Terrain Model created	Aspect, Slope, Altitude (elevation from sea level)	Customization of microclimate in *ex situ*cultivation sites and appropriate positioning in plant displays	
Vegetation zones / Mavromatis (1980)	11 different vegetation types identified for the Greek territory	Selection of the *ex situ*conservation sites for different groups or target species Selection of shading regime for different groups (seasonal, if originating from deciduous vegetation or all year round, if originating from evergreen vegetation) or for target species Creation of species assemblages Selection of sites for specific plant assemblages and displays	
Land cover classes and types/ CORINE Land cover (EC & ETC/LC, 1999)	45 different land-use classes and types		

- Indicate the preferable *ex situ* growing conditions for each target plant (linking the *in situ* natural conditions with the *ex situ* cultivation regimes) and provide guidelines regarding species-specific treatments (Fig. 2, 3, 5, 6),
- Rank or filter included target species in terms of different quantitative variables or group target plants sharing common preferences or requirements (Fig. 7, 8),

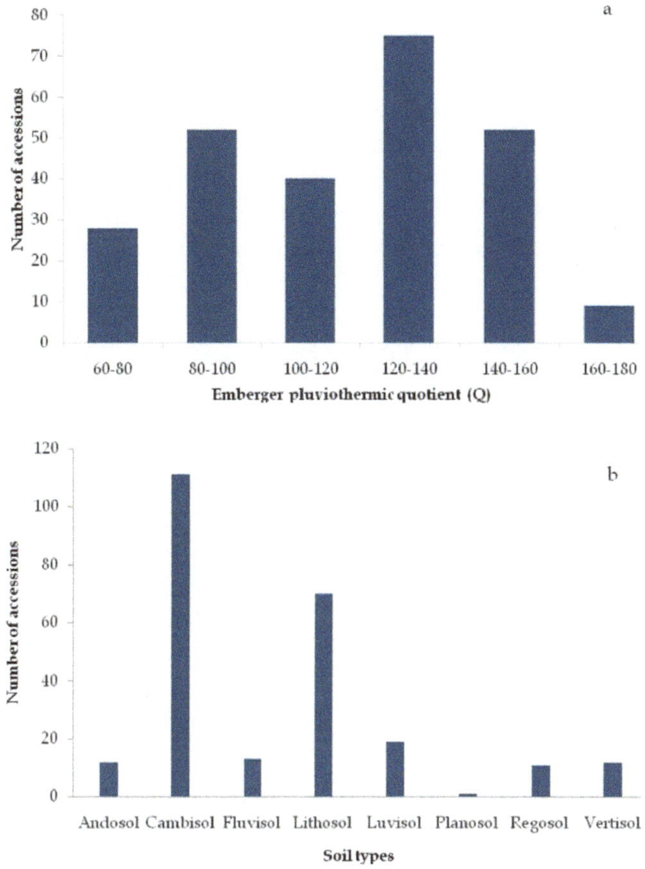

Figure 7: Groupings of target plants based on GIS-derived ecological criteria: Number of accessions of target plants (n=256) originating from the Aegean Archipelago, the Ionian Islands, Crete and Peloponnese, southern Greece grouped into different Emberger's Pluviothermic Quotient classes (a) and according to basic soil types (b). For each class or type specific plant lists can also be generated.

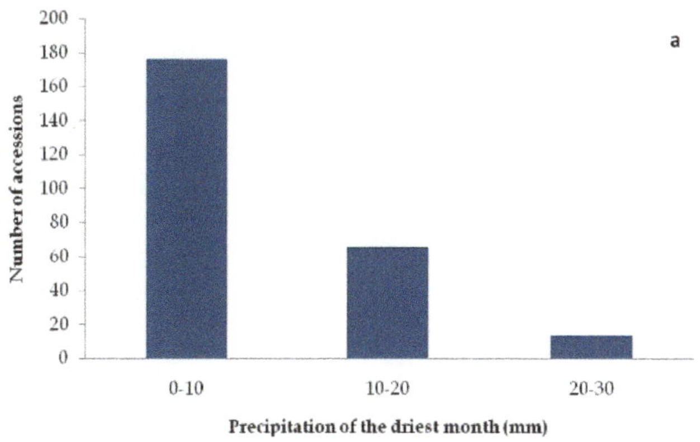

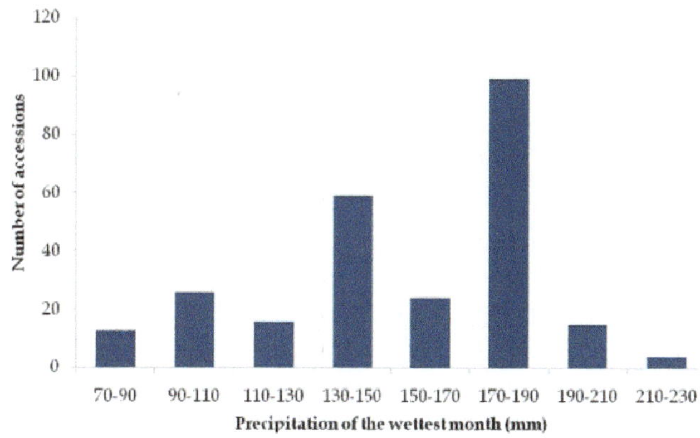

Figure 8: Groupings of target plants based on GIS-derived ecological criteria: Number of accession numbers of target plants (n=256) originating from the Aegean Archipelago, the Ionian Islands, Crete and Peloponnese (Greece) grouped according to precipitation classes during the driest (a) and wettest (b) months of the year. For each class specific plant lists can also be generated.

- Produce lists of species or frequency graphs within specified ranges of environmental variables or combinations thereof (Fig. 7, 8),
- Organize groupings of target plants based on ecological criteria (groups of plants with similar preferences or requirements), thus possibly improving their growing conditions as well as saving human-hours and space in *ex situ* conservation areas (Fig. 7, 8),
- Pinpoint target plants that may or cannot be accommodated in specific areas,
- Reduce trial-and-error losses during the *ex situ* cultivation of target species which seems to be common in the community of botanic gardens, often due to absence of previous experience (Krigas et al. 2010),
- Formulate and establish species-specific guidelines for the *ex situ* cultivation of target plants in botanic gardens,
- Facilitate the gap analysis of the botanic expeditions organized for the collection of plant material (which may also permit better planning of future ones),
- Reveal gaps in the representation of target plants from different altitudes, vegetation zones, habitat types, phytogeographic and climatic regions of specified geographical areas,
- Pinpoint by spatial niche modelling new probable locations where target species may be encountered, permitting the search for new populations in the wild (Jarvis et al., 2005),
- Assess conservation strategies and actions of institutions related to plant conservation.

By exploiting the GIS-derived ecological information for target plant species, the Balkan Botanic Garden of Kroussia (Greece) was the first to initiate this pilot GIS application dedicated to the *ex situ*plant conservation (Krigas et al., 2010). To date the majority of the target plants collected from the wild and propagated at the grounds of the BBGK are able to grow and flower regularly and produce fruits with no problems reported so far (BBGK, pers. comm.). Additionally, their seeds are regularly collected and deposited in a seed bank for future studies. This valuable material of conservation important target plants -in case of catastrophic events and if deemed necessary-could serve as a means to enrich the wild plant populations with individuals raised *ex situ,* actually reducing the risk of their extinction (Bowes, 1999; Bunn et al., 2011).

This novel GIS application described here presents an invaluable (time and money saving) tool with a broad-scale potential in enhancing the prospects of the ex situ plant conservation of target species collected from diverse environmental conditions and transferred to man-made sites such as botanic gardens, nurseries and private gardens.

ACKNOWLEDGEMENT

The authors would like to thank Dr E. Maloupa and the Balkan Botanic Garden of Kroussia (Greece) for funding earlier stages in this research and appreciate access to the plant material conserved ex situ in its grounds. The field work of N. Krigas in the Ionian Islands, SW Greece was partially supported by the Stanley Smith Horticultural Trust (UK) in the frame of the Ionian Island Project (Collections of rare, threatened and endemic plants of the Ionian Islands for their ex situ conservation in Greek and British botanic gardens). The work of A. D. Mazaris was partially supported by the EU FP7 SCALES project ('Securing the conservation of biodiversity across administrative levels and spatial, temporal and ecological Scales'; project #226852).

REFERENCES

1. J. Allison, A. J. Miller, J. H. Knouft, 2006GIS-based characterization of the geographic distributions of wild and cultivated populations of the Mesoamerican fruit tree Spondias purpurea (Anacardiaceae). American Journal of Botany, 93, 175717670002-9122
2. C. C. Baskin, J. M. Baskin, 1988Germination ecophysiology of herbaceous plant species in a temperate region. American Journal of Botany, 75(2), 2863050002-9122
3. E. E. Benson, 1999An introduction to plant conservation biotechnology, In: E.E. Benson (Ed), Plant conservation biotechnology, 310Taylor & Francis, 0-20348-419-3UK & Philadelphia, USA
4. B. G. Bowes, 1999A colour atlas of plant propagation and conservation, Manson Publishing Ltd, 978-1-87454-592-7London, UK
5. N. C. Brady, R. R. Weil, 2002The nature and properties of Soils, Prentice Hall, Upper Saddle River, 130167630Jersey, USA
6. E. Bunn, S. R. Turner, K. W. Dixon, 2011Biotechnology for saving rare and threatened flora in a biodiversity hotspot. In vitro Cellular & Developmental Biology-Plant. 47(1), 1882001054-5476
7. CBD-Convention on Biological Diversity1992United Nations Environment Programme, Rio de Janeiro, Retrieved fromhttp://www.cbd.int/doc/legal/cbd-en.pdf
8. T. Constantinidis, E. Kalpoutzakis, 2009Achillea occulta Constantin. & Kalpoutz., Vulnerable (VU), In: The red data book of rare and threatened plants of Greece, 1A-D), D. Phitos, T. Constantinidis, & G. Kamari (Eds), 4042Hellenic Botanical Society, 978-9-60940-712-0Patras, Greece (in Greek)
9. D. Draper, A. Rosselló-Graella, C. Garcia, Gomes. C. Tauleigne, C. Sérgio, 2003Application of GIS in plant conservation programmes in Portugal. Biological Conservation 113, 3373490960-3115
10. EC & ETC/LC1999CORINE land cover-technical guide, European Community, European Environment Agency, Retrieved fromhttp://www.eea.europa.eu/publications/tech40add
11. EC1985Soil Map of the European Communities (CORINE soil geodatabase) at 1:1000000, The Commission of the European Communities, Directorate General for Agriculture, Coordination of Agricultural Research, Luxembourg, Retrieved from http://www.eea.europa.eu/data-and-maps/data/soil-type
12. EC2004The European Soil Database distribution version 2.0, European Commission and the European Soil Bureau Network (CD-ROM) EUR 19945 EN
13. ESPC2009A sustainable future for Europe: The European Strategy for Plant Conservation 20082014Planta Europa Network, Council of Europe & CBD, Retrieved from http://www.plantaeuropa.org/assets/New%20European%20Strategy%20for%20Plant%20Conservation%20(2008-2014).pdf
14. E. J. Farnsworth, S. Klionsky, W. E. Brumback, K. Havens, 2006A set of simple decision matrices for prioritizing collection of rare plant species for ex situ conservation. Biological Conservation 128, 1120960-3115
15. J. A. Fernández, O. Santana, J. L. Guardiola, R. V. Molina, P. Heslop-Harrison, G. Borbely, F. Branca, S. Argento, E. Maloupa, T. Talou, J. M. Thiercelin, K. Gasimov, H. Vurdu, M. Roldán, M. Santaella, E. Sanchís, A. García-Luis, G. Suranyi, A. Molnár, G. Sramko, G. Gulyas, L. Balazs, O. Horvat, M. F. Rodríguez, R. Sánchez-Vioque, M. A. Escolano, J. V. Reina, N. Krigas, T. Pastor, B. Renau-Morata, C. Raynaud, O. Ibadli, M. Polissiou, M. Z. Tsimidou, A. Tsaftaris, M. Sharaf-Eldin, J. Medina, T. Constantinidis, T. Karamplianis, M. De -Los-Mozos-Pascual, 2011The World Saffron and Crocus collection: strategies for establishment, management, characterization and utilization. Genetic Resources & Crop Evolution 58(1), 1251370925-9864
16. L. Glowka, F. Burhenne-Guilman, H. Synge, J. Mc Neely, L. Güdling, 1994Guide to the convention on biological diversity, Environmental and law paper 30International Union for the Conservation of Nature and Natural Resources, 2-83170-222-4Switzerland & Cambridge, UK
17. S. Godefroid, T. Vanderborght, 2010Seed banking of endangered plants: are we conserving the right species to address climate change? Biodiversity and Conservation 19, 304930580960-3115
18. K. Grigoriadou, N. Krigas, E. Maloupa, 2011GIS-facilitated in vitro propagation and ex situ conservation of Achillea occulta. Plant Cell Tissue & Organ Culture 107(3), 5315400167-6857
19. GSPC-Global Strategy for Plant Conservation2002Secretariat of the convention of

biological diversity in association with Botanic Gardens Conservation International, Montreal, Quebec, Retrieved from http://www.cbd.int/gspc/

20. L. Guarino, A. Jarvis, R. J. Hijmans, N. Maxted, 2002Geographic information systems (GIS) and the conservation and use of plant genetic resources. In: Managing plant genetic diversity, J.M.M. Engels, V. Ramanatha Rao, A.H.D. Brown & M.T. Jackson (Eds), 3874040-85199-522-5Publishing, International Plant Genetic Resources Institute, Rome, Italy
21. R. J. Hijmans, S. E. Cameron, J. L. Parra, P. G. Jones, A. Jarvis, 2005Very high resolution interpolated climate surfaces for global land areas. International Journal of Climatology 25, 196519781097-0088
22. IUCN2001IUCN red list categories and criteria, Version 3.1, IUCN Species Survival Commission, Gland, Switzerland & Cambridge, UK, Retrieved from http://www.iucn.org
23. L. R. Iverson, A. Prasad, 1998Estimating regional plant biodiversity with GIS modeling. Diversity and Distributions 4, 49611472-4642
24. A. Jarvis, K. Williams, D. Williams, L. Guarino, P. J. Caballero, G. Mottram, 2005Use of GIS for optimizing a collecting mission for a rare wild pepper (Capsicum flexuosum Sendtn.) in Paraguay. Genetic Resources & Crop Evolution 52, 6716820925-9864
25. V. Karagianni, G. Kamari, D. Phitos, 2009Silene cephallenia subsp. cephallenia Heldr., Critically Endangered (CR), In: The red data book of rare and threatened plants of Greece, 2E-Z), D. Phitos, T. Constantinidis & G. Kamari (Eds), 319320Hellenic Botanical Society, 978-9-60940-712-0Patras, Greece (in Greek)
26. N. Katsouni, V. Karagianni, G. Kamari, 2009Viola cephalonica Bornm., Critically Endangered (CR), In: The red data book of rare and threatened plants of Greece, 2E-Z), D. Phitos, T. Constantinidis & G. Kamari (Eds), 385387Hellenic Botanical Society, 978-9-60940-712-0Patras, Greece (in Greek)
27. N. Krigas, Th. Karamplianis, 2009Ranunculus cacuminis Strid & Papan., Vulnerable (VU), In: The red data book of rare and threatened plants of Greece, 2E-Z), D. Phitos, T. Constantinidis & G. Kamari (Eds), 269271Hellenic Botanical Society, 978-9-60940-712-0Patras, Greece (in Greek)
28. N. Krigas, E. Maloupa, 2008The Balkan Botanic Garden of Kroussia, Northern Greece: a garden dedicated to the conservation of native plants of Greece and the Balkans. Sibbaldia 6, 9271-87229-144-9
29. N. Krigas, 2009Allium samothracicum Tzanoud., Strid & Kit Tan, Endangered (EN), In: The red data book of rare and threatened plants of Greece, 1A-D), D. Phitos, T. Constantinidis & G. Kamari (Eds), 6869Hellenic Botanical Society, 978-9-60940-712-0Patras, Greece (in Greek)
30. N. Krigas, G. Mouflis, K. Grigoriadou, E. Maloupa, 2010Conservation of important plants from the Ionian Islands at the Balkan Botanic Garden of Kroussia, N Greece: using GIS to link the in situ collection data with plant propagation and ex situ cultivation. Biodiversity & Conservation 19, 358336030960-3115
31. P. Krogstrup, J. I. Find, D. J. Gurkov, M. M. H. Kristensen, 2005Micropropagation of Socotran fig, Dorstenia gigas Schweinf. ex Balf. f.-A threatened species, endemic to the island of Socotra, Yemen. In vitro Cellular & Developmental Biology- Plant 41, 81861054-5476
32. B. A. Loiselle, P. M. Jørgensen, T. Consiglio, I. Jiménez, J. G. Blake, L. G. Lohmann, O. M. Montiel, 2008Predicting species distributions from herbarium collections: does climate bias in collection sampling influence model outcomes? Journal of Biogeography 35, 1051111365-2699
33. E. Maloupa, K. Grigoriadou, K. Papanastassi, N. Krigas, 2008Conservation, propagation, development and utilization of xerophytic species of the native Greek flora towards commercial floriculture. Acta Horticulturae 766, 2052130567-7572
34. M. Maunder, S. Higgens, A. Culham, 2001The effectiveness of botanic garden collections in supporting plant conservation: a European case study. Biodiversity & Conservation 10, 3834010960-3115
35. G. Mavromatis, 1980The Bioclimate of Greece, relationship of climate and natural vegetation, bioclimatic maps, Institute of Forest Research of Athens, Athens, Greece
36. E. S. Minor, S. M. Tessel, K. A. D. Engelhardt, T. R. Lookingbill, 2009The role of landscape connectivity in assembling exotic plant communities: a network analysis. Ecology 90, 180218090012-9658
37. J. Moat, P. P. Smith, 2003Applications of Geographical Information Systems in Seed Conservation, In: Seed Conservation: Turning Science into practice, R.D. Smith, J.B. Dickie, S.H. Linigton, H.W. Pritchard & R. J. Probert (Eds), Chapter IV, 8187Royal Botanic Gardens, 1-84246-052-8UK, Retrieved from http://www.kew.org/science-research-data/kew-in-depth/msbp/publications-data-resources/technical-resources/seed-conservation-science-practice/index.htm
38. Å. Ø. Pedersen, S. Nyhuus, T. Blindheim, Krog. O. M. Wergeland, 2004Implementation of a GIS-based management tool for conservation of biodiversity within the municipality of Oslo, Norway. Landscape & Urban Planning 68, 4294380169-2046
39. M. Powel, A. Accad, A. Shapcott, 2005Geographic information system (GIS) predictions of past, present habitat distribution and areas for re-introduction of the endangered subtropical rainforest shrub Triunia robusta (Proteaceae) from south-east Queensland Australia. Biological Conservation 123, 1651750006-3207
40. J. Ramirez-Villegas, C. Khoury, A. Jarvis, D. G. Debouck, L. Guadrino, 2010A gap analysis methodology for collecting crop genepools: a case study with Phaseolus beans. PloS ONE 5(10), 1181932-6203
41. B. B. Salem, 2003Application of GIS to biodiversity monitoring. Journal of Arid Environments 54, 911140140-1963
42. V. Sarasan, R. Cripps, M. M. Ramsay, C. Atherton, M. Mc Michen, G. Prendergast, J. K. Rowntree,

2006Conservation in vitro of threatened plants-progress in the past decade. In vitro Cellular & Developmental Biology- Plant 42, 2062141054-5476

43. L. Schulman, S. Lehvävirta, 2010Botanic gardens in the age of climate change. Biodiversity & Conservation 20, 217220Editorial note), 0960-3115

44. L. Schulman, T. Toivonen, K. Ruokolainen, 2007Analysing botanical collecting effort in Amazonia and correcting for it in species range estimation. Journal of Biogeography, 34, 138811391365-2699

45. S. Sharrock, M. Jones, 2009Conserving Europe's threatened plants: progress towards Target 8 of the Global Strategy for Plant Conservation, Botanic Garden Conservation International, Richmond, UK, 978-1-90516-430-1Retrieved fromhttp://www.bgci.org/files/Worldwide/Publications/euro_report.pdf

46. R. Store, J. Jokimäki, 2003A GIS-based multi-scale approach to habitat suitability modeling. Ecological Modeling 169, 1150304-3800

47. R. Store, J. Kangas, 2001Integrating spatial multi-criteria evaluation and expert knowledge for GIS-based habitat suitability modeling. Landscape & Urban Planning 55, 79930169-2046

48. N. Turland, 1995Origanum dictamnus L., Vulnerable (VU), In: The red data book of rare and threatened Plants of Greece, D., Phitos, A., Strid, S., Snogerup & W., Greuter (Eds), 394395WWF, 9-60750-604-9Greece

49. USGS-United States Geological Survey2004Shuttle Radar Topography Mission, 30 Arc Second scene SRTM_GTOPO_u30_mosaic, Unfilled Unfinished 2.0, Global Land Cover Facility, University of Maryland, College Park, Maryland, February 2000, Retrieved fromhttp://glcf.umiacs.umd.edu/data/srtm/

50. A. Vanderpoorten, A. Sotiaux, P. Engels, 2006A GIS-based model of the Distribution of the rare liverwort Aneura maxima at the landscape scale for an improved assessment of its conservation status. Biodiversity and Conservation 15, 8298380960-3115

CHAPTER 8

EX SITU CONSERVATION OF BIODIVERSITY WITH PARTICULAR EMPHASIS TO ETHIOPIA

Mohammed Kasso and Mundanthra Balakrishnan

Department of Zoological Sciences, Addis Ababa University, P.O. Box 1176, Addis Ababa, Ethiopia

ABSTRACT

Biodiversity encompasses variety and variability of all forms of life on earth that play a great role in human existence. Its conservation embraces maintenance, sustainable utilization, and restoration, of the lost and degraded biodiversity through two basic and complementary strategies called in situ and ex situ. Ex situconservation is the technique of conservation of all levels of biological diversity outside their natural habitats through different techniques like zoo, captive breeding, aquarium, botanical garden, and gene bank. It plays key roles in communicating the issues, raising awareness, and gaining widespread public and political support for conservation actions and for breeding endangered species in captivity for reintroduction. Limitations of ex situconservation include maintenance of organisms in artificial habitats, deterioration of genetic diversity, inbreeding depression, adaptations to captivity, and accumulation of deleterious alleles. It has many constraints in terms of personnel, costs, and reliance on electric power sources. Ethiopia is considered to be one of the richest centers of genetic resources in the world. Currently, a number of stakeholders/actors are actively working on biodiversity conservation through ex situ conservation strategies by establishing gene banks, botanical garden, and zoo.

INTRODUCTION

According to the Convention on Biological Diversity, biodiversity refers to the variability among living organisms (animals, plants, and microorganisms) including inter alia, terrestrial, marine, and other aquatic ecosystems with their ecological complexes. In another expression, biodiversity encompasses the variety and variability of all forms of life on earth that play a great role in human existence [1, 2]. It also includes the ethnical value of biodiversity such as tradition and traditional knowledge of the indigenous and local communities [2] and the diversity within species (genetics), between species and of ecosystems [3].

Genetic diversity refers to the variation within species of any plant, animal or microbes in the functional units of heredity. Species diversity refers to the variety of species within a geographical area, which become central in the evaluation of diversity, and used as a point of reference in biodiversity conservation. Finally, ecosystem diversity refers to the variety of life forms in a given territory or area with all its functional ecological processes, which is often evaluated based on the diversity of all of its components [1].

Citation: Mohammed Kasso and Mundanthra Balakrishnan, "Ex Situ Conservation of Biodiversity with Particular Emphasis to Ethiopia," ISRN Biodiversity, vol. 2013, Article ID 985037, 11 pages, 2013. doi:10.1155/2013/985037.

Biodiversity is important for the maintenance of a healthy environment and used for direct human benefits like food, medicine, and energy. It is also used for recycling of different essential elements, for mitigation of pollution, for protection of watersheds, to mitigate soil erosion and to control excessive variations in climate and catastrophic events. For example, biodiversity provides different services free of charge worth of billions dollar every year for crucial well-being of the society. Some of these services are providing clean water and air, soil formation and protection, pollination, pest control, food, fuel, fibers, medicine, and construction and industry raw materials [4]. Agricultural biodiversity is another important component of biodiversity, which has a more direct link to the well-being and livelihood of mankind than other forms of biodiversity. Food plant and animal species have been collected, used, domesticated, and improved through traditional systems of selection over many generations [5].

However, today much of the lines of evidences are increasingly pointing out a significant global decline in biodiversity by numerous, varied, and interacting drivers [6]. More than half of the habitable surface of the earth has already been significantly altered by human activities. As a consequence, biodiversity of our planet is on the verge of decline and extinction despite our limited and incomplete knowledge on them [7]. Biodiversity loss and extinction processes can occur in two phases. The first phase is known as deterministic and often resulted from human threats such as habitat loss, fragmentation and degradation, direct exploitation of the species, competition from exotic and domestic species, and persecution and killing due to human animal conflicts. The second phase is known as deterministic that resulted from failures in mitigating threats that eventually result in very small, fragmented, and isolated remnant populations. Then these small remnant populations become vulnerable to a number of other, nonhuman caused threats mainly stochastic, genetic (genetic drift and inbreeding), and demographic events [8]. Thus, small, fragmented, isolated populations can find themselves being dragged into an extinction vortex whereby genetic and demographic stochastic events can cause the species to go extinct. During this second phase of the extinction process, very intensive management of populations and individuals is often necessary to prevent extinction [9].

Several human induced impacts are leading to a mass extinction process affecting global biodiversity. The major reasons for rapid diminishing of biodiversity are attributed to conversion of land for agriculture, wild fires, poor management of available land, over-exploitation for food, fuel-wood, medicine, construction, overgrazing by cattle, displacement and loss of landraces, lower yielding varieties, pests and diseases, global climate change, pollution (e.g., acid rain), and gap of scientific knowledge on some of the biological resources [1, 7, 10]. Human beings are destroying biodiversity, particularly during livelihood activities with or without knowledge of the consequences of their actions [6]. Agriculture is one of the most important land-use that results in detrimental environmental consequences from increased use of fertilizers and biocides, land draining, irrigation, and the loss of many biodiversity-rich landscape features [6]. There are many threats to biodiversity as a result of agricultural practice through changes in land-use, replacement of traditional varieties by modern cultivars, agricultural intensification, increased population, poverty, land degradation, and environmental changes (including climate change) [5]. Recent estimates indicate that humans use more than 40% of the terrestrial components and significantly modified global biodiversity [7]. As a consequence, many species of living organisms are classified as threatened today and this has become a central concern for conservation [11].

Conserving biodiversity has economic, social, and cultural values. Conservation of biodiversity is integral to the biological and cultural inheritance of many people and the critical components of healthy ecosystems that are used to support economic and social developments. Moreover, it is used to maintain the earth's genetic library from which society has derived the basis of its agriculture and medicine [5, 12]. The twenty-first century is predicted to be an era of bioeconomy driven by advances of bioscience and biotechnology. Bio-economy may become the fourth economy form after agricultural, industrial, and information technology economies, having far-reaching impacts on sustainable development in agriculture, forestry, environmental protection, industry, food supply, health care and other micro-economy aspects. Thus, a strategic vision for conservation and sustainable use of biodiversity in the 21st century is of far-reaching significance for sustainable development economy and society [13].

Biodiversity conservation refers to the management of human use of biodiversity in order to get the greatest sustainable benefit to present and future generations. Thus, conservation of biodiversity embraces the protection, maintenance, sustainable utilization, restoration, and enhancement of biodiversity [1]. Biodiversity conservation mainly focuses on genetic conservation with its diverse life-support systems (ecosystems) for the connotation of human well-being [3].

Conservation techniques can be grouped into two basic, complementary strategies: in situ and ex situ [14]. As also outlined in the articles 8 and 9 of the Convention of Biological Diversity (CBD), biodiversity is conserved by two major methods called in situ and ex situ. The conservation efforts, either in situ or ex situ, involve the establishment and management of protected areas and relevant research institutes or academic institutions, which establish and manage arboreta, botanical or zoological gardens, tissue culture, and gene banks [1]. The concept of ex situ conservation is fundamentally different from that of in situ conservation; however, both are important complementary methods for conservation of biodiversity. The principal difference (and hence the reason for the complementarities) between the two lies in the fact that ex situ conservation implies the maintenance of genetic materials outside of the "normal" environment where the species has evolved and aims to maintain the genetic integrity of the material at the time of collection, whereas in situ conservation (maintenance of viable populations in their natural surroundings) is a dynamic system, which allows the biological resources to evolve and change over time through natural or human-driven selection processes [5].

In situ conservation is defined as conservation of ecosystems and natural habitats, the maintenance of viable populations of the species in their natural surroundings and, in the case of the cultivated species, in the surroundings where they have developed their distinctive properties. In situ conservation can be done in farmers' fields, in pasture lands, and in protected areas [15]. For cultivated species, in situ conservation concerns the maintenance of the local intra- and inter-population diversity available in various ecological and geographical sites [1, 16]. Thus, it allows ongoing host-parasite coevolution, which is likely to provide material resistance to pests and diseases, and CBD recognized it as a primary approach to conserve biodiversity [4]. However, in situ conservation has certain limitations like more difficult access to breeders requiring the application of its complimentary technique. For example, some of the natural habitats or wild habitats are very risky when compared to relatively safe captive environment [9].

The second biodiversity conservation technique receiving the most attention to conserve biodiversity is ex situ.Ex situ conservation techniques are mostly used to be applied to species with one or some of the following characteristics: endangered species, species with a past, present or future local importance, species of ethno-botanical interest, species of interest for the restoration of local ecosystems, symbolic local species, taxonomically isolated species, and monotypic or oligotypic genera [17]. Intensive conservation and management of populations and individuals can come in many different forms, like translocation, breeding in a fenced wild habitat, supplementary feeding, captive hand rearing of young of wild parents to become pregnant sooner, and captive breeding [9].

EX SITU CONSERVATION

Ex situ conservation is a technique of conservation of biological diversity outside its natural habitats, targeting all levels of biodiversity such as genetic, species, and ecosystems [1, 2, 16]. Its concept was developed earlier before its official adoption under the Convention on Biological Diversity signed in 1992 in Rio de Janeiro [2]. In general, ex situ conservation is applied as an additional measure to supplement in situ conservation, which refers to conservation of biological diversity in its natural habitats [16]. In some cases, ex situ management will be central to a conservation strategy and in others it will be of secondary importance [18]. Broadly, ex situconservation includes a variety of activities, from managing captive populations, education and raising awareness, supporting research initiatives and collaborating with in situ efforts [19]. It is used as valuable tools in studying and conserving biological resources (plants, animals, and microorganisms) for different purposes [2] through different techniques such as zoos, captive breeding, aquarium, botanical gardens, and gene banks [16].

Types of Ex Situ Conservation

Zoos

Zoos or zoological gardens or zoological parks in which animals are confined within enclosures or semi-natural and open areas, displayed to the public, and in which they may also breed. They are considered by universal thinkers and environmentalists as important means of conserving biodiversity [19–21]. Zoos attract as many as 450 million visitors each year and so are uniquely placed to have very large educational and economic values [22]. Zoos not only act as places of entertainment and observing animal behavior, but are also as institutions, museums, research laboratories, and information banks of rare animals [23]. Although some people dislike zoos, many people enjoy them. Over the last several decades, zoos have made significant progress in its cooperative management of ex situ populations of a variety of biodiversity [9].

Zoos breed many endangered species to increase their numbers. Such captive breeding in zoos has helped to save several species from extinction [19]. Management of animals in zoos includes animal identification, housing, husbandry, health, nutrition as well as addressing and ways of interaction with the public [20]. There are various processes and mechanisms used to determine whether a species or taxon is included within a zoo's collection plan. The frequently used criteria include how the species is valued, according to its uniqueness, contribution to research or education, and conservation status [19]. Zoos help the animal to secure food, shelter, social contact and mates, and to be motivated by desire (appetitive behavior), which is reinforced by pleasure (consummative behavior) [21].

In the past, some zoos paid little attention to the welfare of the animals, and some zoos today have poor environments for animals [24]. They were also once reliant on harvest from the wild to populate their exhibits and reliance on continued wild collection to breeding closed populations [23]. Many zoo animals also became endangered or extinct due to visitor disturbances, unfavorable climate and due to insufficient space [20]. From this aspect, many scholars state on the negative features of keeping animals in zoo as it causes pain, stress, distress, sufferings and evolutionary impacts [21]. Animal welfare, education, conservation, research, and entertainment are major goals of modern zoos, but these can be in conflict. For example, visitors enjoy learning about and observing behavior in captive animals, but visitors often want to observe and interact with the animals in close proximity. Unfortunately, proximity to and interactions with humans induce stress for many species [25]. The same is true for Addis Ababa Lion Zoo Park.

However, progressive zoos are engaged in education, research, and conservation, with the aim of maintaining healthy animals, which behave as if they are in their natural habitats [24]. The current paradigm for managing essential populations is to minimize the rate of genetic decay, slow adaptation to the captive environment, and retain typical behaviors [23, 26]. It is widely accepted that the more generations a population spends in captive breeding, the less suitable it is for attempted restorations in the wild. Hence, population management is designed not to deplete too quickly the resource obtained from the founders [23]. Thus, for true sustainability of the species for the purpose of conservation, display, education, and research, constant refreshing of populations is required [9, 23]. Majority of the current breeding programs base on the genetic management of populations by the analysis of individual pedigrees in order to minimize kinship [9].

Captive Breeding

Captive breeding is an integral part of the overall conservation action plan for a species that helps to prevent extinction of species, subspecies, or population. It is an intensive management practice for threatened individuals, populations, and species by anthropogenic and natural factors [9]. In small and fragmented populations, even if the human caused threats could be magically reversed, the species would still have a high probability of extinction by random demographic and genetic events, environmental variations, and catastrophes.

Thus, under sufficient knowledge on the biology and husbandry of the species, captive breeding helps individuals in the relative safety of captivity, under expert care and sound management by providing an insurance against extinction [9]. Stock for reintroduction or reinforcement efforts, opportunities for education, raising of awareness, scientific and husbandry research, and other contributions to conservation are also possible through captive breeding [9, 27].

Environmental enrichment strategies are used to improve both physiological and psychological welfares of captive animals, which can be achieved by increasing the expression of natural behavior and decreasing abnormal behaviors. Successful environmental enrichment includes the improvement of enclosure design and the provision of feeding devices, novel objects, appropriate social groupings, and other sensory stimuli [27]. The minimum requirement for successful ex situ management, particularly in the captive populations, is the inclusion of as much of the genetic diversity present in wild populations. Genetic sustainability (retention of 90% of the gentic diversity of the wild population for 100 years) in captive breeding is maintained if consideration is given on number of founders, population growth rate, effective population size, and duration of the captive program [19]. However, even if at least 30 founders in captive breeding are recommended to ensure the representation of large enough proportion of the genetic diversity of the wild population, for critically endangered species, actively removing individuals from the wild population to serve as founders may compromise the survival of the wild population [9]. For example, the Arabian Oryx captive breeding program was based on fewer founders and grew to a couple of thousand individuals through breeding management, which helped to reduce risks.

However, there are several challenges (biological and environmental) that are limiting factors to the attainment of the goal of captive breeding for many species [19]. One of the major challenges is a circular consequence of small-population management that has inherent genetic and demographic problems due to genetic diversity loss and demographic stochasticity [19]. In addition, individuals that are well adapted to the circumstances in captivity may also be less well adapted to the circumstances in the wild and may show lower fitness upon reintroduction [28]. Most notably, within the captive environment, housing and husbandry will also have significant impacts on birth and death rates [19].

Aquarium

An aquarium is an artificial habitat for water-dwelling animals. It can also be used to house amphibians or large marine mammals and plant species for tourist attractions. It is usually found in zoos or marine parks with different size. The 15,750 described species of freshwater fish comprise around 25% of living vertebrate species diversity and a key for global economic and nutritional resources of which more than 11% is threatened (60-extinct, 8-extinct in the wild and 1679-threatened) [22]. Fresh waters (0.3%) of available global surface water support 47–53% of all extant fish species that are threatened by overfishing, pollution, habitat loss, damming, alien invasive species, and climate change. This requires world's zoos and aquariums to identify the potential targets (species or areas) for in situ and ex situ conservation program [18, 22]. Aquarium is used to admire at home by hobbyists, to portray as public exhibits, to provide large quantities of human food and animal fodder [18].

Fishes are often overlooked within the development of conservation priorities. This leads to the low focus on meaningful conservation efforts rather than giving more attention for their importance to food supply and livelihoods. For example, it provides job opportunity for over 60 million people, as source of food for over 200 million people in Africa, for US$1.5 billion income from trade of 4000 species global ornamental fish industry, and many are displayed in the world's public zoos and aquaria to a global audience of as many as 450 million people per year [22, 29]. However, despite the clear value of freshwater fish diversity, wetland habitats and their associated freshwater-fish species continue to be lost or degraded at an alarming rate [30]. One recommendation is for aquariums to set up sustainable breeding program that prioritizes threatened species (VU, EN, and CR) and those classified as EW to support species conservation in situ and aid the recovery of species via collaborative reintroduction or translocation efforts when appropriate [22].

Botanical Gardens

Botanical gardens consist of living plants, grown out of doors or under glass in greenhouses and conservatories. They are used to grow and display plants primarily for scientific and educational purposes. They also include herbarium, lecture rooms, laboratories, libraries, museum and experimental or research plantings. It can be taxonomic collection of a particular family, genus or group of cultivars, native plants, wild relatives, medicinal, aromatic, or textile plants [4]. There are over 2,000 botanic gardens, holding 80,000 plant species in their living collections and receiving hundreds of millions of visitors per annum [14, 31]. Furthermore, they have valuable and distinctive mix of officials dedicated to plant research, systematics, conservation education, and public awareness [31]. They are now extremely well networked both among themselves and with other professionals, conservation organizations, and nongovernmental organizations (NGOs) [31]. They provide different services for sectors that utilize and conserve plant diversity like agriculture, forestry, pharmaceutical and biofuel industries, protected area management, and ecotourism. They have a unique opportunity as visitor attraction places and scientific institutions for documentation and conservation of plant diversity by shaping and mobilizing citizens to the current environmental challenges [31]. They also play a great role in attaining target of the Global Strategy for Plant Conservation for 2020 to cultivate 75% of the world's threatened plant species inex situ [14]. Botanical gardens give opportunity for arable plants to be grown under relatively modified environmental conditions (intense cultivation, relatively high fertility, and high levels of disturbance) [14].

However, most of the cultivated taxa are held in a small number of collections and mostly only in small populations. Lack of genetic exchange and stochastic processes in small populations make them susceptible to detrimental genetic effects [14]. The low number of ex situ populations in most botanical gardens poses a fundamental problem for conservation. The total ex situ breeding collection is therefore very small with respect to the stated aim of conserving regional gene pools

[14]. The striking lack of information on source populations casts doubt on the value of using such ex situ populations for potential reintroductions. They also require testing for fitness and similarity to wild populations before they are brought to the field [14]. Thus, conservation actions of botanic gardens such as training and capacity building, needs to be better understood and better coordinated [31].

Gene Banks

Genome resource banking is another management technique used for biodiversity conservation. Different types of gene banks have been established for the storage of biodiversity, depending on the type of materials conserved. These include seed banks (for seeds), field gene banks (for live plants), in vitro gene banks (for plant tissues and cells), pollen, chromosome, and deoxyribonucleic acid (DNA) banks for animals (living sperm, eggs, embryos, tissues, chromosomes, and DNA) that are held in short term or long term laboratory storage; usually cryopreserved or freeze-dried [32]. Currently, there are about 7.4 million PGRFA accessions conserved in over 1750 gene banks [5]. The Genome 10K (G10K) project aims to sequence the genomes of 10,000 vertebrates, of which 4,000 will be fish. To date, 60 fish genomes have been sequenced in laboratories worldwide and added to the database, and a further 100 are targeted. The Frozen Ark database holds details of 28,060 frozen DNA samples. Among these 6,997 are from species listed in the IUCN Red List [18]. The principal aim of gene bank conservation is to maintain genetic diversity alive as long as possible and to reduce the frequency of regeneration that may cause the loss of genetic diversity [5].

With the rapid development in the field of molecular genetics and genomics, DNA is becoming more and more in demand for molecular studies and is one of the most requested materials from gene banks. Establishing DNA storage facility as a complementary "backup" to traditional ex situ collections has been suggested [5]. Some efforts have been made to establish DNA banks for endangered animals [33], and a few plant DNA banks in different parts of the world such as the Missouri Botanic Garden, Kew Royal Botanic Garden and Australian Plant DNA Bank. Many research groups are already developing their own archives of extracted genomic DNA. Recently, the Global Biodiversity Information Facility in Germany has established a DNA bank network, which provides DNA samples of complementary collections (microorganisms, protists, plants, algae, fungi, and animals) [5].

Seeds are usually the most convenient and easiest material to collect and to maintain in a viable state for long periods of time and that makes it preferred for conservation in gene banks [14, 34]. Seed banking techniques rely on the storage of dried seeds of threatened or other plants at low temperatures as the most important factors influencing seed longevity are temperature, seed moisture content, and relative humidity [14, 35, 36]. Seeds are typically conserved at moisture content between 3 and 7 percent and stored at 4 degrees Celsius for short-term conservation, and between −18 and −20 degrees Celsius for long-term conservation [5, 34]. Current research is showing that there exists variability in seed longevity for different species being conserved under similar conditions. In addition, it has been found that the type of seed (endospermic or non-endospermic) and intraspecific variation may also affect accessions longevity [5]. In addition, high initial quality seeds are a major prerequisite for ensuring seed longevity in seed banks [5].

Plants that cannot be conserved as seeds because of their recalcitrant nature (i.e. seeds that are desiccation and/or cold sensitive) or are clonally propagated are traditionally conserved as live plants in ex situ field gene banks. But, field gene banks present real logistical challenges; they require large areas and are costly, they are vulnerable to pests and diseases, natural disasters, political unrest, extreme weather, fire, vandalism, and theft, and they often are at risk due to policy changes on land use [5].

In vitro conservation refers to one type of gene bank known as slow-growth conservation method. It involves culturing of different parts of the plant (meristem, tissues, and cells) into pathogen-free sterile culture in a synthetic medium with growth retardants, which has been cited as a good way of complementing and providing backup to field collections [5].

The other genome conservation technique is cryopreservation, in which living tissues are conserved at very low temperatures (−196°C) in liquid nitrogen to arrest mitotic and metabolic activities [5]. It is now realized that cryopreservation method can offer greater security for long-term, cost effective conservation of plant genetic resources, including orthodox seeds [5]. The storage in liquid nitrogen clearly prolonged shelf life of lettuce seeds with half-lives projected as 500 and 3400 years for fresh lettuce seeds stored in the vapor and liquid phases of liquid nitrogen, respectively [5].

Advantages of Ex Situ Conservation

It is generally preferred to conserve threatened species in situ, because evolutionary processes are more likely to remain dynamic in natural habitats [14, 19]. However, considering the rate of habitat loss worldwide, ex situcultivation is becoming increasingly important [14]. Further more, as many of the taxa are located outside natural parks or reserves, in situ measures are not enough to assure their conservation. Translocation, introduction, reintroduction, and assisted migrations are species conservation strategies that are attracting increasing attention, especially in the face of climate change [37].

As approximately 450 million people per year visit zoos and aquaria globally, their education and marketing services play a key role in communicating the issues, raising awareness, changing behavior, and gaining widespread public and political support for conservation actions. Zoos support conservation by educating the public, raising money for conservation programs, developing technology that can be used to track wild populations, conducting scientific research, advancing veterinary medicine, and developing animal handling techniques [22]. By studying animals in captivity and applying that knowledge to their husbandry, zoos can provide valuable and practical information that may be difficult or impossible to gather from the wild [24]. Zoos and aquaria have significant roles to play in improveing public awareness of the issue facing species and their habitats; for example, through presentation of maps and photographs of species recently extinct as a result of anthropogenic impacts. A similar display of threatened species, even if not currently in the collections of the zoo, would help convey to the public the magnitude of the threat facing the species [18, 22]. It also reaches a wide cross-section of the society, because zoo audiences are not limited to those who are already passionately interested in wildlife and because many zoo visitors are children. Some of these children may become committed conservationists. Some may grow up to be oil company tycoons, politicians, or movie stars, with great potential influence. Some may even live next door to a poacher or wildlife dealer. Thus, instilling an interest in conservation of wildlife in people from all walks of life while they are young is one vital role zoos can play [38]. It is often claimed that zoos perform valuable conservation work by breeding endangered species and returning them to the wild. Zoos can also be used for businesses that make money. This means that animals are often bred for commercial purposes because the public like to see new-born animals. Such breeding leads to a surplus of animals, and in order to keep numbers down sold to private collectors, circuses, or even research laboratories. A zoo with good and attractive entertainments encourages initial visits and subsequent returns to the zoo, which is used to get more revenue for conservation efforts, research, and general animal care and welfare and also to develop more positive perceptions of animals in zoos and become more supportive of conservation efforts [25, 39].

Disadvantages of Ex Situ Conservation

Some ex situ conserved collections showed lower resistance levels, although still others showed higher resistance levels than their in situ conserved counterparts mainly due to the high evolutionary drive and complex nature of evolutionary scenario [40].

The behavior of animals in the zoo may be affected by the frequent arrival of large number of people, who are unfamiliar to the animals [41]. Animals housed in artificial habitats are confronted by a wide range of potentially offensive environmental challenges such as artificial lighting, exposure to loud or aversive sound, arousing odors, and uncomfortable temperatures or substrates. In addition, confinement-specific stressors such as restricted movement, reduced retreat space, forced proximity to humans, reduced feeding opportunities, maintenance in abnormal social groups, and other restrictions of behavioral opportunity are considered [42]. However, over the course of the twentieth century, as knowledge of wildlife biology improved, zoo animals began to be kept in more natural surroundings and social groupings, and diets and veterinary care began to improve. Thus, survival and breeding rates of captive populations improved [38]. Evidence mainly from studies of rodents and primates strongly indicates that prenatal stress can impair stress-coping ability and is able to cause a disruption of behavior in aversive or conflict-inducing situations. Prenatally stressed animals show retarded motor development, reduced exploratory and play behavior, and impairments of learning ability, social behavior, and sexual and maternal behavior. Prenatal stress may also affect the sex ratio at birth and the reproductive success [43].

Although populations of some species managed in ex situ may have the best hope for their long-term survival, they might be challenged if not properly managed during translocation and reintroduction with the effects of climate change [22]. Some species may lose their biological integrity particularly on morphology. For example, an experimental study on black-footed ferrets (Mustela nigripes) in ex situ indicated a decrease of 5–10% body size than pre-captive, in situ animals [44]. In other words, the small cage size and environmental homogeneity inhibit mechanical stimuli necessary for long bone development. Thus, in the absence of such an environment, "unnatural" morphologies can result that may contribute to poor fitness or perhaps even for domestication and reintroduction and relocation [44]. It would be very difficult to reintroduce some zoo-reared animals to their natural habitats because, after generations of captivity, many have lost the necessary skills to survive in their original habitats [22].

For naturally out-breeding species, the high levels of inbreeding in captivity often have negative effects on life history traits related to reproduction and survival [11]. It makes the population in captivity deteriorate due to loss of genetic diversity, inbreeding depression, genetic adaptations to captivity, and accumulation of deleterious alleles [17]. For plants, ecological shifts, small population size, genetic drift, inbreeding, and gardener-induced selection may negatively affect population structure after several generations of ex situcultivation [16, 45, 46]. These factors could seriously put at risk the success of ex situ conservation [17].

Captive breeding of threatened species has used increasingly sophisticated technologies and protocols in recent years [47]. Although, this has blurred the dichotomy between in situ and ex situ species management, the value of captive breeding as a conservation tool remains controversial [48]. It is recognized that ex situ conservation has many constraints in terms of personnel, costs, and reliance on electric power sources (especially in many developing countries where electricity power can be unreliable) for gene banks. It requires high facilities and financial investments. It cannot also conserve all of the thousands of plant and animal species that make up complex ecosystems such as tropical rainforests [49]. Capture of individuals from the wild for captive breeding or translocation some times can have detrimental effects on the survival prospects of the species as a whole through disease infection [50].

Even though the management of irreplaceable animal populations in zoos and aquarium has focused primarily on minimizing genetic decay with the use of advanced technologies, recent analyses have shown that as most zoo programs are not projected to meet the stated goals due to lack of achieving "sustainability" of the populations [23]. Thus, managing zoo populations as comprehensive conservation strategies for the species requires research on determinants of various kinds of genetic, physiological, behavioral, and morphological variations, and their roles in population viability, development of an array of management techniques, tools, and training of managers [23].

Challenges to Ex Situ Conservation

Ex situ conservation requires different kinds and levels of intensity of management, and a multistakeholder approach like the input from experts on aquarium and zoo husbandry, ex situ breeding, gene-banking, reintroduction, and habitat restoration [51]. Other expert input may include taxonomy, ecology and conservation, ethnography, and sociology. For outreach program, there is a need to liaise with local communities and national government fisheries and wildlife departments; with international (nongovernmental and intergovernmental) conservation bodies [18].

The most important challenges of applying ex situ conservation (captive breading) are the difficulty in recognizing the right time, identifying the precise role of the conservation efforts within the overall conservation action plan, and setting realistic targets in terms of required time span, population size, founder numbers, resources, insurance of sound management and cooperation, and the development of much needed new technical methods and tools [9]. In captive breeding to achieve the retention of 90% of the wild gentic diversity, it is necessary to incorporate sufficient number of founders, careful pair combinations and management [9]. Evidence also exists, which demonstrates that manipulation of housing and husbandry variables can also have significant positive influence on animal reproduction in captivity [19].

In many cases, ex situ populations are founded from only a few individuals, which cause genetic bottlenecks. Small populations are exposed to threats such as stochastic demographic events as well as genetic effects, including loss of genetic diversity, inbreeding depression or accumulation of new, potentially deleterious mutations [11]. More specific problems in garden populations include poorly documented or even unknown sources of material, accidental hybridization of material from various localities, and or unintended selection for traits more suited to garden conditions [14]. In every region, most of the cooperatively managed breeding programs have too few animals, too few animals in appropriate situations for breeding, too few successful breeders, too few founders, and too many animals with undocumented ancestries and/or too little cooperation with scientifically designated breeding recommendations. These deficiencies are resulting in declining populations or declining gentic diversity or both [23, 52].

Problems associated with small founder populations such as inbreeding depression, removal of natural selection, and rapid adaptation to captivity pose considerable challenges for managers of captive populations of threatened species [48]. Equally, reintroduction of captive-bred stock to the wild may require implementation of rigorous protocols that embrace acclimation, pre- and post-release training, health screening, genetic management, long-term monitoring, and involvement of local stakeholders [53, 54]. Shortfalls in implementing such protocols may jeopardize the likelihood of achieving success [47].

Inbreeding due to the mating between two related individuals is unavoidable in small, fragmented, or isolated populations typical of many threatened species, and it can lead to a significant reduction in fitness. The deleterious effects of inbreeding on individual fitness can be large and may be an important factor contributing to population extinction. Inbreeding depression has potential significance for the management and conservation of endangered species [55]. As populations get smaller, the probability increases for all offspring in a given generation are of the same sex [19].

Evaluating the long-term efficiency of ex situ conservation is important, but is complicated because of the difficulty of finding more than one sample of a documented (origin and cultivation) ex situ population and its corresponding still-existing in situ source population [46].

Animal translocations are usually risky and expensive, and a number of biological and nonbiological factors can influence success. Biological considerations include knowledge of genetics, demography, behavior, disease, and habitat requirements. It also includes legal framework, fiscal and intellectual resources, monitoring capacity, goal of the translocation, logistic challenges, and organizational structure of decision making [56].

The regeneration process is one of the most critical steps and a major challenge in gene bank management, during which there is the highest probability for genetic erosion [57]. It is equally important to understand how different conservation methods (seed, field, and cryopreservation) and their management can affect or change the genetic make up, thereby reducing the effective population size (Ne). This will also contribute to decision-making process for determining which methods to use for conservation of the wide diversity [5].

If people are discouraged or prevented from interacting with the resident animals, fewer visitors attend, decreasing public financial support. The visitors' noise and crowding become a source of stress for many species that affects both their welfare and the enjoyment of the visitor [25].

EX SITU CONSERVATION PRACTICE IN ETHIOPIA

Ethiopia is considered to be one of the richest centers of genetic resources in the world. It is believed that indigenous crops such as teff (Eragrostis tef), Noug (Guizotia abyssinica), and Enset (Ensete ventricosum) were first domesticated in Ethiopia. Numerous major crop species including durum wheat (Triticum durum), barley (Hordeum vulgare), sorghum (Sorghum bicolor), sesame (Sesamum indicum), castor (Ricinus communis), and coffee (Coffea arabica) are also known to show significant diversity in the Ethiopian region [58]. Almost 85% of the populations of Ethiopia live in rural areas and most of this population depends directly or indirectly on biodiversity. Biodiversity also plays a crucial role in the different sectors like energy, agriculture, forestry, fisheries, wildlife, industry, health, tourism, commerce, irrigation, and power [15].

The records on biodiversity conservation efforts in Ethiopia date back to the days of Emperor Zera-Yakob (1434–1468 E.C.). The Emperor brought juniper seedlings from Wof Washa of North Shewa and planted in Managesha-Suba area. Modern conservation intervention began by Emperor Menilik in 1908 E.C. and eventually evolved to the establishment of protected areas in the 1960s [4]. Currently, a number of stakeholders are actively working on biodiversity related issues at the federal government level. These include Institute of Biodiversity Conservation (IBC), Ethiopian Institute of Agricultural Research (EIAR), Ethiopian Wildlife Conservation Authority (EWCA), Ministry of Agriculture and Rural Development (MoARD), Ministry of Science and Technology (MoST), Higher Learning Institutions (HLIs), particularly Addis Ababa University (AAU), and offices in various regional states of Ethiopia [59]. Ex situ conservation as complementary to the rehabilitation and restoration of degraded ecosystems and the recovery of threatened species was started 1976 with the establishment of IBC [4].

Ex situ conservation activities mostly focus on high socioeconomic value and internationally important crop types that are considered to be facing immediate danger of genetic erosion [15]. The collections held at IBC are mostly of indigenous landraces some of which are not seen today in farmlands. The collections of root crops, medicinal plants, weedy species, and wild relatives of cultivated species are still relatively inadequate within the existing ex situ collections [60]. However, appropriate emphasis is being placed on conservation and sustainable use of all forms of plant biological resources [60]. Since the establishment of IBC, systematic crop germplasm exploration and collection operations have been undertaken in the different administrative regions of the country, covering a wide range of agroecological zones. Collection priorities were set based on factors like economic importance, degree of genetic erosion and diversity, researchers' needs, the rate of diffusion of improved varieties, clearing of natural vegetation, agricultural policies, natural disasters, and resettlement program [4].

Currently about 68,014 seed accessions of 200 plant species, 6,704 accessions of 205 species of forestry, medicinal, forage and pasture plants (in field gene bank), 290 species of microbial genetic resources, and three semen of threatened breed of domestic animals are conserved by IBC. About 90% of the total germplasm holdings in the gene bank consist of field crops. The total collection is composed of cereal seeds, pulses, oil crops, spices, and species of medicinal and industrial value. Aside from the crop collections, the gene bank also holds 650 collections of micro-organisms. Over 9,000 accessions of horticultural crops, medicinal plants, and herbs are maintained in field gene banks. The type and nature of collection missions and number and lists of plant species and landraces collected have been documented in manuals and reports. Regular monitoring activities are performed for seed viability [15].

For plant species with recalcitrant and intermediate storage behavior, there are ten field gene banks under IBC control and small sized fields in the various research stations of the Ethiopian Institute of Agricultural Research (EIAR) and at universities. The plan for the immediate future is to increase the number of field gene banks in different agroecological zones. Community gardens, backyards, and holy places are being considered for inclusion in the future plan. Spices, vegetables and medicinal plants require management on a large scale and with the full involvement of the local communities [15]. The initiative at national level is still in its infancy and there is currently no well-established national botanical garden in Ethiopia including the Gulele Botanical Garden Center [4, 59]. The Gulele Botanic Garden Center was established through the Proclamation No. 18/2005 E.C. in October 30/2002 E.C. in a 705 hectare land at Gulele and Kolfe-Kernayo subcities. It was established with a vision to see the center to be developed as an exemplary garden in terms of education, ecotourism attraction and center for originality of the Ethiopian plant species, and to be a place of research and nurturing of plant species. The center also has a mission to provide persistent ecotourism services to tourists by taking care of plant species and carrying out educational and research works [59].

Although there is no well-established zoo or zoological garden in Ethiopia, the Addis Ababa Lion Zoo Park can be dominantly cited [59]. The Addis Ababa Lion Zoo Park was established in 1948 with five founder lions presented to Emperor Haile Selassie as gifts. The park accommodates lions with cubs, tortoises, baboons, monkeys, apes, and ducks. A team of international researchers has provided the first comprehensive DNA evidence from 15 (eight males and seven females) samples of Addis Ababa lion indicating the genetically unique samples that requires immediate conservation action. Both microsatellite and mitochondrial DNA data suggest that the zoo lions are genetically distinct from all existing lion populations for which comparative data exist.

Desiccation-intolerant seeds and species that do not readily produce seeds are conserved ex situ in field gene banks. For example, accessions of coffee (Coffea arabica), root crops such as yam (Dioscorea bulbifera) and "Oromo dinich" (Coleus edulis), and spices like ginger (Zingiber officinale) and Ethiopian cardamom (Aframomum corrorima) are conserved in agro-ecological zones in field gene banks [60].

The need for action for global biodiversity conservation is now well understood, and government agencies, nongovernmental organizations, and botanic gardens have all been working in various ways to promote environmental sustainability and reduce species and habitat loss [31].

Seed banking is the major ex situ conservation method employed in Ethiopia. There are three major seed banks operating in Ethiopia. The National Tree Seed Project processes seeds from a narrow range of tree species and uses short-term storage facilities. It aims to cater for the annual seed demand from commercial and small-scale forestry enterprises. Of the 70 species regularly collected and processed, 20 are indigenous. The Forage Genetic Resources Centre maintained by the Consultative Group on International Agricultural Research at the International Livestock Research Institute maintains long-term conservation of a wide range of native and exotic forage species. The Institute for Biodiversity Conservation and Research holds active collections of seeds mainly for research and distribution and as a base collection for long-term conservation [60].

For security reasons, the collected and stored germplasm need to be conserved in duplicate gene banks. However, except for the limited samples of the Ethiopian germplasm held by the Consultative Group on International Agricultural Research, United States Agency for International Development, and the Nordic Gene Bank, majority of the Ethiopian collections are still kept in a single copy at the National Gene Bank. Greater efforts need to be made to store duplicate collections to avoid future genetic erosion [15]. For the continuing power supply, the Ethiopian Gene Bank has independent power supply in the form of a stand by generator to overcome power cuts [15].

Lack of adequate knowledge with respect to collection, handling, and treatment of seeds often impedes the planting of indigenous trees and shrubs. Inadequate work has been done on establishing the seed storage behavior of native species resulting in only limited availability of ex situ conservation seed collections especially with respect to native forest species [60] and lack of alternative storage facilities for the existing conventional cold rooms (e.g., in vitro and cryo-preservation methods) [4]. The current holdings of the IBC gene bank reach over 60,000 accessions of plant species. Some collections are in the medium-term storage mainly due to insufficient seed samples [15].

CONCLUSION

Biodiversity plays a great role in human existence and in healthy function of natural systems although it is on the way of depletion dominantly due to anthropogenic activities. This requires conservation of biodiversity either in in situ or ex situ or both methods in combination based on the conservation situation and its objective. Although in situ conservation is more encouraged to be used for biodiversity conservation, ex situ conservation is recommended as it complements through different techniques like zoo, captive breeding, aquarium, botanical garden, and gene bank. Ex situ conservation has its own advantages, disadvantages, and challenges making decision on its application by evaluating advantages, disadvantages and challenges. Although, Ethiopia is rich in biodiversity resources, more people depend on it for their livelihood directly or indirectly causing a great loss. Even if the conservation of biodiversity in Ethiopia has long-time history, its progress, coverage, and enforcement of the rule for conservation seem to be weak. Despite of good progress made in gene bank conservation, it is yet to be developed. In the same way, attention should be given for developing a National Zoological Park and a Botanical Garden.

ACKNOWLEDGMENTS

The authors are extremely grateful to Dr. Habte Jebessa for his valuable comments, suggestions and corrections on the draft of this paper. Our appreciation also goes to Professor. Afework Bekele for his remarkable encouragements and critical comments on the draft of this paper.

REFERENCES

1. T. I. Borokini, A. U. Okere, A. O. Giwa, B. O. Daramola, and W. T. Odofin, "Biodiversity and conservation of plant genetic resources in Field Gene-bank of the National Centre for Genetic Resources and Biotechnology, Ibadan, Nigeria," The International Journal of Biodiversity and Conservation, vol. 2, pp. 37–50, 2010.
2. M. Antofie, "Current political commitments' challenges for ex situ conservation of plant genetic resources for food and agriculture," Analele Universităţii din Oradea—Fascicula Biologie, vol. 18, pp. 157–163, 2011.
3. C. A. Tisdell, "Core issues in the economics of biodiversity conservation," Annals of the New York Academy of Sciences, vol. 1219, no. 1, pp. 99–112, 2011.
4. Institute of Biodiversity Conservation, National Biodiversity Strategy and Action Plan, Institute of Biodiversity Conservation, Federal Democratic Republic of Ethiopia, Addis Ababa, Ethiopia, 2005.
5. M. E. Dulloo, D. Hunter, and T. Borelli, "Ex situ and in situ conservation of agricultural biodiversity: major advances and research needs," Notulae Botanicae Horti Agrobotanici Cluj-Napoca, vol. 38, no. 2, pp. 123–135, 2010.
6. J. Young, C. Richards, A. Fischer et al., "Conflicts between biodiversity conservation and human activities in the central and eastern European countries," Ambio, vol. 36, no. 7, pp. 545–550, 2007.
7. H. Debela, "Human influence and threat to biodiversity and sustainable living," Ethiopian Journal of Education and Sciences, vol. 3, no. 1, pp. 85–95, 2007.
8. R. C. Lacy, "Considering threats to the viability of small populations using individual-based models,"Ecological Bulletin, vol. 48, pp. 39–51, 2000.
9. K. Leus, "Captive breeding and conservation," Zoology in the Middle East, vol. 54, supplement 3, pp. 151–158, 2011.
10. K. A. Wilson, J. Carwardine, and H. P. Possingham, "Setting conservation priorities," Annals of the New York Academy of Sciences, vol. 1162, pp. 237–264, 2009. ·
11. R. Frankham, J. D. Ballou, and D. A. Briscoe, Introduction to Conservation Genetics, Cambridge University Press, Cambridge, UK, 2002.
12. T. Getachew, "Biodiversity hotspots: pitfalls and prospects," IBC News Letter, vol. 1, pp. 14–16, 2012.
13. H. Huang, "Plant diversity and conservation in China: planning a strategic bioresource for a sustainable future," Botanical Journal of the Linnean Society, vol. 166, no. 3, pp. 282–300, 2011. ·
14. C. Brutting, I. Hensen, and K. Wesche, "Ex situ cultivation affects genetic structure and diversity in arable plants," Plant Biology, vol. 15, pp. 505–513, 2013.
15. Institute of Biodiversity Conservation, Ethiopia: Second Country Report on the State of PGRFA to FAO, FAO, Rome, Italy, 2007.
16. E. D. Kjaer, L. Graudal, and I. Nathan, Ex Situ Conservation of Commercial Tropical Trees: Strategies, Options and Constraints, Danida Forest Seed Centre, Humlebaek, Denmark, 2001.
17. J. Hakansson, Genetic Aspects of Ex Situ Conservation: Introductory Paper, Department of Biology, Linköping University, 2004.
18. G. M. Reid, T. C. Macbeath, and K. Csatadi, "Global challenges in freshwater-fish conservation related to public aquariums and the aquarium industry," International Zoo Yearbook, vol. 47, pp. 6–45, 2013.
19. V. A. Melfi, "Ex situ gibbon conservation: status, management and birth sex ratios," International Zoo Yearbook, vol. 46, no. 1, pp. 241–251, 2012.
20. C. Ratledge, "Towards conceptual framework for wildlife Tourism," Tourism Management, vol. 22, pp. 31–40, 2001.
21. J. Balcombe, "Animal pleasure and its moral significance," Applied Animal Behaviour Science, vol. 118, no. 3-4, pp. 208–216, 2009.
22. S. F. Carrizo, K. G. Smith, and W. R. T. Darwall, "Progress towards a global assessment of the status of freshwater fishes (Pisces) for the IUCN Red List: application to conservation programmes in zoos and aquariums," International Zoo Yearbook, vol. 47, pp. 46–64, 2013.
23. R. C. Lacy, "Achieving true sustainability of zoo populations," Zoo Biology, vol. 32, pp. 19–26, 2013.
24. R. L. Eaton, "An overview of zoo goals and exhibition

principles," International Journal for the Study of Animal Problems, vol. 2, pp. 295–299, 1981.

25. E. J. Fernandez, M. A. Tamborski, S. R. Pickens, and W. Timberlake, "Animal-visitor interactions in the modern zoo: conflicts and interventions," Applied Animal Behaviour Science, vol. 120, no. 1-2, pp. 1–8, 2009.

26. L. A. Dickie, "The sustainable zoo: an introduction," International Zoo Yearbook, vol. 43, no. 1, pp. 1–5, 2009.

27. A. M. Claxton, "The potential of the human-animal relationship as an environmental enrichment for the welfare of zoo-housed animals," Applied Animal Behaviour Science, vol. 133, no. 1-2, pp. 1–10, 2011.

28. S. E. Williams and E. A. Hoffman, "Minimizing genetic adaptation in captive breeding programs: a review," Biological Conservation, vol. 142, no. 11, pp. 2388–2400, 2009.

29. R. J. Whittington and R. Chong, "Global trade in ornamental fish from an Australian perspective: the case for revised import risk analysis and management strategies," Preventive Veterinary Medicine, vol. 81, no. 1–3, pp. 92–116, 2007.

30. D. Dudgeon, A. H. Arthington, M. O. Gessner et al., "Freshwater biodiversity: importance, threats, status and conservation challenges," Biological Reviews of the Cambridge Philosophical Society, vol. 81, no. 2, pp. 163–182, 2006.

31. S. Blackmore, M. Gibby, and D. Rae, "Strengthening the scientific contribution of botanic gardens to the second phase of the Global Strategy for Plant Conservation," Botanical Journal of the Linnean Society, vol. 166, no. 3, pp. 267–281, 2011.

32. A. G. Clarke, "The Frozen Ark Project: the role of zoos and aquariums in preserving the genetic material of threatened animals," International Zoo Yearbook, vol. 43, no. 1, pp. 222–230, 2009. ·

33. O. A. Ryder, A. McLaren, S. Brenner, Y.-P. Zhang, and K. Benirschke, "DNA banks for endangered animal species," Science, vol. 288, no. 5464, pp. 275–277, 2000.

34. C. W. Vertuccj, "Predicting the optimum storage conditions for seeds using thermodynamic principles,"Journal of Seed Technology, vol. 17, pp. 41–52, 1993.

35. R. H. Ellis and E. H. Roberts, "Improved equations for the prediction of seed longevity," Annals of Botany, vol. 45, no. 1, pp. 13–30, 1980.

36. J. B. Dickie, R. H. Ellis, H. L. Kraak, K. Ryder, and P. B. Tompsett, "Temperature and seed storage longevity," Annals of Botany, vol. 65, no. 2, pp. 197–204, 1990.

37. M. L. Moir, P. A. Vesk, K. E. C. Brennan et al., "Considering extinction of dependent species during translocation, ex situ conservation, and assisted migration of threatened hosts," Conservation Biology, vol. 26, no. 2, pp. 199–207, 2012.

38. S. Christie, "Why keep tigers in zoos?" in Tigers of the World: The Science, Politics and Conservation of Panthera Tigris, R. Tilson and P. Nyhus, Eds., pp. 205–214, Elsevier Inc., Amsterdam, The Netherlands, 1998.

39. U. S. Anderson, A. S. Kelling, R. Pressley-Keough, M. A. Bloomsmith, and T. L. Maple, "Enhancing the zoo visitor's experience by public animal training and oral interpretation at an otter exhibit,"Environment and Behavior, vol. 35, no. 6, pp. 826–841, 2003. ·

40. H. R. Jensen, A. Dreiseitl, M. Sadiki, and D. J. Schoen, "The Red Queen and the seed bank: pathogen resistance of ex situ and in situ conserved barley," Evolutionary Applications, vol. 5, no. 4, pp. 353–367, 2012.

41. G. R. Hosey, "How does the zoo environment affect the behaviour of captive primates?" Applied Animal Behaviour Science, vol. 90, no. 2, pp. 107–129, 2005.

42. K. N. Morgan and C. T. Tromborg, "Sources of stress in captivity," Applied Animal Behaviour Science, vol. 102, no. 3-4, pp. 262–302, 2007.

43. B. O. Braastad, "Effects of prenatal stress on behaviour of offspring of laboratory and farmed mammals,"Applied Animal Behaviour Science, vol. 61, no. 2, pp. 159–180, 1998.

44. S. M. Wisely, R. M. Santymire, T. M. Livieri et al., "Environment influences morphology and development for in situ and ex situ populations of the black-footed ferret (Mustela nigripes)," Animal Conservation, vol. 8, no. 3, pp. 321–328, 2005.

45. K. Helenurm and L. S. Parsons, "Genetic variation and the reintroduction of Cordylanthus maritimus ssp. maritimus to Sweetwater Marsh, California," Restoration Ecology, vol. 5, no. 3, pp. 236–244, 1997.

46. D. Lauterbach, M. Burkart, and B. Gemeinholzer, "Rapid genetic differentiation between ex situ and theirin situ source populations: an example of the endangered Silene otites (Caryophyllaceae)," Botanical Journal of the Linnean Society, vol. 168, no. 1, pp. 64–75, 2012. ·

47. A. Balmford, G. M. Mace, and N. Leader-Williams, "Designing the ark: setting priorities for captive breeding," Conservation Biology, vol. 10, no. 3, pp. 719–727, 1996.

48. R. A. Griffiths and L. Pavajeau, "Captive breeding, reintroduction, and the conservation of amphibians,"Conservation Biology, vol. 22, no. 4, pp. 852–861, 2008.

49. R. J. Probert, M. I. Daws, and F. R. Hay, "Ecological correlates of ex situ seed longevity: a comparative study on 195 species," Annals of Botany, vol. 104, no. 1, pp. 57–69, 2009.

50. L. M. Clayton, E. J. Milner-Gulland, D. W. Sinaga, and A. H. Mustari, "Effects of a proposed ex situconservation program on in situ conservation of the babirusa, an endangered suid," Conservation Biology, vol. 14, no. 2, pp. 382–385, 2000.

51. W. G. Conway, "Buying time for wild animals with zoos," Zoo Biology, vol. 30, no. 1, pp. 1–8, 2011.

52. N. C. Ellstrand and D. R. Elam, "Population genetic conseqences of small population size: implications for plant conservation," Annual Review of Ecology and Systematics, vol. 24, pp. 217–242, 1993.

53. A. A. Cunningham, "Disease risks of wildlife

translocations," Conservation Biology, vol. 10, no. 2, pp. 349–353, 1996.

54. R. P. Reading, T. W. Clark, and B. Griffith, "The influence of valuational and organizational considerations on the success of rare species translocations," Biological Conservation, vol. 79, no. 2-3, pp. 217–225, 1997.

55. L. I. Wright, T. Tregenza, and D. J. Hosken, "Inbreeding, inbreeding depression and extinction,"Conservation Genetics, vol. 9, no. 4, pp. 833–843, 2008.

56. B. Miller, K. Ralls, R. P. Reading, J. M. Scott, and J. Estes, "Biological and technical considerations of carnivore translocation: a review," Animal Conservation, vol. 2, no. 1, pp. 59–68, 1999.

57. L. Laikre, L. C. Larsson, A. Palmé, J. Charlier, M. Josefsson, and N. Ryman, "Potentials for monitoring gene level biodiversity: using Sweden as an example," Biodiversity and Conservation, vol. 17, no. 4, pp. 893–910, 2008.

58. N. I. Vavilov, "The origin, variation, immunity and breeding of cultivated plants," Chronica Botanica, vol. 13, pp. 1–366, 1951.

59. S. Demissew, "How has government policy post-global strategy for plant conservation impacted on science? The Ethiopian perspective," Botanical Journal of the Linnean Society, vol. 166, no. 3, pp. 310–325, 2011.

60. B. Girma, T. Pearce, and D. Abebe, "Biological diversity and current ex situ conservation practices in Ethiopia," in Seed Conservation Turning Science into Practice, R. D. Smith, J. B. Dickie, S. H. Linington, H. W. Pritchard, and R. J. Probert, Eds., pp. 849–856, Kew Publishing, Kew, UK, 2003.

CHAPTER 9 DISTRIBUTIONS, EX SITU CONSERVATION PRIORITIES, AND GENETIC RESOURCE POTENTIAL OF CROP WILD RELATIVES OF SWEETPOTATO [IPOMOEA BATATAS(L.) LAM., I. SERIES BATATAS]

Colin K. Khoury[1,2], Bettina Heider[3], Nora P. Castañeda-Álvarez[1,4], Harold A. Achicanoy[1], Chrystian C. Sosa[1], Richard E. Miller[5], Robert W. Scotland[6], John R. I. Wood[6], Genoveva Rossel[3], Lauren A. Eserman[7], Robert L. Jarret[8], G. C. Yencho[9], Vivian Bernau[1,10], Henry Juarez[3], Steven Sotelo[1], Stef de Haan[1,3] and Paul C. Struik[2]

[1] International Center for Tropical Agriculture, Cali, Colombia

[2] Centre for Crop Systems Analysis, Wageningen University, Wageningen, Netherlands

[3] International Potato Center, CGIAR Research Program on Roots, Tubers and Bananas, Lima, Peru

[4] School of Biosciences, University of Birmingham, Birmingham, UK

[5] Department of Biological Sciences, Southeastern Louisiana University, Hammond, LA, USA

[6] Department of Plant Sciences, University of Oxford, Oxford, UK

[7] Department of Plant Biology, University of Georgia, Athens, GA, USA

[8] Plant Genetic Resources Conservation Unit, United States Department of Agriculture – Agricultural Research Service, Griffin, GA, USA

[9] Department of Horticultural Science, North Carolina State University, Raleigh, NC, USA

[10] Department of Horticulture and Crop Science, The Ohio State University, Columbus, OH, USA

ABSTRACT

Crop wild relatives of sweetpotato [Ipomoea batatas (L.) Lam., I. series Batatas] have the potential to contribute to breeding objectives for this important root crop. Uncertainty in regard to species boundaries and their phylogenetic relationships, the limited availability of germplasm with which to perform crosses, and the difficulty of introgression of genes from wild species has constrained their utilization. Here, we compile geographic occurrence data on relevant sweetpotato wild relatives and produce potential distribution models for the species. We then assess the comprehensiveness of ex situ germplasm collections, contextualize these results with research and breeding priorities, and use ecogeographic information to identify species with the potential to contribute desirable agronomic traits. The fourteen species that are considered the closest wild relatives of sweetpotato generally occur from the central United States to Argentina, with richness concentrated in Mesoamerica and in the extreme Southeastern United States. Currently designated

Citation: Colin K. Khoury, Bettina Heider, Nora P. Castañeda-Álvarez, Harold A. Achicanoy, Chrystian C. Sosa, Richard E. Miller, Robert W. Scotland, John R. I. Wood, Genoveva Rossel, Lauren A. Eserman, Robert L. Jarret, G. C. Yencho, Vivian Bernau, Henry Juarez, Steven Sotelo, Stef de Haan and Paul C. Struik, Distributions, ex situ conservation priorities, and genetic resource potential of crop wild relatives of sweetpotato [Ipomoea batatas (L.) Lam., I. series Batatas], http://dx.doi.org/10.3389/fpls.2015.00251.

species differ among themselves and in comparison to the crop in their adaptations to temperature, precipitation, and edaphic characteristics and most species also show considerable intraspecific variation. With 79% of species identified as high priority for further collecting, we find that these crop genetic resources are highly under-represented in ex situ conservation systems and thus their availability to breeders and researchers is inadequate. We prioritize taxa and specific geographic locations for further collecting in order to improve the completeness of germplasm collections. In concert with enhanced conservation of sweetpotato wild relatives, further taxonomic research, characterization and evaluation of germplasm, and improving the techniques to overcome barriers to introgression with wild species are needed in order to mobilize these genetic resources for crop breeding.

INTRODUCTION

Sweetpotato [*Ipomoea batatas* (L.) Lam.] counts among the world's most important root crops, grown on at least eight million hectares in 114 countries worldwide, with particular significance to food supplies in the tropics and subtropics of East and Southeast Asia, and Sub-Saharan Africa. The success of the crop in these regions is due to its adaptability to a wide range of agro-ecological conditions, ease of propagation from cuttings, low input cultivation requirements, and high productivity as well as nutritive value (Woolfe, 1992). The crop can be cultivated from humid to semi-arid conditions, from sea level to 3000 m.a.s.l. (Huaman, 1987), and can translocate photosynthetic products to the storage roots throughout the growing season, thereby mitigating negative effects of temporary adverse conditions (Kays, 1985). Sweetpotato produces among the highest amounts of edible energy per hectare of all major food crops (De Vries et al., 1967), and is an important source of vitamin A and C, calcium, iron, and a number of essential amino acids (Kays and Kays, 1998; Tumwegamire et al., 2011). In addition to human consumption of the storage roots and young leaves, the crop is utilized for animal feed, fuel, and starch production.

In Sub-Saharan Africa, sweetpotato is predominantly cultivated in small plots characterized by low fertility and drought-prone soils, producing relatively good yields with low inputs and minimal labor costs. The crop has recently become the focus of targeted bio-fortification for enhanced vitamin A. Orange-fleshed varieties have been bred with 50-fold more β-carotene than standard varieties and these newly released varieties rank first among roots and tubers in Sub-Saharan Africa for their nutritional quality (Low et al., 2007; Hotz et al., 2012). Given its adaptability, low-external input requirements, nutritional quality, and improvement potential, it is not surprising that sweetpotato has become a priority in crop based strategies for enhancing food security in the tropics (Pfeiffer and McClafferty, 2007; Bill and Melinda Gates Foundation, 2011; Bouis and Islam, 2012).

The full potential of sweetpotato is far from realized, with particularly large yield gaps (ca. 20 t ha^{-1}) remaining across rain-fed Sub-Saharan Africa due to a range of biotic and abiotic constraints, especially sweetpotato virus disease (SPVD) and sweetpotato weevils (SPW), as well as susceptibility to drought (Sutherland, 1986; Valverde et al., 2007; Ngailo et al., 2013; Kivuva et al., 2015a,b). SPVD is a severe constraint on the continent, caused by the synergistic interaction of two viruses transmitted by whiteflies and aphids, causing yield losses of up to 98% under severe infections (Ngailo et al., 2013). SPW (*Cylas*spp.) are the most devastating insect pests of the crop. *Cylas formicarius elegantulus* Summers, *C. puncticollis* Boheman and *C. brunneus* Fabricius can cause yield losses of between 67 and 100% in Sub-Saharan Africa (Smit, 1997). The concealed feeding behavior, oviposition, and larval development of the weevils in the storage roots make their control very difficult, necessitating the development of improved management options, in particular via enhanced genetic resistance. Drought, and the compounding effect of increasing heat on drought, is a rising concern particularly in regions undergoing significant climatic change, both due to its direct effect on productivity (Low et al., 2009; Schafleitner et al., 2010) as well as to its association with increased severity of damage from SPW and SPVD (Munyiza et al., 2007; Mwololo et al., 2007). Lack of drought tolerance in high β-carotene sweetpotato varieties has led to constraints in their adoption (Mwanga and Ssemakula, 2011).

Crop wild relatives (CWR) are increasingly being recognized for their potential to contribute valuable traits to breeding programs (Feuillet et al., 2008; Guarino and Lobell, 2011; Dempewolf et al., 2014). CWR have provided breeders with genes for pest and disease resistance, abiotic stress tolerance, and quality traits in an ever increasing number of food crops, such as banana, barley, bean, cassava, chickpea, maize, lettuce, oat, potato, rice, sugarcane, sunflower, tomato, and wheat, among others (Xiao et al., 1996; Hajjar and Hodgkin, 2007; McCouch et al., 2007; Khoury et al., 2010). Yet despite the successful history of contribution to the improvement of major crops, systematic approaches to the use of CWR in the breeding programs of a number of important staples, including sweetpotato, remain underdeveloped.

The morning glory tribe Ipomoeeae contains ~650–900 species and includes the genus *Ipomoea* and nine other related genera (Wilkin, 1999; Mabberley, 2008). Although many genera, subgenera, and sections of the Ipomoeeae are not monophyletic in phylogenetic analyses, *Ipomoea* series *Batatas* (Choisy) D. F. Austin, which contains sweetpotato and 14 closely related CWR (Austin, 1978; McDonald and Austin, 1990), does form a monophyletic lineage (Miller et al., 1999; McDonald et al., 2011; Eserman et al., 2014). These species include wild *I. batatas* (L.) Lam. [including *I. batatas* var. *apiculata* (Martens and Galeotti) McDonald and Austin], *Ipomoea cordatotriloba* Dennstedt, *Ipomoea cynanchifolia* Meisn., *Ipomoea grandiflora* (Dammer) O'Donell, *Ipomoea lacunosa* L., *Ipomoea leucantha* Jacquin, *Ipomoea littoralis*Blume, *Ipomoea ramosissima* (Poir.) Choisy, *Ipomoea splendor-sylvae* House, *Ipomoea tabascana*McDonald and Austin, *Ipomoea tenuissima* Choisy, *Ipomoea tiliacea* (Willd.) Choisy in D. C., *Ipomoea trifida* (H. B. K.) G. Don, and *Ipomoea triloba* L.

Many sweetpotato CWR can be hybridized with the crop through controlled pollinations, somatic cell, and/or ovule culture techniques (Diaz et al., 1996). Crosses involving *I. tabascana*, *I. trifida*, *I. triloba*, *I.littoralis*, *I. grandifolia*, *I. lacunosa*, *I. leucantha*, and wild *I. batatas* in particular have resulted in relatively viable progeny (Nimmakayala et al.,

2011). The wild conspecific as well as *I. trifida* have been documented for their contribution to increases in protein and starch content, and nematode and SPW resistance (Iwanaga, 1988; Shiotani et al., 1994), although there is uncertainty for some material as to whether they may actually have been feral forms of the cultivar (Nimmakayala et al., 2011). Species that have been explored for potential traits of use in crop improvement include *I. trifida* and *I. littoralis* for yield and SPW, scab [*Elsinoë batatas* (Saw.) Viegas et Jenkins], and black rot disease (*Ceratocystis fimbriata* Ell. et Halst.) resistance; *I. grandifolia* for sweetpotato stem nematode and SPVD resistance; and *I. triloba* for drought tolerance, root rot resistance, and foliar fungal disease resistances (Iwanaga, 1988; Jarret et al., 1992; Komaki, 2004; Zhang and Liu, 2005; Nimmakayala et al., 2011). Challenges in the creation of viable progeny between the CWR and the cultivated species are not insignificant, though, due to differences in ploidy and interspecific incompatibility (Martin, 1970, 1982; Teramura, 1979; Lu and Li, 1992; Shiotani et al., 1994; Diaz et al., 1996; Komaki, 2004; Nimmakayala et al., 2011).

A lack of basic knowledge about boundaries between species within *I.* series *Batatas* and a dearth of diagnostic characters enabling differentiation of taxa – to facilitate reliable and accurate species identification – is a fundamental stumbling block constraining the utilization of sweetpotato CWR (Austin, 1978, 1988; Jarret et al., 1992; Diaz et al., 1996; McDonald et al., 2011; Nimmakayala et al., 2011;Eserman et al., 2014). Needed studies have been delayed in part due to the absence of plant materials for research, in particular the availability of specimens with flowers and ripe fruits. Studies that have been performed have often been based upon limited sampling (e.g., single accessions for *I. littoralis* and *I.tabascana*).

Many unanswered questions regarding the relationships of CWR to sweetpotato potentially impact the efficiency of breeding strategies for the crop. For example, do species such as *I. tabascana* and *I.tenuissima* represent distinct taxa (hybrid or otherwise) or rather rare variants of the crop? What is the range and genetic diversity present in truly wild forms of the crop conspecific, compared to feral escapees? How accurate are the classifications of species with highly disjunct distributions (e.g., *I. cordatotriloba*)? Are there as yet unrecognized cryptic species within *I.* series *Batatas* [e.g., *I.* '*austinii*' (Duncan and Rausher, 2013)]? What are the lineages and genetic resources potential of purported hybrid species (i.e., *I.leucantha* and *I. grandifolia*)? What are the geographic locations of new variation being generated through hybridization among sweetpotato CWR?

The investigation, conservation, and availability of genetic resources of sweetpotato provide a foundation for the crop's long term viability and for its potential for improvement. To contribute to these objectives, we analyzed the comprehensiveness of *ex situ* conservation of sweetpotato CWR through a series of questions: (a) what constitutes a potentially useful wild relative of sweetpotato?, (b) where are these species encountered?, (c) what is the state of conservation and availability of these species to researchers, and what are the highest taxonomic and ecogeographic priorities for further collecting? And finally, (d) what traits do sweetpotato CWR possess that may be valuable to crop improvement?

MATERIALS AND METHODS

Identification of Target CWR Species and Occurrence Data Compilation

The CWR of sweetpotato analyzed in this study were selected based upon recent and historical taxonomic and molecular phylogenetic research (Austin, 1978, 1997; Austin and Huamán, 1996; Miller et al., 1999;McDonald et al., 2011; Germplasm Resources Information Network [GRIN], 2013; Eserman et al., 2014), identifying those wild species with a relatively close phylogenetic relationship to the crop (i.e., members of*I.* series *Batatas*). We included all 14 wild species comprising the series in the analysis.

Domesticated sweetpotato *I. batatas* (6x) has been proposed as originating from interspecific hybridization involving *I. trifida* (2x, 4x, 6x), *I. littoralis* (2x), and/or *I. leucantha* (2x) (Nishiyama, 1971,1982; Austin, 1988). The species most closely related to the crop have been posited to be *I. trifida* followed by *I. tabascana* (4x) (Srisuwan et al., 2006; Nimmakayala et al., 2011). Following the genepool concept ofHarlan and de Wet (1971) and due to ploidy incompatibility with the cultivated species, the putative closest related taxa to sweetpotato are placed in the secondary genepool: wild forms of *I. batatas* (4x), *I. trifida*, *I. littoralis*, and *I. tabascana* (Jarret et al., 1992; Jarret and Austin, 1994; Rajapakse et al., 2004;Germplasm Resources Information Network [GRIN], 2013). Species classified as tertiary wild relatives include: *I. cordatotriloba* (syn. *Ipomoea trichocarpa* Elliott), *I. cynanchifolia*, *I. grandifolia*, *I. lacunosa*,*I. leucantha*, *I. ramosissima*, *I. splendor-sylvae* (syn. *Ipomoea umbraticola* House), *I. tenuissima*, *I.tiliacea*, and *I. triloba* (Jarret and Austin, 1994; Huang et al., 2002; Rajapakse et al., 2004; Germplasm Resources Information Network [GRIN], 2013).

Occurrence records for these species were acquired from online biodiversity, herbaria, and germplasm databases; through communications with herbaria and genebank managers, and other crop researchers; and via direct recording of provenance data during visits to selected herbaria (Supplementary Table S1). Germplasm data were obtained from repositories that provide straightforward access to genetic resources and associated data to the global research community through online information systems. The occurrence data were then compiled in a standardized format, nomenclature was checked againstGermplasm Resources Information Network [GRIN] Taxonomy for Plants (2013) and The Plant List (2010), and duplicate records were eliminated. Existing coordinates were cross-checked to country and being on land (Hijmans et al., 1999), and records with locality information but no coordinates were geo-referenced using the Google Maps Geocoder (2013) v.3 application programming interface. Occurrence data were mapped, iteratively evaluated for correctness, and further processed in order to form a final dataset of improved taxonomic and spatial accuracy.

Challenges in using and in improving the large quantities of occurrence data now available from online resources such as the Global Biodiversity Information Facility (GBIF) have been noted (Gaiji et al., 2013), including geographic and nomenclatural data quality and the slow speed with which aggregated datasets are updated (Mesibov, 2013; Otegui et al., 2013; Hjarding et al., 2014). In addition, particular caution must be applied to the occurrence records used in the current paper as ongoing work (unpublished data) indicates that many *Ipomoea* occurrence records in such online databases are identified as synonyms, excluded or invalid names, and that many valid names were applied to specimens well outside of species known ranges. We have identified some of these obvious errors but until all specimen records are correctly identified and checked against an accurate taxonomy these data must be treated with caution.

A total of 5,614 occurrence records for the 14 taxa were included in potential distribution modeling and/or in the conservation analysis, including 749 germplasm records sourced from four genebanks, and 4,865 herbarium and other occurrence reference records sourced from 42 providers. Records per species ranged from eight (*I. tabascana*) to 1,409 (*I. trifida*). Of these, 3,650 records containing unique cross-checked coordinates were used to model species potential distributions and to locate the original collecting site of existing germplasm accessions.

Species Potential Distribution Modeling

A potential distribution model for each species was calculated using the maximum entropy (Maxent) algorithm (Phillips et al., 2006), with a set of ecogeographic variables and unique species presence records as inputs. We chose Maxent due to its extensive application in predicting species distributions (Elith et al., 2006; Phillips and Dudik, 2008; Costa et al., 2010), including those for wild relatives (Ramírez-Villegas et al., 2010; Conolly et al., 2012; Khoury et al., 2015). We performed modeling at a resolution of 2.5 arc-minutes (~5 km^2 cell size at the equator), employing 10,000 background points for model training over the combined distributional range of the sweetpotato CWR. Ecogeographic inputs included altitude and 19 bioclimatic variables from the WorldClim database (Hijmans et al., 2005), and seven major edaphic drivers of plant species distributions with consistent data coverage throughout the range of the sweetpotato CWR species, obtained from ISRIC-World Soil Information (Hengl et al., 2014; Supplementary Table S2). For the edaphic variables we calculated a weighted mean across 0–5, 5–15, 15–30, 30–60, and 60–100 cm soil depth values in order to derive a single data value for 0–100 cm. We then resampled the 30 arc-seconds resolution data to form 2.5 arc-minutes inputs aligned with the WorldClim datasets.

Potential distribution models were produced by calculating the mean of replicates ($k = 5$), and clipped by measuring the shortest distance between the receiver operating characteristic curve (ROC-curve) and the top-left corner of the plot (Liu et al., 2005). Models were constrained per species by a native range defined at the country level as given in GRIN Taxonomy for Plants (Germplasm Resources Information Network [GRIN], 2013), in order to focus prioritization recommendations on those regions with populations exhibiting long-term adaptations to local ecogeographic conditions. We further cross-validated and refined occurrence data based upon our knowledge of native distributions, constraining localities for wild*I. batatas* to Mexico south to Peru, and not in the Caribbean; *I. leucantha* to the USA and Mexico, and *I. littoralis* to points within 100 km of the ocean. In order to derive robust distribution models for each species, we analyzed Maxent results across three groups of ecogeographic inputs: (a) the full set of 19 bioclimatic variables (Ramírez-Villegas et al., 2010); (b) the bioclimatic variables, altitude, and the additional set of seven edaphic variables, totaling 27 input variables; and (c) a species-specific derivation of the most important drivers of distribution based upon presence data, further refined by removing highly correlated variables. For the ecogeographic variables in the species-specific subset method, we utilized a non-linear iterative partial least squares (NIPALS) algorithm to perform a principal-component analysis (PCA), as NIPALS has the potential to handle data arrays in which the number of observations is less than the number of input variables. We then identified those variables with the greatest contribution (>0.7 or < -0.7) to the first two principal components per species. Finally, we used a variance inflation factor (VIF) to identify the variables with a low degree of collinearity (see Supplementary Table S3 for a list of variables identified per species). A calibrated area under the ROC curve (cAUC) was obtained to assess the predictive performance of each model (Hijmans, 2012; Khoury et al., 2015).

The three modeling methods were compared against null models, and the species-specific subset variables method showed the least overall spatial sorting bias among methods (Spearman's p for the 19 variables was 0.65, for 27 variables it was 0.72, and for the subset method it was 0.25), although the differences in median AUC distributions across species for each method were not significant ($p = 0.095$) through a Kruskal–Wallis non-parametric analysis of variance test. Maxent models were subjected to a four-fold assessment process including: (a) the fivefold average area under the ROC curve of test data (ATAUC), b) the standard deviation of the test AUC of the five different folds (STAUC), (c) the proportion of the potential distribution coverage with SD above 0.15 (ASD15), and (d) the cAUC value. Models with ATAUC above 0.7, STAUC below 0.15, ASD15 below 10%, and cAUC exceeding 0.40 were considered accurate and stable (Ramírez-Villegas et al., 2010; Khoury et al., 2015).

The potential distribution models of the sweetpotato CWR generally performed well in regard to the modeling assessment process. Four species demonstrated low cAUC values and one of these an elevated ASD15 value, indicating greater uncertainty in the models (Supplementary Table S3). Species-specific subset model outputs for taxa with relatively few distinct occurrence points (<20; *I. tabascana*, *I.tenuissima*, and *I. cynanchifolia*) lacked sufficient discriminatory power, leading to highly inflated spatial models in comparison to recorded distributions. Potential distribution models for these species were resolved by deriving an ensemble (i.e., overlap) of the three input variation methods, verified by researchers knowledgeable in the distribution of the species as more accurately representing the true known distributions. Potential distribution models based upon the species-specific subset variables method were therefore utilized in subsequent analyses for all species with adequate distinct occurrence points (>20). The ensemble method was used for the three species with limited occurrence data.

Analysis of Current *Ex Situ* Conservation and Further Collecting Needs for CWR

We adapted a gap analysis methodology proposed by

Ramírez-Villegas et al. (2010), combining three metrics used to assess the urgency of further collecting in order to fill gaps in *ex situ* conservation of CWR. The total sample representation of each species in genebank collections was estimated via a sampling representativeness score (SRS), calculated as the number of germplasm samples (G) divided by the total number of samples [G + herbarium samples (H); i.e., all other records aside from available genebank accessions].

The sufficiency of geographic coverage of germplasm collections of each species was estimated through a geographic representativeness score (GRS), calculated as the share of the combined total area of circular buffers of 50 km (CA50) placed around existing germplasm collection points compared with the overall potential geographic distribution of the species.

The comprehensiveness of ecological coverage of germplasm collections of each species was estimated through an ecological representativeness score (ERS), calculated by estimating the distinct ecosystem classifications (Olson et al., 2001) represented in the CA50 of existing germplasm collection points compared with the diversity of ecosystems in which the overall potential geographic distribution model of the species occurs.

A final priority score (FPS) for further collecting for *ex situ* conservation was assigned to each species by averaging the three gap analysis metrics (SRS, GRS, and ERS). FPS scores were further classified into four categories of urgency for collecting: high priority species (HPS) for taxa whose $0 < FPS \leq 2.5$ or when no germplasm accessions are currently conserved, medium priority species (MPS) when $2.5 < FPS \leq 5$, low priority species (LPS) when $5 < FPS \leq 7.5$, and 'no further collecting recommended' (NFCR) when $7.5 < FPS \leq 10$. We produced collecting priorities maps for all species, displaying the geographic areas that have not yet been collected from within the potential distributions of taxa.

The ecogeographic data preparation, species distribution modeling, and gap analysis were written and performed in R v2.15.1 (R Core Team, 2013), utilizing packages maptools (Bivand and Lewin-Koh, 2014), rgdal (Bivand et al., 2014), SDMTools (van der Wal et al., 2014), raster (Hijmans, 2014), sp (Pebesma and Bivand, 2005; Bivand et al., 2013), dismo (Hijmans et al., 2013), and plsdepot (Sanchez, 2012). Resulting spatial files were mapped in ArcMap v.10 (ESRI, 2011). Collecting priorities spatial files were analyzed using the Zonal Statistics tool in ArcMap to list the countries prioritized for further collecting for *ex situ*conservation.

Expert Evaluation of Conservation Assessment Results

In order to validate and/or expose deficiencies in our findings, we subjected the gap analysis numerical and spatial results to an evaluation performed by seven crop experts with experience in the systematics, distribution, and/or conservation status of CWR of sweetpotato: Richard E. Miller, Southeastern Louisiana University; Robert W. Scotland and John R. I. Wood, University of Oxford; Genoveva Rossel, International Potato Center; Lauren A. Eserman, University of Georgia; Robert L. Jarret, USDA – ARS Griffin; and G. Craig Yencho, North Carolina State University. These experts were first asked to provide an evaluation of the sufficiency of germplasm collections per species based only upon their knowledge of the total number of accessions conserved, and geographic as well as environmental gaps. Such an assessment [comparable expert priority score (EPS)] was considered directly comparable to the FPS of the gap analysis results.

A second evaluation score (contextual EPS) based on the entirety of expert knowledge, including additional factors such as threats to species *in situ* and prioritization by usefulness in crop breeding, was collected in order to provide additional information to collecting prioritization efforts. In both cases, an EPS between 0 and 10, aligned with the gap analysis results prioritization scale, was requested. After these steps, experts were shown the gap analysis data and asked to comment on assessed quantitative results, occurrence data, potential distribution models, and maps of collecting priorities. Following these contributions by experts, input occurrence data were further refined by eliminating clearly incorrect points and adjusting country-level native areas, and the potential distribution modeling and gap analyses were re-run using the refined datasets in order to improve the quantitative and spatial outputs. Expert evaluation metrics displayed in the results pertain to a final evaluation of improved gap analysis outputs, performed by five of the experts.

A multiple factor analysis was used in order to compare the various forms of expert evaluation inputs (i.e., comparable expert priority score, contextual expert priority score, evaluation of gap analysis results score, evaluation of occurrence data, evaluation of potential species distribution models, and evaluation of collecting priorities map) with the gap analysis results. An expert evaluation index was created, which estimated the degree of accord between all experts and the gap analysis results for each species, with a scale from 0 (disagreement) to 100 (agreement). Analyses were performed using R package FactoMineR (Husson et al., 2009).

Identification of Ecogeographic Characteristics of CWR

In order to evaluate the pairwise degree of geographic overlap between sweetpotato CWR distribution models, we calculated an overlap measure equal to the frequency of shared 2.5 arc-minute geographic cells between taxa divided by the sum of the total number of cells in which either of the species are present (Kernohan et al., 2001; Fieberg and Kochanny, 2005). To assess the pairwise degree of ecogeographic niche overlap between species, we used Schoener's index for niche similarity (D) and the adjusted similarity index (I) as outlined in Warren et al. (2008), using species distribution models and the species-specific subset of the 27 ecogeographic layers/ensemble models as inputs. Overlap indices range from 0 (no overlap) to 1 (complete overlap) and were implemented in the R package Phyloclim (Heibl, 2011).

We utilized ecogeographic information in combination with species presence data in order to identify populations of species with the potential for outstanding adaptations to climatic and/or edaphic conditions of interest to sweetpotato breeding objectives. We assessed the relative importance of the 27 ecogeographic variables (Supplementary Table S2) in explaining the total variation through a PCA after confirming the validity of its use through a Bartlett's test performed in R package psych (Revelle, 2015). We created a hierarchical cluster of principal components (HCPC) in order to identify ecogeographic clusters for the sweetpotato wild relatives, driven by those variables demonstrating ≥15% difference (±) from the mean and a reduction of ≥20% from the mean standard deviation exhibited across all species, using R package FactoMineR. Boxplots for each of the 27 ecogeographic variables were created based upon CWR species occurrence data points, displaying the median and variance parameters per species per variable. Comparable ecogeographic variable data for the sweetpotato crop was extracted from area of cultivation maps (Monfreda et al., 2008) at a resolution of 5 arc-minutes, with a random sample of 1,000 points weighted by harvested area, taken from the major cultivation areas in Asia, Africa, and Latin America.

RESULTS

Distributions of the Wild Relatives of Sweetpotato

Sweetpotato CWR were modeled to occur from the central USA to northern Argentina in the Americas, including the Caribbean (Supplementary Figure S1). Species richness was greatest in central Mexico through Central America to the northern Andean region, and in the Southeastern USA, with up to nine species potentially overlapping in Mexico from the states of Veracruz through the Yucatan peninsula (Figure 1). The Mexican and Central American regions of distribution represent one of the posited centers of origin and primary diversity of cultivated sweetpotato (Austin, 1988; Austin and Huamán, 1996; Zhang et al., 2000; Gichuki et al., 2003; Roullier et al., 2013). Northwestern South America, with archeological remains of cultivated sweetpotato from Peru dating back to 8,000 years BP, which are among the oldest recorded domestication events on the continent (Piperno and Pearsall, 1998; Solis et al., 2001), displayed a considerably lesser but still notable degree of CWR species richness. One Old World species (*I. littoralis*;Austin, 1991) was also modeled to occur in coastal areas of Madagascar, South and Southeast Asia, Australia, and the Pacific region.

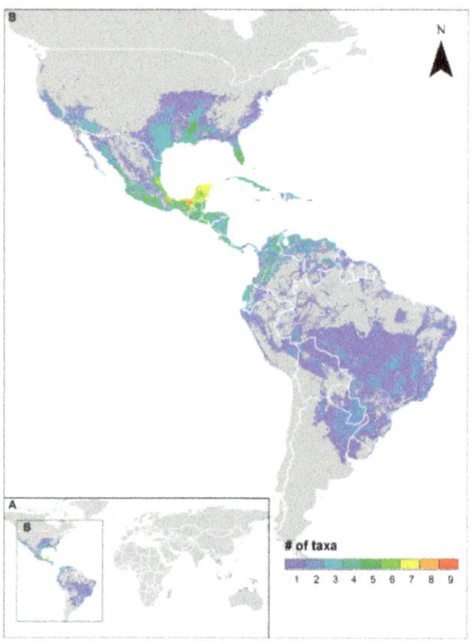

Figure 1: Species richness map for assessed sweetpotato crop wild relative potential distribution models worldwide (A), with concentration of species in the neotropics (B).

Analysis of Current *Ex Situ* Conservation and Further Collecting Needs for CWR

Eleven out of the fourteen CWR species were assigned high priority for further collecting due to insufficient genebank accessions in comparison to the total number of occurrence samples (SRS), and to large geographic (GRS) and ecological (ERS) gaps in *ex situ* germplasm collections in comparison to the full potential distributions of the species (Figure 2, Table 1, Supplementary Figure S2). Six of these taxa are currently represented by ≤10 accessions conserved *ex situ*, moreover, with few exceptions these accessions lack associated geographic occurrence information (Supplementary Table S3). Total sampling representativeness and geographic coverage of species in germplasm collections were particularly lacking for taxa assessed high priority, while gaps in ecological representativeness were less extreme for some species (e.g., *I. cordatotriloba*, *I. triloba*, and *I. splendor-sylvae*). *Ipomoea grandifolia* and *I. trifida* were assessed as relatively well covered in regard to ecosystems represented *ex situ*, which raised their FPS into the medium priority for further collecting category. *Ipomoea tabascana* was assessed as of low priority for further collecting due to existing germplasm collections largely covering its very restricted distribution in southern Mexico, resulting in a high GRS score. The mean FPS across all CWR was 1.75 ± 1.82.

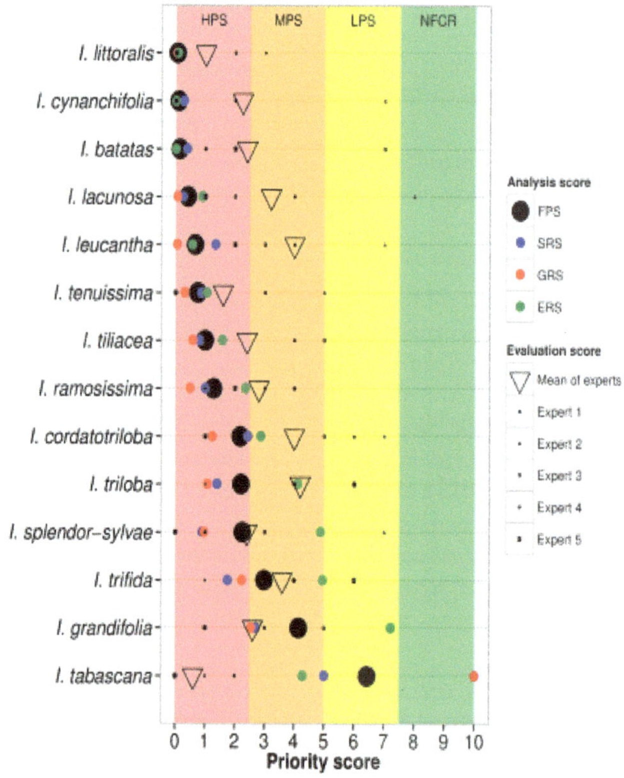

Figure 2: Gap analysis results and comparable expert assessments per sweetpotato crop wild relative. Species are listed by descending priority for further collecting by priority categories [high priority species, HPS (red); medium priority species, MPS (orange); low priority species, LPS (yellow); and no further collecting recommended, NFCR (green)]. The black circle represents the final priority score (FPS) for the species, which is the mean of the sampling representativeness score (SRS), geographic representativeness score (GRS), and ecological representativeness score (ERS).

Table 1: Utilization characteristics, number of germplasm accessions conserved *ex situ*, further collecting priorities, and potential adaptive traits associated with ecogeographic niches of sweetpotato crop wild relatives

Taxon	Gene pool	Ploidy	Germplasm accessions	Gap analysis priority	Mean expert priority	Eco geographic cluster	Potential adaptation to
Ipomoea batatas	2	4x = 60	4 (0)	HPS	HPS	2	Heat, high precipitation, drought, precipitation seasonality, clay soils
I. cordatotriloba	3	2x, 4x	103 (67)	HPS	MPS	1	Cold, temperature variation, clay soils, sandy soils
I. cynanchifolia	3	2x = 30	1 (0)	HPS	HPS	1,2	Drought, precipitation seasonality, sandy soils
I. grandifolia	3	2x = 30	124 (83)	MPS	MPS	1	Cold, temperature variation, clay soils, sandy soils
I. lacunosa	3	2x = 30	10 (1)	HPS	MPS	1	Cold, temperature variation, drought
I. leucantha	3	2x = 30	18 (15)	HPS	MPS	1,2	Heat, drought, precipitation seasonality, sandy soils
I. littoralis	2	2x = 30	2 (2)	HPS	HPS	2	Heat, high precipitation, drought, precipitation seasonality, sandy soils
I. ramosissima	3	2x = 30	34 (30)	HPS	MPS	2,1	Cold, high precipitation, clay soils
I. splendor-sylvae	3	2x = 30	16 (9)	HPS	HPS	2	Heat, high precipitation, drought, precipitation seasonality, clay soils
I. tabascana	2	4x = 60	4 (2)	LPS	HPS	2	Heat, high precipitation, clay soils
I. tenuissima	3	2x = 30	3 (1)	HPS	HPS	1	Heat, cold, temperature variation, sandy soils
I. tiliacea	3	4x = 60	61 (44)	HPS	HPS	2	Heat, high precipitation, clay soils
I. trifida	2	2x,3x,4x,6x	248 (159)	MPS	MPS	2	Heat, high precipitation, drought, precipitation seasonality
I. triloba	3	2x = 30	121 (74)	HPS	MPS	2,1	Heat, drought

Ploidy data adapted from Nimmakayala et al. (2011). Germplasm accessions displays both the total number of accessions recorded in genebanks, as well as the number of accessions with unique geographic coordinates (i.e., unique populations) in parenthesis. HPS = high, MPS = medium, and LPS = low priority species for further collecting.

Paralleling the distribution of species richness of sweetpotato CWR, the regions identified for further collecting of the greatest number of species included central and southern Mexico and the Southeastern USA, with up to seven species prioritized for further collecting occurring in the same area (Figure 3). Further collecting priorities were recognized in a total of 50 countries throughout the range of the genepool (Supplementary Figure S3, Supplementary Table S4). Occurrence data, potential distribution models, and collecting priorities maps for all modeled species are available in an interactive map format athttp://www.cwrdiversity.org/distribution-map/.

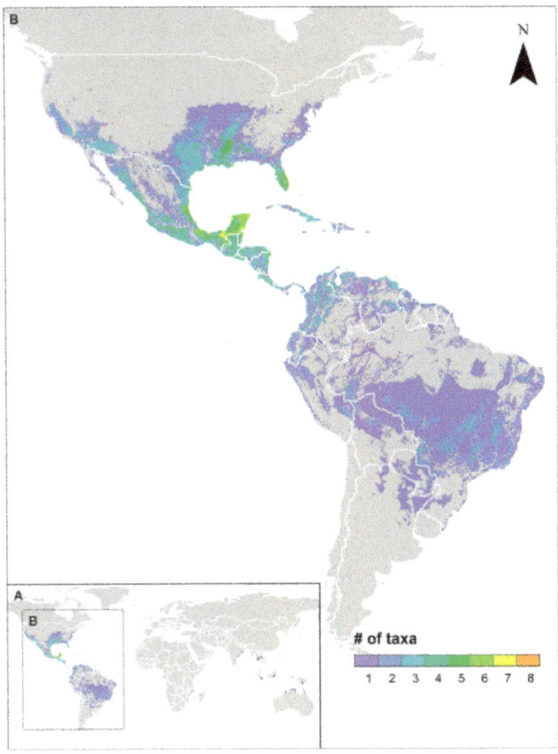

Figure 3: Further collecting priorities hotspots map for high priority species (HPS) sweetpotato crop wild relatives. The map displays areas worldwide (A) within the potential distributions of HPS species that have not been previously collected for *ex situ* conservation, with concentration of species in the neotropics (B).

Expert Evaluation of Conservation Assessment Results

The average of the directly comparable EPS across experts was 2.65 (±1.10) as a mean across species, varying 1.95 points on average from the FPS, with seven taxa designated by the experts as HPS, and seven as MPS (Table 1, Figure 2). For most species, this mean was highly influenced by one or two experts' giving species considerably less priority than the other experts. Species with closest accord between the gap analysis results and the comparable expert analysis included *I. splendor-sylvae*, *I. trifida*, *I. tenuissima*, *I. littoralis*, *I. tileacea*, *I. ramosissima*, and *I. grandifolia*. Taxa with the greatest difference between gap analysis and comparable expert values included *I. tabascana* and *I. leucantha*.

Two species were assessed by the experts as higher priority for further collecting than the results of the gap analysis. Although *ex situ* collections for the highly restricted distribution of *I. tabascana* were determined in the gap analysis to be fairly comprehensive (LPS), the experts assigned the species HPS for further collecting due to its very limited overall number of germplasm holdings, and to the difficulty in producing viable seed in *ex situ* conservation. *Ipomoea grandifolia* was assessed in the gap analysis as reasonably comprehensively represented in regard to ecosystem diversity, and thus assigned medium priority for further collecting, while the experts gave moderately higher priority to the species.

The contextual expert priority score per species, which also included the expert's knowledge of threats *in situ* as well as usefulness for crop improvement, did not vary widely from the comparable score [mean across all experts and species was 2.76 (±0.85); mean difference between comparable and contextual expert scores across all species and experts was 0.11]. Due largely to differences between the perceptions of relative sufficiency in regard to the overall number of germplasm accessions by the experts versus gap analysis results (e.g., for *I. tabascana*, *I. triloba*, and *I. leucantha*), the comparable and contextual assessments did not correlate well with the gap analysis results for the genepool as a whole (Supplementary Figures S4A,B).

The multiple factor analysis combining the comparable expert priority score, contextual expert priority score, evaluation of gap analysis results score, evaluation of occurrence data, evaluation of potential species distribution models, and evaluation of collecting priorities map, revealed sufficient agreement among experts and variables to justify the use of the expert evaluation index, although variation in expert option was notable for many species. Data inputs and resulting distribution models were generally assessed positively as a mean across experts, with eight species receiving very positive index values. Those species with the highest accord between all variables and experts and the gap analysis results included *I. littoralis*, *I. spendor-sylvae*, *I. ramosissima*, *I. tenuissima*, and *I. tiliacea*. Those species with the lowest accord included *I. tabascana*, *I. triloba*, and *I. grandifolia* (Supplementary Figure S4). FPS results were particularly influenced by spotty occurrence records for the majority of species, with gaps recognized by the experts.

Identification of Ecogeographic Characteristics of CWR

The analysis of geographic and ecogeographic accord between pairwise potential species distribution models segregated species well into temperate North American (e.g., *I. lacunosa*, *I. tenuissima*), Mesoamerican (e.g., *I. splendor-sylvae*, *I. tabascana*), widely distributed tropical (e.g., *I. triloba*, *I. trifida*), and South American (e.g., *I. grandifolia*, *I. cynanchifolia*) areas (Supplementary Figure S5).

Strong linear relationships were found between bioclimatic variables within the study area, justifying the application of the PCA, with 75.6% of variance explained through four principal components (Supplementary Figure S6). The first principal component (32% of variation) was correlated with temperature extremes and fluctuation. The second component (21.6% of variation) was most occupied by precipitation variables related to drought. The third component (13.9% of variation), was related with altitude, and the final component (8.1%) with soil texture characteristics.

Species occurrence data segregated into temperate and tropical ecogeographic clusters, with the great majority of species' distributions strongly occurring within a single cluster. Ecogeographic variables most strongly influencing the definition of the temperate cluster (1) included those associated with temperature variation and relatively low precipitation, elevation, and soil organic matter. The most determinant variables in the tropical cluster (2) were related to relatively high and consistent temperatures. Those species displaying a notable proportion of occurrences within both clusters included *I. cynanchifolia* and*I. triloba*, and to a lesser degree *I. leucantha* and *I. ramosissima* (Table 1, Supplementary Figure S6).

Ecogeographic niche assessments of sweetpotato CWR based upon occurrence data points revealed large differences in potential adaptation to temperature, precipitation, and edaphic characteristics (Table 1, Figure 4, Supplementary Figure S7). Such adaptation for many species fell well outside the modeled niche of the cultivated species, particularly for high temperatures both in wet and dry conditions, as well as high precipitation. Species of notable adaptation to high mean annual, monthly, and quarterly temperatures included *I. littoralis*, *I. tabascana*, *I. trifida*, *I. leucantha*, *I. tiliacea*, *I. tenuissima*, *I. triloba*, *I. splendor-sylvae*, and wild *I. batatas*. USA species *I. lacunosa* stood out for adaptation to low temperatures, with *I. grandifolia*, *I. cordatotriloba*, *I. tenuissima*, and *I. ramossissima* also demonstrating cold tolerance. These same species were among those displaying the widest adaptation to temperature fluctuation.

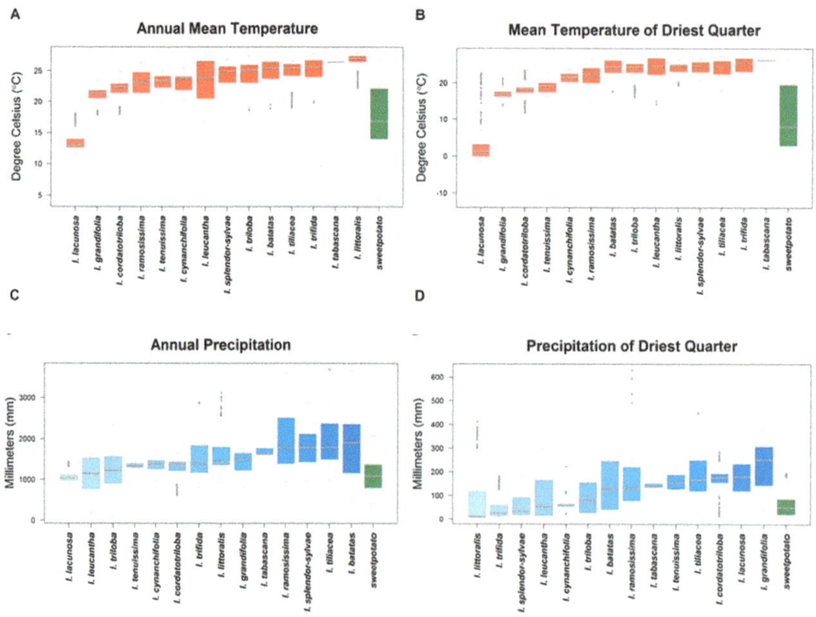

Figure 4: Ecogeographic niches of crop wild relative species based upon their occurrence data presence locations, and the sweetpotato crop, for (A) annual mean temperature, (B) mean temperature of the driest quarter of the year, (C) annual precipitation, and (D) precipitation of the driest quarter of the year. The thick gray line represents median values, boxplots between 25 and 75% variation, and open circles outliers within 90% of total variation. For a principal component analysis of variables see Supplementary Figure S6 and for ecogeographic niches dis-

playing total variation for all variables per species see Supplementary Figure S7.

Crop wild relative of sweetpotato occurring in areas of high precipitation included *I. ramossissima*, *I. littoralis*, *I. splendor-sylvae*, *I. tabascana*, *I. tiliacea*, *I. trifida*, and wild *I. batatas*. *Ipomoea littoralis*, *I. trifida*, *I. splendor-sylvae*, *I. leucantha*, *I. cynanchifolia*, *I. triloba*, *I. lacunosa*, and wild *I. batatas* were distributed in regions with low precipitation. These species were also among those displaying the widest adaptation to precipitation seasonality. Sweetpotato CWR also displayed variation in adaptation to edaphic characteristics. *Ipomoea tabascana*, *I. grandifolia*, *I. tiliacea*, *I. splendor-sylvae*, *I. ramosissima*,*I. cordatotriloba*, and wild *I. batatas* occurred in clay soils, while *I. tenuissima*, *I. littoralis*, *I. cynanchifolia*, *I. grandifolia*, *I. leucantha*, and *I. cordatotriloba* were distributed in sandy soils.

DISCUSSION

This article utilizes the most current knowledge on species concepts within *I.* series *Batatas*. Due to taxonomic uncertainties and to the notable dearth of study material for sweetpotato CWR, the results represent a preliminary understanding of the geography and conservation status of the series, to be further refined following increased levels of collecting sufficient to support the needed taxonomic and crossability research. Further collecting of germplasm thus serves two purposes: (a) conserves genetic resources for the long-term and makes these resources available to breeders; and (b) provides the basic materials needed by researchers to understand the diversity present in the CWR of sweetpotato.

A total of 78.6% of the wild relatives of sweetpotato considered in this study were assessed as high priority for further collecting for *ex situ* conservation. With general agreement from expert evaluators of medium to high importance for all species, it is clear that much remains to be done to safeguard the wild genetic resources of this critically important root crop. Included in this list of priorities are species with very few germplasm accessions accessible to the global community in genebank information systems, including *I.cynanchifolia*, *I. littoralis*, *I. tenuissima*, *I. tabascana*, *I. lacunosa*, *I. leucantha*, *I. splendor-sylvae*, and clearly designated wild forms of the crop conspecific *I. batatas*. Such species represent the highest level of priority for further collecting for use in systematic analyses as well as genetic resources conservation. As the species diversity gaps in genebank collections largely align with the geographic distribution of species richness of sweetpotato CWR, hotspots in central Mexico to Central America, and in the extreme Southeastern USA, represent particularly high priority regions for efficient collecting of the sweetpotato genepool (Figure 3). Additional unrepresented populations of HPS such as *I. littoralis* and *I. cynanchifolia* occur outside those regions, thus targeted collecting throughout the geographic distribution of the genepool is necessary in order to form comprehensive germplasm collections.

Due to having relatively large potential distributions occupying a diversity of ecosystems, species such as*I. triloba*, *I. cordatotriloba*, and *I. tiliacea* were categorized as high priority, and *I. trifida* and *I.grandifolia* as medium priority for further collecting despite having sizable currently existing germplasm collections. As the cost of conserving and investigating germplasm *ex situ* is significant, a further assessment of what constitutes sufficient representation of these species in germplasm collections is warranted. Given adequate resources, further collecting may be worthwhile, as extremely valuable traits sourced from CWR of some crops have been found in only a few populations despite screening of a large number of accessions (Brar and Khush, 1997), and accessions of individual CWR species such as *I. triloba*have been shown to possess notable variation in traits such as tolerance to precipitation (Martin and Jones, 1973; Nimmakayala et al., 2011).

As Maxent models are based upon known presence points for species and are thus subject to sampling bias, they may not fully capture the possibility of occurrence of populations of CWR species in unique climates (Araújo and Guisan, 2006; Loiselle et al., 2007; Kramer-Schadt et al., 2013). Further field exploration of climatic extremes beyond the edges of the distributions created through these methods may therefore lead to the discovery of new populations with particularly valuable adaptations to abiotic stress (Williams et al., 2009). Investigation of non-native populations (e.g., *I. trifida* in Asia) may also yield novel adaptations of use in crop improvement. As techniques for the utilization of distantly related germplasm improve, the exploration of other *Ipomoea* species outside of *I.* series *Batatas* may also result in the identification of beneficial traits [e.g., *Ipomoea purpurea* (L.) Roth, for stem nematode and SPVD resistance (Cao et al., 2009)].

Analysis of geographic overlap and ecogeographic similarities between species, as well as ecogeographic clusters among all species, can supplement morphological and genetic analyses in differentiating useful genetic resources, and can serve as a point of departure for identifying taxonomically problematic populations for further investigation. These analyses may also indicate geographic areas of particular interest in regard to high rates of hybridization, as in the case of *I. cordatotriloba* and *I. lacunosa*(Duncan and Rausher, 2013), which indeed were identified as sharing similar geographies and ecogeographic niches. The purported hybrid descendent of these species, *I. leucantha*, was modeled as containing a much more extended latitudinal gradient in the northern hemisphere than its parents, as well as a differing ecogeographic niche, including potential adaptation to high heat and to drought.

Genetic resistance is essential to efforts to overcome major biotic and abiotic constraints in sweetpotato production. As these constraints are often interrelated, e.g., drought stress with SPW and SPVD damage, enhancement of broad resistance for traits such as drought may improve yield across relatively large geographic areas, without the need to breed for resistance to localized viral strains (Ngailo et al., 2013). Such broad tolerance may also improve adoption rates for sweetpotato varieties with other desirable characteristics, such as high β-carotene content.

Reliable funding for germplasm collections is paramount in order to safeguard sweetpotato CWR genetic resources in the long-term and to continue to make *ex situ* collections available to the global community. Further investment in genebank information systems, *ex situ* conservation technologies (i.e., storage, testing, and regeneration), safety duplication of unique germplasm, and ensuring facilitated access to genetic diversity is equally essential (Food and Agriculture Organization of the United Nations [FAO], 2002, 2010; Khoury et al., 2010). In order to maximize the usefulness of conserved germplasm, characterization and evaluation for traits of interest, and increased breeding research, which have been limited for CWR of sweetpotato, are also needed. Further research combining morphological studies, trait evaluations, and genetic diversity analyses is likewise critically needed for elucidating species boundaries and highlighting accessions of particular value for use in breeding. Recent focused research has produced quick gains, including the identification of new species (Duncan and Rausher, 2013; Wood et al., in press). Through these actions the crop research community will contribute to ensuring the long term viability of this important root crop.

AUTHOR CONTRIBUTIONS

CK, BH, NC-A, HA, CS, HJ, and SdH conceived and designed the study. CK, NC-A, HA, CS, and VB gathered data and performed the analyses. CK, BH, NC-A, HA, CS, RM, RS, JW, GR, LE, RJ, CY, SdH, and PS analyzed the results. CK and BH wrote the manuscript. CK, BH, NC-A, HA, CS, RM, RS, JW, GR, LE, RJ, CY, VB, HJ, SdH, and PS edited the manuscript.

ACKNOWLEDGMENTS

We dedicate this research to the late Dan F. Austin, who provided the foundation for our understanding of the taxonomy of *Ipomoea* series *Batatas*. We thank the taxonomic and geographic data providers whose contributions made this analysis possible, and S. Calderón, A. Castañeda, O. Rivera, and I. Vanegas for gathering and organizing occurrence data. Data gathering and analyses were undertaken as part of the initiative "Adapting Agriculture to Climate Change: Collecting, Protecting and Preparing Crop Wild Relatives" which is supported by the Government of Norway. The project is managed by the Global Crop Diversity Trust with the Millennium Seed Bank of the Royal Botanic Gardens, Kew, UK, and implemented in partnership with national and international genebanks and plant breeding institutes around the world. For further information, see Dempewolf et al. (2014), and http://www.cwrdiversity.org/. Support from the CGIAR Research Program on Roots, Tubers, and Bananas (RTB) is kindly acknowledged.

REFERENCES

1. Araújo, M. B., and Guisan, A. (2006). Five (or so) challenges for species distribution modelling. *J. Biogeogr.* 33, 1677–1688. doi: 10.1111/j.1365-2699.2006.01584.x
2. Austin, D. F. (1978). The *Ipomoea batatas* complex – 1. Taxonomy. *Bull. Torrey Bot. Club* 105, 114–129. doi: 10.2307/2484429
3. Austin, D. F. (1988). "The taxonomy, evolution and genetic diversity of sweetpotatoes and related wild species," in *Exploration, Maintenance, and Utilization of Sweetpotato Genetic Resources*, ed. P. Gregory (Lima: International Potato Center), 27–60.
4. Austin, D. F. (1991). *Ipomoea* littoralis (Convolvulaceae) - Taxonomy, distribution, and ethnobotany. *Econ. Bot.* 45, 251–256. doi: 10.1007/BF02862052
5. Austin, D. F. (1997). Dissolution of *Ipomoea* series Anisomerae (Convolvulaceae). *Taxon* 28, 359–361.
6. Austin, D. F., and Huamán, Z. (1996). A synopsis of *Ipomoea* (Convolvulaceae) in the Americas. *Taxon* 45, 3–38. doi: 10.2307/1222581
7. Bill and Melinda Gates Foundation. (2011). *Agricultural Development Strategy Overview*. Available at:https://docs.gatesfoundation.org/Documents/agricultural-development-strategy-overview.pdf
8. Bivand, R., and Lewin-Koh, N. (2014). *Maptools: Tools For Reading and Handling Spatial Objects. R Package version 0.8-30*. Available at: http://CRAN.R-project.org/package=maptools
9. Bivand, R., Keitt, T., and Rowlingson, B. (2014). *Rgdal: Bindings for the Geospatial Data Abstraction Library. R Package version 0.8-16*. Available at: http://CRAN.R-project.org/package=rgdal
10. Bivand, R., Pebesma, E., and Gomez-Rubio, V. (2013). *Applied Spatial Data Analysis With R*, 2nd Edn. New York, NY: Springer. Available at: http://www.asdar-book.org/
11. Bouis, H., and Islam, Y. (2012). *Delivering Nutrients Widely through Biofortification: Building on Orange Sweet Potato - Scaling Up in Agriculture, Rural Development and Nutrition. Focus 19, Brief 11.* Washington, DC: International Food Policy Research Institute.
12. Brar, D. S., and Khush, G. S. (1997). Alien introgression in rice. *Plant Mol. Biol.* 35, 35–47. doi: 10.1023/A:1005825519998
13. Cao, Q., Zhang, A., Ma, D., Li, H., Li, Q., and Li, P. (2009). Novel interspecific hybridization between sweetpotato (*Ipomoea batatas* (L.) Lam.) and its two diploid wild relatives. *Euphytica* 169, 345–352. doi: 10.1007/s10681-009-9967-7
14. Conolly, J., Manning, K., Colledge, S., Dobney, K., and Shennan, S. (2012). Species distribution modelling of ancient cattle from early neolithic sites in SW Asia and Europe. *Holocene* 22, 997–1010. doi: 10.1177/0959683612437871
15. Costa, G. C., Nogueira, C., Machado, R. B., and Colli, G. R. (2010). Sampling bias and the use of ecological niche modeling in conservation planning: a field evaluation in a biodiversity hotspot. *Biodivers. Conserv.* 19, 883–899. doi: 10.1007/s10531-009-9746-8
16. De Vries, C. A., Ferwerda, J. D., and Flach, M. (1967). Choice of food crops in relation to actual and potential production in the tropics.*Neth. J. Agr. Sci.* 15, 241–248.
17. Dempewolf, H., Eastwood, R. J., Guarino, L., Khoury, C. K., Müller, J. V., and Toll, J. (2014). Adapting agriculture to climate change: a global initiative to collect, conserve, and use crop wild relatives. *Agroecol. Sust. Food* 38, 369–377. doi: 10.1080/21683565.2013.870629
18. Diaz, J., Schmiediche, P., and Austin, D. F. (1996). Polygon of crossability between eleven species of *Ipomoea*: section batatas (Convolvulaceae). *Euphytica* 88, 189–200. doi: 10.1007/BF00023890
19. Duncan, T. M., and Rausher, M. D. (2013). Evolution of the selfing syndrome in *Ipomoea*. *Front. Plant Sci.*

4:301. doi: 10.3389/fpls.2013.00301

20. Elith, J., Graham, C. H., Anderson, R. P., Dudik, M., Ferrier, S., Guisan, A., et al. (2006). Novel methods improve prediction of species' distributions from occurrence data. *Ecography* 29, 129–151. doi: 10.1111/j.2006.0906-7590.04596.x

21. Eserman, L. A., Tiley, G. P., Jarret, R. L., Leebens-Mack, J. H., and Miller, R. E. (2014). Phylogenetics and diversification of morning glories (tribe Ipomoeeae, Convolvulaceae) based on whole plastome sequences. *Am. J. Bot.* 101, 92–103. doi: 10.3732/ajb.1300207

22. ESRI. (2011).*ArcGIS Desktop: Release* 10.1. Available at: http://www.esri.com/software/arcgis/arcgis-for-desktop

23. Food and Agriculture Organization of the United Nations [FAO]. (2002). *International Treaty on Plant Genetic Resources for Food and Agriculture*. Rome: FAO.

24. Food and Agriculture Organization of the United Nations [FAO]. (2010). *The Second Report on the State of the World's Plant Genetic Resources for Food and Agriculture*. Rome: FAO.

25. Feuillet, C., Langridge, P., and Waugh, R. (2008). Cereal breeding takes a walk on the wild side. *Trends Genet.* 24, 24–32. doi: 10.1016/j.tig.2007.11.001

26. Fieberg, J., and Kochanny, C. O. (2005). Quantifying home-range overlap: the importance of the utilization distribution. *J. Wildl. Manag.* 69, 1346–1359. doi: 10.2193/0022-541X(2005)69[1346:QHOTIO]2.0.CO;2

27. Gaiji, S., Chavan, V., Ariño, A. H., Otegui, J., Hobern, D., Sood, R., et al. (2013). Content assessment of the primary biodiversity data published through GBIF network: status, challenges and potentials. *Biodivers. Inform.* 8, 94–172.

28. Germplasm Resources Information Network [GRIN]. (2013). *USDA, ARS, National Genetic Resources Program*. Available at:http://www.ars-grin.gov/cgi-bin/npgs/html/wep.pl

29. Gichuki, S. T., Berenyi, M., Zhang, D., Hermann, M., Schmidt, J., Glössl, J., et al. (2003). Genetic diversity in sweetpotato [*Ipomoea batatas* (L.) Lam.] in relationship to geographic sources as assessed with RAPD markers. *Genet. Resour. Crop Evol.* 50, 429–437. doi: 10.1023/A:1023998522845

30. Google Maps Geocoder. (2013).*Version 3*. Available at: https://developers.google.com/maps/documentation/geocoding/?hl=pl

31. Guarino, L., and Lobell, D. B. (2011). A walk on the wild side. *Nat. Clim. Change* 1, 374–375. doi: 10.1038/nclimate1272

32. Hajjar, R., and Hodgkin, T. (2007). The use of wild relatives for crop improvement: a survey of developments over the past 20 years.*Euphytica* 156, 1–13. doi: 10.1007/s10681-007-9363-0

33. Harlan, J. R., and de Wet, J. M. J. (1971). Toward a rational classification of cultivated plants. *Taxon* 20, 509–517. doi: 10.2307/1218252

34. Heibl, C. (2011). *Phyloclim: Integrating Phylogenetics and Climatic Niche Modeling. v.0.9-4*. Available at: http://cran.r-project.org/web/packages/phyloclim/index.html

35. Hengl, T., de Jesus, J. M., MacMillan, R. A., Batjes, N. H., Heuvelink, G. B. M., Ribeiro, E., et al. (2014). SoilGrids1km – global soil information based on automated mapping. *PLoS ONE* 9:e105992. doi: 10.1371/journal.pone.0105992

36. Hijmans, R. J. (2012). Cross-validation of species distribution models: removing spatial sorting bias and calibration with a null model.*Ecology* 93, 679–688. doi: 10.1890/11-0826.1

37. Hijmans, R. J. (2014). *Raster: Geographic Data Analysis and Modeling. R package version 2.2-31*. Available at: http://CRAN.R-project.org/package=raster

38. Hijmans, R. J., Cameron, S., Parra, J., Jones, P. G., and Jarvis, A. (2005). *WorldClim, version 1.3*. University of California, Berkeley.

39. Hijmans, R. J., Phillips, S., Leathwick, J., and Elith, J. (2013). *Dismo: Species Distribution Modeling. R package version 0.9-3*. Available at: http://CRAN.R-project.org/package=dismo

40. Hijmans, R. J., Schreuder, M., De La Cruz, J., and Guarino, L. (1999). Using GIS to check co-ordinates of genebank accessions. *Genet. Resour. Crop Evol.* 46, 291–296. doi: 10.1023/A:1008628005016

41. Hjarding, A., Tolley, K. A., and Burgess, N. D. (2014). Red List assessments of East African chameleons: a case study of why we need experts. *Oryx* doi: 10.1017/s0030605313001427

42. Hotz, C., Loechl, C., Lubowa, A., Tumwine, J. K., Ndeezi, G., Nandutu Masawi, A., et al. (2012). Introduction of β-carotene-rich orange sweet potato in rural Uganda resulted in increased vitamin a intakes among children and women and improved vitamin a status among children. *J. Nutr.* 142, 1871–1880. doi: 10.3945/jn.111.151829

43. Huaman, Z. (1987). "Current status on maintenance of sweetpotato genetic resources at CIP," in *Proceedings of the Exploration, Maintenance and Utilization of Sweetpotato Genetic Resources. Report of the First Sweetpotato Planning Conference 1987*. (Lima, Peru: CIP), 101–120.

44. Huang, J., Corke, H., and Sun, M. (2002). Highly polymorphic AFLP markers as a complementary tool to ITS sequences in assessing genetic diversity and phylogenetic relationships of sweetpotato (*Ipomoea batatas* (L.) Lam.) and its wild relatives. *Genet. Resour. Crop Evol.* 49, 541–550. doi: 10.1023/A:1021290927362

45. Husson, F., Josse, J., Le, S., and Mazet, J. (2009). *FactoMineR: Multivariate Exploratory Data Analysis and Dmining with R. V.1.26*. Available at: http://factominer.free.fr

46. Iwanaga, M. (1988). "Use of wild germplasm for sweetpotato breeding," in *Exploration, Maintenance, and Utilization of Sweetpotato Genetic Resources*, ed. P. Gregory (Lima: International Potato Center), 199–210.

47. Jarret, R. L., and Austin, D. F. (1994). Genetic diversity and systematic relationships in sweetpotato (*Ipomoea*

batatas (L.) Lam.) and related species as revealed by RAPD analysis. *Genet. Resour. Crop Evol.* 41, 165–173. doi: 10.1007/BF00051633

48. Jarret, R. L., Gawel, N., and Whittemore, A. (1992). Relationships of the sweetpotato [*Ipomoea batatas* (L.) Lam.]. *J. Am. Soc. Hortic. Sci.* 117, 633–637.

49. Kays, S. J. (1985). "The physiology of yield in sweet potato," in *Sweet Potato Products: A Natural Resource of the Tropics*, ed. J. Bouwkamp (Boca Raton, FL: CRC Press), 19–133.

50. Kays, S. J., and Kays, S. E. (1998). "Sweet potato chemistry in relation to health," in *Proceedings of the Sweet Potato Production System Towards the 21st Century*, Miyakonojo: Kyushu National Agricultural Experimental Station, 231–272.

51. Kernohan, B. J., Girzen, R. A., and Millspaugh, J. J. (2001). "Analysis of animal space use and movements," in *Radio Tracking Animal Populations*. eds J. J. Millspaugh and J. M. Marzluff (San Diego: Academic Press), 125–166. doi: 10.1016/B978-012497781-5/50006-2

52. Khoury, C., Laliberté, B., and Guarino, L. (2010). Trends in ex situ conservation of plant genetic resources: a review of global crop and regional conservation strategies. *Genet. Resour. Crop Evol.* 57, 625–639. doi: 10.1007/s10722-010-9534-z

53. Khoury, C. K., Castañeda Álvarez, N. P., Achicanoy, H., Sosa, C. C., Bernau, V., Kassa, M. T., et al. (2015). Crop wild relatives of pigeonpea [*Cajanus cajan* (L.) Millsp.]: distributions, ex situ conservation status, and potential genetic resources for abiotic stress tolerance. *Biol. Conserv.* 184, 259–270. doi: 10.1016/j.biocon.2015.01.032

54. Kivuva, B. M., Githiri, S. M., Yencho, C., and Sibiya, J. (2015a). Combining ability and heterosis for yield and drought tolerance traits under managed drought stress in sweetpotato. *Euphytica* 201, 423–440. doi: 10.1007/s10681-014-1230-1

55. Kivuva, B. M., Githiri, S. M., Yencho, C., and Sibiya, J. (2015b). Screening sweetpotato genotypes for tolerance to drought stress. *Field Crop Res.* 171, 11–22. doi: 10.1016/j.fcr.2014.10.018

56. Komaki, K. (2004). "Breeding value of wild species closely related to sweetpotato," in *Proceedings of international workshop on production, utilization and development of sweetpotato*, Korea, 164–172.

57. Kramer-Schadt, S., Niedballa, J., Pilgrim, J. D., Schroder, B., Lindenborn, J., Reinfelder, V., et al. (2013). The importance of correcting for sampling bias in MaxEnt species distribution models. *Divers. Distrib.* 19, 1366–1379. doi: 10.1111/ddi.12096

58. Liu, C., Berry, P. M., Dawson, T. P., and Pearson, R. G. (2005). Selecting thresholds of occurrence in the prediction of species distributions. *Ecography* 3, 385–393. doi: 10.1111/j.0906-7590.2005.03957.x

59. Loiselle, B., Jørgensen, P. M., Consiglio, T., Jiménez, I., Blake, J. G., Lohmann, L. G., et al. (2007). Predicting species distributions from herbarium collections: does climate bias in collection sampling influence model outcomes? *J. Biogeogr.* 35, 105–116. doi: 10.1111/j.1365-2699.2007.01779.x

60. Low, J. W., Arimond, M., Osman, N., Cunguara, B., Zano, F., and Tschirley, D. (2007). A food-based approach introducing orange-fleshed sweet potatoes increased vitamin a intake and serum retinol concentrations in young children in rural Mozambique. *J. Nutr.* 137, 1320–1327.

61. Low, J., Lynam, J., Lemaga, B., Crissman, C., Barker, I., Thiele, G., et al. (2009). "Sweetpotato in Sub-Saharan Africa," in *The Sweet Potato*, eds G. Loebenstein and G. Thottappilly (Amsterdam: Springer), 359–390.

62. Lu, S., and Li, T. (1992). Study on the characterization of intra and inter-specific incompatibility of Batatas section. *Acta Agron. Sin.* 18, 161–168.

63. Mabberley, D. J. (2008). *Mabberley's Plant Book: A Portable Dictionary of Plants, Their Classifications, and Uses*, 3rd Edn. Cambridge: Cambridge University Press.

64. Martin, F. W. (1970). Self- and interspecific incompatibility in the Convolvulaceae. *Bot. Gaz.* 131, 139–144. doi: 10.1086/336524

65. Martin, F. W. (1982). "Analysis of the incompatibility and sterility of sweetpotato," in *Proceedings of 1st International Symposium on Sweetpotato*, eds R. L. Villareal and T. D. Griggs (Tainan: AVRDC), 275–283.

66. Martin, F. W., and Jones, A. (1973). The species of *Ipomoea* closely related to the sweetpotato. *Econ. Bot.* 26, 201–215. doi: 10.1007/BF02861033

67. McCouch, S. R., Sweeney, M., Li, J., Jiang, H., Thomson, M., Septinginsih, E., et al. (2007). Through the genetic bottleneck: *O. rufipogon*as a source of trait-enhancing alleles for *O. sativa. Euphytica* 154, 317–339. doi: 10.1007/s10681-006-9210-8

68. McDonald, J. A., and Austin, D. F. (1990). Changes and additions in *Ipomoea* section Batatas (Convolvulaceae). *Brittonia* 42, 116–120. doi: 10.2307/2807625

69. McDonald, J. A., Hansen, D. R., McDill, J. R., and Simpson, B. B. (2011). A phylogenetic assessment of breeding systems and floral morphology of North American *Ipomoea* (Convolvulaceae). *J. Bot. Res. Inst. Texas* 5, 159–177. doi: 10.1007/s00606-011-0550-z.1.15

70. Mesibov, R. (2013). A specialist's audit of aggregated occurrence records. *ZooKeys* 293, 1–18. doi: 10.3897/zookeys.293.5111

71. Miller, R. E., Rausher, M. D., and Manos, P. S. (1999). Phylogenetic systematics of *Ipomoea* (Convolvulaceae) based on ITS and waxy sequences. *Syst. Bot.* 24, 209–227. doi: 10.2307/2419549

72. Monfreda, C., Ramankutty, N., and Foley, J. A. (2008). *Farming the planet: 2. Geographic distribution of crop areas, yields, physiological types, and net primary production in the year 2000. Global biogeochemical cycles 22: GB1022.* Available at:http://www.geog.mcgill.ca/landuse/pub/Data/175crops2000/ doi: 10.1029/2007GB002947

73. Munyiza, H., Stevenson, P., Mwanga, R., Talwana, H., Murumu, J., and Odongo, B. (2007). The relationship between stem base and root damage by *Cylas* spp. on

sweet potato. *Afr. Crop Sci. J.* 8, 955–957.

74. Mwanga, R., and Ssemakula, G. (2011). Orange-fleshed sweet potatoes for food, health and wealth in Uganda. *IJSA* 9, 42–49. doi: 10.3763/ijas.2010.0546

75. Mwololo, J. K., Mburu, M. W. K., Njeru, R. W., Ateka, E. M., Kiarie, N., Munyua, J. K., et al. (2007). Resistance of sweet potato genotypes to sweet potato virus disease in Coastal Kenya. *Proc. ACSS Conf.* 8, 2083–2086.

76. Ngailo, S., Shimelis, H., Sibiya, J., and Mtunda, K. (2013). Sweet potato breeding for resistance to sweet potato virus disease and improved yield: progress and challenges. *Afr. J. Agric. Res.* 8, 3202–3215.

77. Nimmakayala, P., Vajja, G., and Reddy, U. K. (2011). "*Ipomoea*," in *Wild Crop Relatives: Genomic and Breeding Resources*, ed. C. Kole (Berlin: Springer), 123–132.

78. Nishiyama, I. (1971). Evaluation and domestication of sweetpotato. *Bot. Magazine Tokyo* 84, 377–387. doi: 10.15281/jplantres1887.84.377

79. Nishiyama, I. (1982). "Autohexaploid evolution of the sweetpotato," in *Proceedings of 1st International Symposium on Sweetpotato*, eds R. L. Villareal and T. D. Griggs (Shanhua: Asian Vegetable Research and Development Center), 263–274.

80. Olson, D. M., Dinerstein, E., Wikramanayake, E. D., Burgess, N. D., Powell, G. V. N., Underwood, E. C., et al. (2001). Terrestrial ecoregions of the world: a new map of life on earth. *BioScience* 51, 933–938. doi: 10.1641/0006-3568(2001)051[0933:TEOTWA]2.0.CO;2

81. Otegui, J., Ariño, A. H., Encinas, M. A., and Pando, F. (2013). Assessing the primary data hosted by the Spanish node of the global biodiversity information facility (GBIF). *PLoS ONE* 8:e55144. doi: 10.1371/journal.pone.0055144

82. Pebesma, E. J., and Bivand, R. S. (2005). Classes and methods for spatial data in R. *R News 5* (2). Available at: http://cran.r-project.org/doc/Rnews/

83. Pfeiffer, W. H., and McClafferty, B. (2007). HarvestPlus: breeding crops for better nutrition. *Crop Sci.* 47, 88–105. doi: 10.2135/cropsci2007.09.0020IPBS

84. Phillips, S. J., Anderson, R. P., and Schapire, R. E. (2006). Maximum entropy modeling of species geographic distributions. *Ecol. Model.* 190, 231–259. doi: 10.1016/j.ecolmodel.2005.03.026

85. Phillips, S. J., and Dudik, M. (2008). Modeling of species distributions with maxent: new extensions and comprehensive evaluation. *Ecography* 31, 161–175. doi: 10.1111/j.0906-7590.2008.5203.x

86. Piperno, D. R., and Pearsall, D. M. (1998). *The Origins of Agriculture in the Lowland Neotropics*. San Diego: Academic Press.

87. R Core Team. (2013). *R: A Language and Environment for Statistical Computing*. Available at: http://www.r-project.org/

88. Rajapakse, S., Nilmalgoda, S. D., Molnar, M., Ballard, R. E., Austin, D. F., and Bohac, J. R. (2004). Phylogenetic relationships of the sweetpotato in *Ipomoea* series Batatas (Convolvulaceae) based on nuclear β-amylase gene sequences. *Mol. Phylogenet. Evol.* 30, 623–632. doi: 10.1016/S1055-7903(03)00249-5

89. Ramírez-Villegas, J., Khoury, C., Jarvis, A., Debouck, D. G., and Guarino, L. (2010). A gap analysis methodology for collecting crop genepools: a case study with Phaseolus beans. *PLoS ONE* 5:e13497. doi: 10.1371/journal.pone.0013497

90. Revelle, W. (2015). *Psych: Procedures for Phychological, Psychometric, and Personality Research, v.* 1.5.1. Available at: http://cran.r-project.org/web/packages/psych/index.html

91. Roullier, C., Duputié, A., Wennekes, P., Benoit, L., Fernández Bringas, V. M., Rossel, G., et al. (2013). Disentangling the origins of cultivated sweet potato (*Ipomoea batatas* (L.) Lam.). *PLoS ONE* 8:e62707. doi: 10.1371/journal.pone.0062707

92. Sanchez, G. (2012). *Package 'Plsdepot'. Partial Least Squares (PLS) Data Analysis Methods, v.* 0.1.17. Available at: http://cran.r-project.org/web/packages/plsdepot/index.html

93. Schafleitner, R., Tincopa, L. R., Palomino, O., Rossel, G., Robles, R. F., Alagon, R., et al. (2010). A sweet potato gene index established by de novo assembly of pyrosequencing and Sanger sequences and mining for gene-based microsatellite markers. *BMC Genomics* 11:604. doi: 10.1186/1471-2164-11-604

94. Solis, R. S., Haas, J., and Creamer, W. (2001). Dating caral, a preceramic site in the supe valley on the central coast of Peru. *Science* 292, 723–726. doi: 10.1126/science.1059519

95. Shiotani, I., Huang, Z. Z., Sakamoto, S., and Miyazaki, T. (1994). The role of the wild *Ipomoea* trifida germplasm in sweet potato breeding. *Acta Hort.* (*ISHS*) 380, 388–398.

96. Smit, N. E. J. M. (1997). *Integrated Pest Management for Sweetpotato in Eastern Africa*. Ph.D. thesis, Wageningen University, Wageningen.

97. Srisuwan, S., Sihachakr, D., and Siljak-Yakovlev, S. (2006). The origin and evolution of sweet potato (*Ipomoea batatas* Lam.) and its wild relatives through the cytogenetic approaches. *Plant Sci.* 171, 424–433. doi: 10.1016/j.plantsci.2006.05.007

98. Sutherland, J. A. (1986). A review of the biology and control of the sweetpotato weevil, cylas formicarius elegaluntus (F). *Trop. Pest Manag.* 32, 304–315. doi: 10.1080/09670878609371084

99. Teramura, T. (1979). Phylogenetic study of *Ipomoea* species in the section Batatas. *Mem. Coll. Agric. Kyoto Univ.* 114, 29–48.

100. The Plant List. (2010). *Version 1*. Available at: http://www.theplantlist.org/

101. Tumwegamire, S., Kapinga, R., Rubaihayo, P. R., LaBonte, D. R., Grüneberg, W. J., Burgos, G., et al. (2011). Evaluation of dry matter: protein, starch, β-carotene, iron, zinc, calcium, and magnesium in East African sweetpotato [*Ipomoea batatas* (L.) *Lam*] germplasm. *Hort. Science* 46, 348–357.

102. Valverde, R. A., Clark, C. A., and Valkonen, J. P. T. (2007). Viruses and virus disease complexes of sweetpotato. *Plant Viruses* 1, 116–126.

103. van der Wal, J., Falconi, L., Januchowski, S., Shoo, L., and Storlie, C. (2014). *SDMTools: Species Distribution Modelling Tools: Tools for Processing Data Associated with Species Distribution Modelling Exercises. R Package version 1.1-20*. Available at: http://CRAN.R-project.org/package=SDMTools

104. Warren, D. L., Glor, R. E., and Turelli, M. (2008). Environmental niche equivalency versus conservatism: quantitative approaches to niche evolution. *Evolution* 62, 2868–2883. doi: 10.1111/j.1558-5646.2008.00482.x

105. Wilkin, P. (1999). A morphological cladistic analysis of the Ipomoeeae (Convolvulaceae). *Kew Bull.* 54, 853–876. doi: 10.2307/4111164

106. Williams, J. N., Seo, C., Thorne, J., Nelson, J. K., Erwin, S., O'Brian, J. M., et al. (2009). Using species distribution models to predict new occurrences for rare plants. *Divers. Distrib.* 15, 565–576. doi: 10.1111/j.1472-4642.2009.00567.x

107. Wood, J. R. I., Carine, M. A., Harris, D., Wilkin, P., Williams, B., and Scotland, R. W. (in press). *Ipomoea* (Convolvulaceae) in Bolivia.*Kew Bull.*

108. Woolfe, J. A. (1992). Sweetpotato, an untapped food source. New York, NY: Cambridge University Press.

109. Xiao, J., Grandillo, S., Ahn, S. N., McCouch, S. R., and Tanksley, S. D. (1996). Genes from wild rice improve yield. *Nature* 384, 223–224. doi: 10.1038/384223a0

110. Zhang, D. P., Cervantes, J., Huamán, Z., Carey, E., and Ghislain, M. (2000). Assessing the genetic diversity of sweetpotato (*Ipomoea batatas* (L.) La.) cultivars from tropical America using AFLP. *Genet. Resour. Crop Evol.* 47, 659–665. doi: 10.1023/A:1026520507223

111. Zhang, S. S., and Liu, L. F. (2005). "Utilization of the wild relatives in sweetpotato breeding," in *Sweetpotato Breeding and Industrialization in China*, ed. Q. C. Liu (Beijing: China Agricultural University Press),29–32.

CHAPTER 10

MODELLING SOCIOECONOMIC DETERMINANTS FOR CULTIVATION AND IN-SITU CONSERVATION OF VITEX DONIANA SWEET (BLACK PLUM), A WILD HARVESTED ECONOMIC PLANT IN BENIN

Sognigbe N'Danikou[1,2], Enoch G Achigan-Dako[1], Dedeou A Tchokponhoue[1], Chaldia OA Agossou[1], Carlos A Houdegbe[1], Raymond S Vodouhe[2] and Adam Ahanchede[1]

[1]Horticulture and Genetics Unit, Faculty of Agronomic Sciences, University of Abomey-Calavi, BP: 2549 Abomey-Calavi, Republic of Benin

[2]Bioversity International, West and Central Africa office, 08 BP 0932 Cotonou, Republic of Benin

ABSTRACT

Background

Cultivation is the most appropriate management option when both demand and harvesting of wild plant species increase beyond natural production levels. In the current study we made the assumption that, besides the intrinsic biological and ecological characteristics of the species, the decision to cultivate and/or to conserve an overharvested wild plant species is triggered by the socioeconomic factors such as land tenure and size, origin of respondents, gender, and users' knowledge of the plant phenology.

Methods

We carried out semi-structured inters with 178 informants involved in V. doniana exploitation. The data collected were related to socio-demographic characteristics of informants' household situation, knowledge of the biology and propagation of the species, willingness to cultivate the species, in-situ maintenance of populations, and costs associated with management of the species. According to data types we used Student's t, Spearman correlation, Kruskal-Wallis, Fisher's exact and $\chi 2$ tests to test the effects of land tenure, origin of respondents, gender and users' knowledge of plant phenology on the decision making process. Conditional inference tree models and generalized additive models were also used to identify variables which were significantly determinant in the decision to cultivate and/or to conserve the species in-situ.

Citation: Sognigbe N'Danikou, Enoch G Achigan-Dako, Dedeou A Tchokponhoue, Chaldia OA Agossou, Carlos A Houdegbe, Raymond S Vodouhe and Adam Ahanchede, Modelling socioeconomic determinants for cultivation and in-situ conservation of Vitex doniana Sweet (Black plum), a wild harvested economic plant in Benin, DOI: 10.1186/s13002-015-0017-3.

Results

We found that men were more willing to cultivate the species than women and this is conditioned by land area available. The willingness to conserve the species in-situ depends mainly on the total land area available, the number of trees within the landscape, accessibility of the trees, land tenure, gender, location, seedling cost, and trade-off cost for conservation. People who offered more than one US dollar to acquire a seedling ofV. doniana, landowners, and those who own a total land area in excess of 6.5 ha were most willing to conserve the species in-situ.

Conclusions

From our findings we conclude that future management and conservation initiatives for V. doniana should first target specific user groups for sustainable exploitation of the species. Also, the Cultivation Opportunity Ratio is an important indicator for quick determination of the likelihood of farmers to engage into cultivation and conservation of the species.

BACKGROUND

The decision to conserve wild-harvested species is based on factors that are presumed to be community and resource specific. When the value of a non-wood forest product (NWFP) and quantities harvested are both high, it will be under high risk of local extinction, and cultivation is the most effective management option to sustain both the species and livelihoods of the different user groups [1]. The continuous selection and cultivation of variants of useful species to humans, most often, result in domesticated crops [2,3]. Gepts *et al.*[4] indicated that the increase in human demand represents new push for domestication. The main advantages of domesticating intensively wild-harvested edible plant species include: (a) enhanced local livelihoods and food security, (b) increased supply of market commodities and maintenance of food diversity, and (c) increased ecosystem services and resilience. Overall, the economic, social and environmental rewards of domestication are obvious, and domesticated plants should repay better human investments than collection of resources from the wild [5]. For Clement [6] plant domestication is a co-evolutionary process whereby humans select phenotypes of most utility to them and bring them into cultivation. Domestication implies a process in which wild plants are taken from their natural habitats and grown in cultivation systems where growth conditions can be manipulated to enhance production of the desired product (e.g. fruits, leaves, roots, seeds). In simple terms it is the human-mediated adaptation of the wild plant to cultivation in plantation or farmlands [5]. This involves socioeconomic and biophysical processes [7], which lead to both changes to human culture and plant phenology, and begins when species become objects of ownership, inheritance, purchase and exchange [8]. These processes put local communities at the centre of the domestication process, and Clement [6] noted that "*for plant domestication to take place, there must be selection and management*". Selection of candidates at species and variety levels begins with local communities and is often continued with support of formal breeding and genetic improvement by science. It is important to indicate that domestication is a gradual transition which starts from wild-harvesting through to pre-agriculture (selective harvesting, sparing, etc.), proto-agriculture or incipient domestication (sowing, transplanting, irrigation, etc.), and full domestication characterized by small-scale and/or large-scale agriculture [4]. Success of domestication relies on management of domesticates, which is carried out by farmers. In this context, the recruitment of potential growers and users and their engagement in large scale production of the domesticated species is an important question which needs to be figured out [4]. From the biological point of, domestication is a continuous endeavour, the result of intensive selection and cultivation of variants with useful traits to humans, with no clear demarcated end points, whereby new phenotypes are created and continuously improved based on new social and economic needs [2,3,6,9]. It then appears that cultivation is the ground for domestication of useful plants to humans. Therefore, in this paper we evaluated the motives for cultivation of wild-harvested economic plants.

Although the scope of ethnobotanical studies has expanded and now covers a wide range of topics related to human - plant interactions and ably demonstrated the importance of wild harvesting to livelihoods of the poor, there is now need to move further to provide insights and guidance for conservation professionals to enhance the sustainable use of highly valued species. For the success of conservation programmes designed by natural resource managers, there is need to understand and integrate drivers of active engagement of local communities, individually and collectively. From the literature, there is very limited knowledge of factors affecting people's decisions related to plant selection for cultivation [10-12]. Thus, several variables remained unknown for predicting the success or failure of conservation initiatives aimed at bringing wild plants into cultivation. Here, we used *Vitex doniana* Sweet (Verbenaceae) as a case study to investigate and pinpoint local drivers that favour both active cultivation and conservation of wild populations of the species by local communities.

Vitex doniana (black plum) is one of the most important wild-harvested multipurpose trees and leafy vegetables of tropical Africa [13-15] particularly in Benin (where it is sold in almost all markets [12,16]), Burkina Faso [17] and Nigeria [18,19] to provide income and support livelihoods of many households. It is one of the most widespread of the 150 species of the *Vitex* genus, with several uses [14,20]. *Vitex doniana* is found in savanna woodlands and gallery forests, and is also indicated to be present in about 50% of traditional agroforestry systems in southern Benin [21], where it is just spared during land clearance [12,22,23]. The species is deciduous and its leaves renewed annually, during the dry season. Intensive harvesting occurs after bush burning associated with farm clearance, with the development of new shoots that are used as vegetable. Leaves, bark and roots also have many applications in indigenous medicine [20]. The fruit is eaten raw or used for making candies, jams and syrups [18,19]. However, the species is currently perceived as the most threatened of edible wild-harvested plants by local communities in Benin [12,24]. Field inventories indicated that the natural populations are continuously eroded and densities of trees and regeneration are becoming low in natural habitats [16,22-24]. Agossou [22], N'Danikou [23] and N'Danikou *et al.* [12] reported that harvesting practices represent the most important threat to the species, contrasting with Oumorou *et al.*[24]'s findings. In fact, harvesting of *V. doniana* vegetable by local communities mainly consists in removing all the new shoots, most often with twigs. Furthermore, some harvesters just cut the whole tree down to make harvesting easier [22,23]. These forms of harvesting reduce fruit production, and subsequently sexual propagation. Meanwhile, the suckers that regenerate from root stock are subject to bush fire in the natural habitat, jeopardizing the natural regeneration of the species. Thus, threat is mainly attributed to overharvesting of wild populations and habitat loss. Therefore, understanding and considering the complex interplay of local valuation and perception of threats would give an important ground for policy and decision making that will sustain the use of the species and also the livelihoods of various user groups.

This situation triggers our study knowing that *V. doniana* is under high harvesting pressure and belongs to a complex market chain [12,13,16], with increased price (authors' personal observation) whereas natural stands are subjected to private ownership [22]. In this condition which socioeconomic and cultural factors would favour cultivation of *V. doniana*? Is gender an important issue for the success of cultivation? How the access to resources and local knowledge shape the decision-making to cultivate and/or conserve *V. doniana in-situ*?

To answer these questions, we based this investigation on the assumption that, besides the intrinsic biological and ecological characteristics of the species, the decision to cultivate and/or to conserve an overharvested wild plant species is determined by the socioeconomic factors such as land tenure and size, origin of respondents, gender and users' knowledge of the plant phenology. In fact, in Africa and particularly in Benin, there are societal rules and other factors that affect farmers' decision to manage useful plants. These factors include gender, age, occupation, wealth status, importance of wild-harvesting in household food security, scarcity of resources, and land holding status [10,11,25,26]. For instance, male farmers and young people were more likely to accept production of non-wood forest products in their farms in Nigeria, because men are more likely to invest in the future than women [10], and this may be because women are uncertain about the security of their tenure due to possible change in their marital status [26]. Also, generally the local customary laws do put restrictions on plantation of perennial crops, particularly tree species which a migrant cannot grow on a rented or borrowed land, unless the land is purchased. This is simply because trees are used to mark human presence and dating land occupation. It is also shown that farmers' knowledge of the reproductive biology, including phenology, of species is taken into account for selecting candidate species for conservation [3,12].

MATERIALS AND METHODS

Study area

This study was carried out in Djidja municipality (Figure 1), one of the most important areas where *V. donianais* harvested and sold. The species is a highly valued one which is under high human pressure in this area [12,22]. Djidja is the biggest municipality in size (out of nine) in the Zou region (41.66%) in southern Benin. According to the population census by DED-INSAE [27], Djidja holds the second highest population (84,590 inhabitants), representing 14.50% in the entire population of the Zou region. About 78.00% of the district population live in rural areas. The study area is located on flat land at around 250 m altitude in the Sudano-Guinean transition zone. This is a transitional zone between the Guinean forests and the Sudanian woodlands and savannas, and is characterized by a vegetation mosaic of forest islands, gallery forests, and savannas. The annual rainfall varies from 900 to 1200 mm with two rainy seasons (March to July and September to October). Soils are predominantly hydromorphic and sandy-clay. The natural vegetation is made up of savanna woodlands, fallows, and a mosaic of secondary and gallery forests. This landscape provides a range of forest resources from which many edible wild plants are collected, including black plum. However, the natural vegetation is being depleted by land clearance for agriculture, logging for charcoal production, and grazing [12]. As presented below, a total of 12 villages were investigated in this municipality.

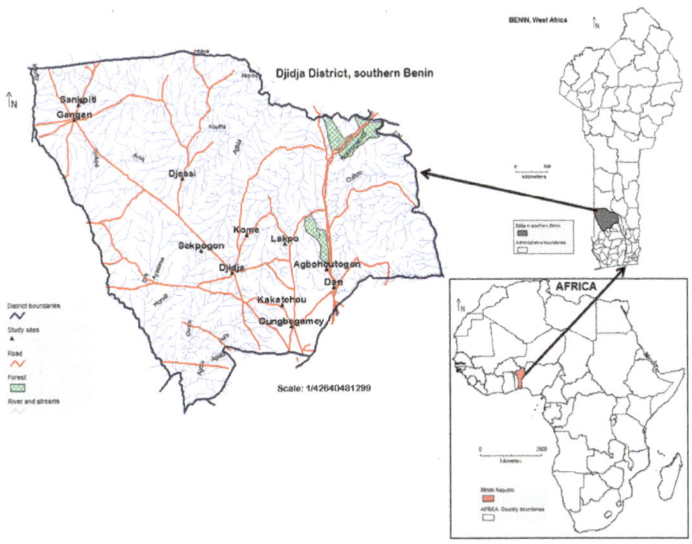

Figure 1: Study area showing surveyed villages in Djidja Ditrict, southern Benin.

Selection of study sites

Prior to selecting survey villages a market investigation was conducted in local markets in Bohicon and Djidja municipalities which are the major townships where *V. doniana* vegetable is sold. In Bohicon where the regional market is located, twelve vegetable sellers were randomly selected and intered. This market survey helped establish the origin of products, the abundance periods and villages supplying the regional market in Bohicon. The interees ranked the supplier villages according to the importance of their contributions to the supply. The ranking exercise revealed that the highest share of supply comes from Djidja municipality. In Djidja, a second market investigation was undertaken in five local markets. A total of 42 vendors of black plum products (fruits and/or vegetable) at the local markets were randomly intered on the most important supplier villages. Interees ranked villages according to the importance of primary collection and sale of the target products. The top 12 villages with higher contribution to the supply were targeted for final survey.

Informants sampling and data collection

A total of 178 key informants (persons having knowledge of uses and users of *V. doniana* in their community) were identified for inters using a snowball sampling approach. The number of key informants was deemed representative when a saturation limit was attained. This saturation limit was defined as the point at which there are no new names of key informants being mentioned by the last interee. At this point, the plot of the number of key informants becomes asymptotic with the addition of any new interee. The data collected include three categories of information: a) the socio-demographic characteristics of informants (gender, age, ethnic background, marital status, size of household, land ownership, origin, education of the head of household, occupation and revenue sources, social group membership, social responsibilities, and experience with management of wild species), b) ranking of the five most important and valued tree species to the respondent, and c) informant's knowledge of black plum (ownership status, uses and use frequency, knowledge of the phenology, threats and drivers, individual conservation initiatives, trade-off and cultivation opportunity costs and ratio, and willingness to cultivate and to maintain the species *in-situ*).

We obtained the permission from the chief of each village involved in the study before inters were conducted, following the ethical guidelines of the International Society of Ethnobiology [28]. Participants were selected and included in the study after obtaining their verbal prior informed consent.

No ethical approval was needed for this study.

Data analysis

Determination of the importance and use value of Vitex doniana in the surveyed communities

Before testing the research hypotheses, we first determined the relative importance of *V. doniana* within the diversity of useful local trees, against a number of socioeconomic characteristics. Based on value scores attributed by each respondent for the five most important tree species to them (55 species mentioned in all), we calculated the Species Use Value (SUV) for each species as the sum of the species value for all 178 respondents as:

$$SUV_i = \sum_{j=1}^{n} S_{ij}$$

with S_{ij} being the score given to the species *i* by the informant *j*.

Ranks were converted into scores and SUV calculated using the method by Lawrence *et al.* [29]. The species that was ranked first by a respondent received the score 5 whereas the fifth received the score 1. A species which was not mentioned by a given informant was given the score zero. The SUV of each species was disaggregated by gender, origin and schooling of respondents.

We also determined the Relative Use Value (RUV) of *V. doniana* for each category of informant (for e.g. farmers versus vegetable sellers). This helped identify user groups with specific knowledge on the use of the species and the relative importance to them.

$$RUV_k = \frac{1}{n}\sum_{j=1}^{n} S_{kj},$$

adapted from Phillips *et al.* [30].

S_{kj} is the informant *j* of user group *k*'s use value for the species; *n* is the total number of informants in the user group *k*.

To reveal a minimum consensus on the use of the species in the community only species cited by at least two respondents were included in further analyses. We used Student's *t*-test to examine the influence of gender and origin and Kruskall-Wallis' statistic to test the influence of schooling on the perceived value attributed to *V. doniana.* We also determined the Informant Agreement Ratio (IAR) following Bakwaye *et al.* [31] as follow:

$$IAR_i = \frac{nr_i - na_i}{nr_i - 1}$$

where nr_i is the total number of citations recorded for a given use category *i*; and na_i the total number of different plant parts that are employed in this specific use category *i*. This index measures if there are specific plant parts employed for particular uses. Five different use categories were identified, viz. food, income, medicine, construction, and others (handicraft, ink, stools).

Test of research hypothesis using classical non-parametric tests

We used Fisher's exact test to test if resource management depends on socioeconomic variables such as tree ownership, origin of respondent, and accessibility to resources. The relationships respectively between gender, land tenure and land holding, schooling and literacy, origin, participation in social organizations, level of income earned from *V. doniana* sales, technical training and management of *V. doniana* were evaluated using Spearman's correlation tests.

As cultivation of the species was not yet evident in Djidja, we assessed the economic value and respondent's willingness to cultivate and also to conserve the species based on five parameters, namely the Cultivation Opportunity Ratio (COR), the Land Allocation Ratio (LAR), the Trade-off Value (TOV) which represents the minimum amount the informant is ready to accept for the maintenance of one "*Cantin*" (equivalent 400 m^2) of the species, the willingness to conserve the species in the wild, and the willingness to experiment with cultivation of the species. These are new parameters that we introduced to determine socioeconomic drivers for plant cultivation by local communities. COR and LAR are calculated as follow:

$$COR = \frac{S_p - A_c}{S_p} * 100 \ ; \ and \ LAR = \frac{S_{\max}}{S_t} * 100$$

S_p is the estimated price at which the respondent would sell a seedling of black plum, A_c is the estimated cost that the respondent would be willing to pay to acquire a seedling of black plum, S_{max} is the maximum size of land an informant is ready to allocate to black plum within his/her farm and S_t the total land size s/he holds. We tested differences in COR, LAR, and TOV using Mann–Whitney and Kruskal-Wallis' tests. Relationships between household socioeconomic characteristics and willingness to maintain the species in its natural habitat (*in-situ* conservation) and/or to invest in black plum cultivation were tested using Chi-square or Fisher exact tests as appropriate.

Test of research hypotheses using statistical models

Due to the large number of explanatory variables (69 in total), we used the conditional inference tree model to select variables with most significant effects on the decision to cultivate and/or to conserve the species *in-situ*. The tree model is appropriate in cases where there is large number of explanatory variables and these can be a mix of continuous and categorical variables [32]. The tree-based models have a number of advantages in estimating a non-parametric regression relationship by binary recursive partitioning using conditional distributions. Such models are not based on any kind of distribution assumptions, and most importantly complex interactions with a large number of explanatory variables and non-linear relationships that are difficult to examine with traditional statistical methods can be modelled. Further details on tree models and conditional inference tree-based classification can be found in Hothorn *et al.* [33] and Crawley [32]. Further analyses of the relationships between the selected variables with the response variables (willingness to cultivate and willingness to conserve wild populations *in-situ*) were tested with generalized additive models (GAMs) which are appropriate for binary responses with at least one continuous explanatory variable.

Definition of terms

In-situ conservation

In Article 2 of the Convention on Biological Diversity (CBD) *in-situ* conservation is defined as "*the conservation of ecosystems and natural habitats and the maintenance and recovery of viable populations of species in their natural surroundings and, in the case of domesticated or cultivated species, in the surroundings where they have developed their distinctive properties*" [34]. On-farm conservation of useful species by farmers has been recognized as an *in-situ* or *circa-situ* conservation strategy [4,35].

Ex-situ conservation

In the same Article 2 of the CBD, *ex-situ* conservation means "*the conservation of components of biological diversity outside their natural habitats*". The conservation of germplasms in genebanks is recognized as an *ex-situ* measure as is conservation in botanical gardens [4].

RESULTS

Household characteristics of informants

Women represented 60.00% of informants, and the Fon ethnic group dominated the sample (89.00%). Other ethnic groups include, by importance, Agouna, Mahi, Nagot and Adja. *Vitex doniana* resource users were between 12 to 66 years old, and 42.73±16.12 years in average. The sample was characterized by a low proportion (17.98%) who has attended primary school and the main activities were farming, and trade of agricultural products. About 83.00% of respondents are owners of their land (with an average of 12.05 ha per household) whereas those who rent their land represented 16.29% with an average of 4.91 ha/household. Those who were landowners but also rented additional land for farming represented a very small proportion (0.71%). About 38.40% of respondents participated in social groups; the most common of these included self-help groups who provide agricultural product storage, groups formed to undertake sales and processing of agricultural products. Women participated more than men in such groups. Autochthons represented 68.00% of the sample against 32.00% migrants. The size of households greatly varied. Households with less than five members represented 28.65% of the sample; a high proportion of households had between 5–10 members (52.81%). Those with 11–15 members represented 16.85%, and households with more than 15 members represented 1.69%. All the households were involved with black plum.

Diversity of most valued tree species in the communities

From the individual respondents' lists of their five most important tree species, we recorded a total of 55 species belonging to 24 families. The determinants of the value given by respondents included the nutritional value, the market value, the medicinal value, the cultural functions and use in construction. The families most often mentioned were Leguminosae-Caesalpinioideae and Anacardiaceae. The species mentioned in the Leguminosae-Caesalpinioideae family included *Daniellia oliveri* (Rolfe) Hutch. & Dalziel, *Afzelia africana* Pers.,*Detarium microcarpum* Guill. & Perr., *Bauhinia thonningii* Schum., and *Tamarindus indica* L.; those in the Anacardiaceae included *Anacardium occidentale* L., *Mangifera indica* L., *Lannea schimperi* (Hochst. ex A.Rich.) Engl., and *Spondias mombin* L. The results also indicated that wild species were dominant (60.00%), especially*V. doniana* and *Parkia biglobosa* (Jacq.) G.Don which were respectively ranked the second and the third species, behind *A. occidentale* which was a cash crop in the survey area. The seven most frequently mentioned species have high nutritional and market values, whereas the next eight species are mainly used in folk medicine and construction (Table 1).

Table 1: Top 15 priority tree species based on species use values (SUV); wild species are in bold

Species	**Instruction level**			**Rank$_{inst}$**	**Origin**		**Ran-k$_{Orig}$**	**Gender**		**Rank-$_{gend}$**
	No schooling	**Primary**	**Secondary**		**Indigenous**	**Migrants**		**Men**	**Women**	
Anacardium occidentale L.	2.16	1.92	1.50	1	2.40	1.98	1	2.33	2.17	1
***Vitex doniana*Sweet**	2.38	1.67	0.25	3	2.15	2.53	2	1.52	2.61	2
***Parkia biglobosa*(Jacq.) G.Don**	2.15	1.68	1.25	2	1.94	2.09	3	1.75	2.26	3
*Mangifera indica*L.	1.26	1.27	0.88	4	1.09	1.49	4	0.82	1.44	5
Elaeis guineensis-Jacq.	1.15	1.00	1.88	5	1.49	0.82	5	1.94	0.69	4
*Tectona grandis*L.f.	1.36	1.03	1.38	6	1.02	0.94	7	1.42	0.84	6
Citrus sinensis (L.) Osbeck	1.53	0.48	0.50	7	1.14	1.12	6	1.28	0.97	6
***Pterocarpus erinaceus* Poir.**	0.51	1.19	0.00	8	0.70	0.48	8	1.00	0.34	8
***Vitellaria paradoxa*C.F.Gaertn.**	0.56	0.58	1.00	9	0.47	0.50	9	0.43	0.53	9
***Pseudocedrela kotschyi*(Schweinf.) Harms**	0.24	0.35	0.00	10	0.19	0.25	11	0.09	0.30	11
***Ficus sur* Forssk.**	0.28	0.20	0.25	12	0.20	0.24	10	0.03	0.31	12
***Prosopis africana*(Guill. & Perr.) Taub.**	0.17	0.38	0.00	11	0.24	0.02	15	0.27	0.16	10
*Azadirachta indica*A. Juss.	0.12	0.33	0.00	13	0.08	0.25	12	0.22	0.12	12
***Daniellia oliveri*(Rolfe) Hutch. & Dalziel**	0.11	0.32	0.00	14	0.16	0.18	14	0.18	0.12	14
***Lannea schimperi*(Hochst. ex A.Rich.) Engl.**	0.15	0.18	0.00	15	0.15	0.13	13	0.16	0.15	14

We also found that among the wild species *V. doniana* was the most valued. This perception was not affected by whether the informant was indigenous or migrant ($t=0.47$; $p=0.64$). However, gender ($t=6.48$; $p<0.001$) and schooling ($H=8.62$; $p=0.013$) significantly affected the perceived value ascribed to the species. Women valued the species significantly higher (mean SUV = 2.61) than men (mean SUV=1.52). *Vitex doniana*seemed to be the most valued species by people who have not attended school (SUV = 2.38) and this value decreased with schooling (SUV=1.67 for primary and SUV=0.25 for secondary and higher education). Gender, origin and schooling also showed strong and positive correlations with each other ($r\approx0.99$; $p<0.001$).

Variations in *Vitex doniana* use patterns

A total of 34 different uses were recorded for six parts of *V. doniana*, namely fruits, leaves, bark, trunk, branches and roots (Table 2). Those uses were classified into five use categories such as household food consumption, folk medicine, sales, construction material and fuel, and others (stools, handcrafts, ink making). The values of Informant Agreement Ratio (IAR) calculated for each use category revealed a consensus on the use of the different plant parts (Table 3). All plant parts are used but roots are only used in folk medicine. Leaves and fruits are used for human food and also sold in markets to generate income. All plant parts had applications in traditional medicine, except trunk and branches; the most commonly treated ailment being haemorrhoids.

Table 2: *Vitex doniana* plant parts and their uses in Djidja district

Plant parts used	Use categories	Uses
Roots	Folk medicine	Against snake and scorpion bites, haemorrhoids, malaria, ulcers, diabetes, sexual impotence, stomach-ache, wounds, constipation, eye pains, intestine worms, sprain
Bark	Folk medicine	Sterility, haemorrhoids, snake and scorpion bites, intestine worms, ulcers, stomach-ache, diabetes, constipation, malaria, wounds, sexual impotence, eye pain, cough, amenorrhea, dysmenorrhea, strokes
Trunk and branches	Construction and fuel	Fuel wood, charcoal, planks, beam
	Other uses	Drums, stool of the king
Leaves	Food	Household use for consumption as cooked vegetable
	Sales	Sale as cooked vegetable
	Folk medicine	Haemorrhoids, snake bites, mouth candidiasis, whitlow, annulet, malaria, ulcers, sexual impotence, stomach-ache
	Other uses	Ink
Fruits	Food	Direct consumption of fruits, juice making
	Sales	Sale of fruits for direct consumption
	Folk medicine	Haemorrhoids

Table 3: *Vitex doniana* major use types and Informant Agreement Ratio (IAR)

Use categories	Number of use citations (nr)	Plant parts involved	Number of parts involved (na)	IAR
Construction and fuel	3	Trunk and branches	2	0.500
Food	298	Fruits and leaves	2	0.996
Sales	120	Fruits and leaves	2	0.992
Folk medicine	58	Fruits, leaves, bark, roots	4	0.947
others	11	Trunk, branches, and leaves	3	0.800

The relative use value (RUV) calculated for each user group indicated that traders of black plum products ascribed higher use value (RUV=3.26) to the species than other stakeholders (Table 4). The average profit from sales of *V. doniana* vegetable during the last season was estimated at XOF 38,100 (US $79.16) per user. The results also indicated that women realized more profit from sales than men ($p<0.001$). Although youths realized more profit than teenagers and elderly people the differences were not statistically significant between age groups (Table 5). The values of the coefficient of variation were greater than 1 for all age categories and for women. This indicates a big variability in the profit earned by these groups out of the sale of *V. doniana* products. It is noteworthy to mention that the profit did not follow a normal distribution. Overall, there is an uneven distribution of the profit along the value chain of *V. doniana* (this was examined further in another study and not discussed here).

Table 4: The relative use values (RUV) of *Vitex doniana* per user group

Occupations	Total scores	n	RUV
Traders of *V. doniana* products and other small businesses	137	42	3.26
Farming only	182	79	2.30
Artisans, forest product businesses	33	16	2.06
Transport, part time farming, singing, traditional healer	20	10	2.00
Other traders of household goods, plus other activities	46	31	1.48

Table 5: Profits from *Vitex doniana* products sale by different age categories and gender in Djidja district

Variable	Groups	Profit earned	Standard deviation	Coefficient of variation	Significance
	<20 years	22,787.5	38,399.7	1.7	Age (p=0.206)
Age	20 – 55 years	41,067.2	41,704.7	1.0	
	>55 years	29,576.9	34,720.8	1.2	
	Men	12,916.7	9,769.4	0.8	Gender (p=0.000)
Gender	Women	40,290.6	41,268.2	1.0	

Management of *Vitex doniana*

Vitex doniana is under both individual and collective management in the study area, with collective management being dominant. About 92.00% of informants maintained trees of the species within their farmlands. The species is harvested from different habitats. By decreased order of importance, *V. doniana* is harvested from farms and fallows (100% of informants), savannas (87.00%), and woodlands and gallery forests (83.00%). An important proportion of informants (55.60%) reported that the species was either very scarce or scarce. According to informants three main constraints were encountered during harvesting, namely: difficulties accessing some sites due to thickets, the personal risks of climbing trees for harvesting, and privatization of some trees (Figure 2). Trees occurring in savannas and woodlands were more difficult to climb while privatization was more prominent for trees occurring on farms and fallows. Felling of taller trees to harvest new shoots has become a common practice.

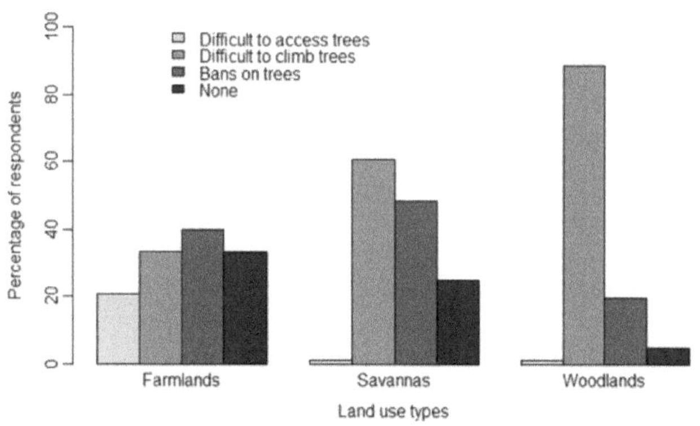

Figure 2: Perception of constraints to *V. doniana* exploitation in different land use types.

The people's knowledge of species' phenology was highly variable. For instance, knowledge of the fruiting period was dependent on gender and origin of respondents ($p<0.001$). Men (average score=5.19) had better knowledge of fruiting period than women (average score=3.51); and autochthones (average score=5.23) had better knowledge of the fruiting period than migrants (average score = 2). In addition, men had a significantly better knowledge of the period of appearance of new shoots than women ($p=0.02$). Land tenure also significantly affected this knowledge ($p<0.001$), with the best knowledge of shooting period registered among autochthon landlords. Moreover, schooling seemed to be negatively correlated with knowledge and we reason that attendance at school deprives children of the chance to be in contact with the species, hence elite people in the surveyed communities had less knowledge on the biology compared to those who have not gone to school ($p=0.01$). Results also indicated that there was a minimum growth required before a specific plant part can be harvested and the timing of this varies with the plant parts harvested (Figure 3). Except for leaves, harvesting of all other plant parts (e.g. bark, branches, stems, and roots) starts at the earliest at the shrub-like stage. Harvesting of leaves can occur on juvenile subjects, while fruits harvesting occurs only on mature trees.

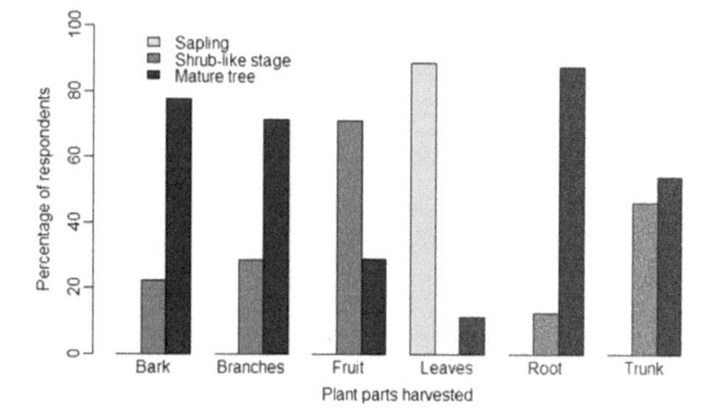

Figure 3: Perception of the minimum growth required before harvesting the different parts of *Vitex doniana.*

Local perception on status and conservation strategies

Most informants (55.00%) considered *V. doniana* populations as declining in the local environment and this perception was equally shared among men and women farmers ($p>0.05$). According to our informants, the decline in *V. doniana*'s populations is triggered by human pressure that is characterized by intense harvesting of new shoots, sometimes involving cutting of whole tree that are difficult to climb. The species is also used in charcoal for which mature trees are cut down. About 50.00% of informants reported that cutting of *V. doniana*tree to harvest leaves was the main threat to the species. Informants also reported low regeneration rates which they attributed to low production of fruits and low germination rates (7.92% of informants), natural death of stands, and other drivers such as bushfires (2.97%), agricultural expansion and urbanization (2.97%).

Respondents recognized propagation through seeds (81.47%) as the major route to regenerate the species. Only 6.17% mentioned the possibility of using stem cuttings. However, attempts to propagate the species were very rare and only 5.61% of informants (five men and five women in a sample of 178) have tried to regenerate it. Among respondents, only owners of land (acquired either through inheritance or gift) regenerated the species. Among those who have tried, 30.00% of them experimented regeneration with stem cuttings; the other 70.00% used seeds. The success rate is 33.33% (only one success over three cases reported) for stem cuttings and 57.14% (four successes over seven cases reported) for the regeneration by seeds. When disaggregated by gender, this success rate was 60.00% for women and 40.00% for men. In the cases reported no additional care was provided to the transplanted seedlings or the regenerated plants. Successful results were obtained for seedlings transplanted in the wet season; seedlings were rain-fed. Trees produced by these means were found on-farm and were intensively harvested by their owners. No consideration of selective sowing of seeds or planting of stem cuttings was reported.

Readiness of communities to engage in cultivation of *Vitex doniana*

Exploration of relationships between socio-economic variables and willingness to actively conserve the species as wild and cultivated resource, using Fisher's test

In order to better promote cultivation of *V. doniana* in the surveyed communities, we evaluated informants' willingness to engage in its active conservation (i.e. both planting and caring of the trees on-farm but also maintenance of wild trees *in-situ*). Overall, 76.50% of informants were ready to invest in agroforestry of the species, whereas 74.40% said they are willing to maintain woodlots of the species. The Fisher test on individual variables indicated that the willingness to plant the species seemed independent of gender, origin, land ownership, literacy level, schooling, and household consumption or sale of black plum products ($p>0.05$). Plant resources management was also independent of social group membership and participation in technical training ($p>0.05$). However, the willingness to maintain wild trees *in-situ* was significantly affected by land tenure ($p<0.05$). Landowners were the most willing to conserve stands of the species. However, this does not significantly affect the trade-off value (TOV). The TOV was only significantly affected by schooling ($p<0.05$). We observed that people reaching secondary and higher formal education reported TOVs more than three times higher (XOF 270,000; equivalent USD 600) to maintain a 400 m^2 stands of black plum, compared to those who stopped at primary level (XOF 78,000; equivalent USD 173). The average land size that respondents were ready to allocate for *V. doniana* agroforestry is 0.98 ha, representing an average land allocation ratio (LAR) of 8.33%. The LAR was significantly affected by gender ($p<0.05$). Women were ready to allocate an average of 23.66% of their land to black plum while men were only willing to allocate an average of 14.04% of their agricultural land to *V. doniana*. However, this did not imply that women allocated more land than men. In absolute terms men's land holding (18.9 ha in average) is far larger than that of women (5.6 ha in average). These ratios imply that men allocated 2.7 ha while women allocated 1.3 ha.

The *in-situ* conservation cost was not evaluated for 8.42% of the respondents, as they were not able to evaluate the costs related to this. For the other 91.58%, there was no significant difference in the Cultivation Opportunity Ratio (COR) with respect to gender, origin, access, and schooling. However, the COR had significant effect on the willingness to cultivate (Table 6) or to conserve the species *in-situ* (Table 7). Negative COR values indicated the informants were ready to be actively engaged ($p<0.05$). The average acquisition cost an informant was ready to pay for a *V. doniana* seedling was disaggregated by negative and positive COR values. These values indicated a great variation among the amounts proposed by users. The negative COR was associated with a high value ascribed to the seedling by the informant in a way that s/he was ready to offer higher price to acquire than s/he would sell his/her own seedlings. Interestingly, informants who were most willing to cultivate or to conserve wild plants of *V. doniana* (negative COR values) offered the highest prices to acquire seedlings. Therefore, the sign of the COR is a good indicator of the willingness to cultivate, but also the willingness to conserve the species *in-situ*. Among those with negative COR, all are willing to cultivate and also willing to conserve, excepted one informant. This was more prominent with other respondents who indicated they "*will never sell seedling of black plum*" due to its importance for them. Those informants were not included in the calculations, as they did not suggest any price. In the case of positive values of COR, there seemed to be additional variables that are taken into account in the decision making.

Table 6: Willingness to cultivate and mean seedling price and acquisition cost (in XOF) for positive and negative cultivation opportunity ratio (COR) value groups, N=101

	Negative COR		Positive COR		Significance
Mean±sd	−1.1±2.1		21.7±82.5		p<0.05
	cultivation		cultivation		
	Yes	No	Yes	No	
	n=23	n=0	n=61	n=17	
Selling price	115.2±424.6	-	10,300.0±29,655.6	42,297.1±122,605.6	p<0.05
Acquisition cost	958.7±1,677.8	-	753.8±1,928.0	370.6±1,197.6	p<0.05

Table 7: Willingness to conserve species *in-situ* and mean seedling price and acquisition cost (in XOF) for positive and negative cultivation opportunity ratio (COR) value groups, N=101

	Negative COR		Positive COR		Significance
Mean±sd	−1.1±2.1		21.7±82.5		p<0.05
	Conservation		Conservation		
	Yes	No	Yes	No	
	n=22	n=1	n=56	n=22	
Selling price	139.5±465.6	0.0±0	10,879.2±31,640.4	36,702.4±11,0351.1	p<0.05
Acquisition cost	1,001.1±1704.6	25.0±0	970.0±2140.2	83.6±116.3	p<0.05

Determinants of the willingness to cultivate Vitex doniana, using the conditional inference tree approach

The conditional inference tree classification using all socioeconomic and biological variables (Figure 4) indicated that the maximum land size that people are ready to allocate for the agroforestry of *V. doniana* was the main indicator of readiness to cultivate. Among people whose maximum allocated land was less than 0.0002 ha, gender was key in the decision making process, and men were more (or the most) willing to cultivate. For the women informants in this category, educational level determined the decision. Non-schooled women and school drop-outs were not willing to cultivate *V. doniana*, but those women who have attended secondary or higher education were more willing to cultivate the species.

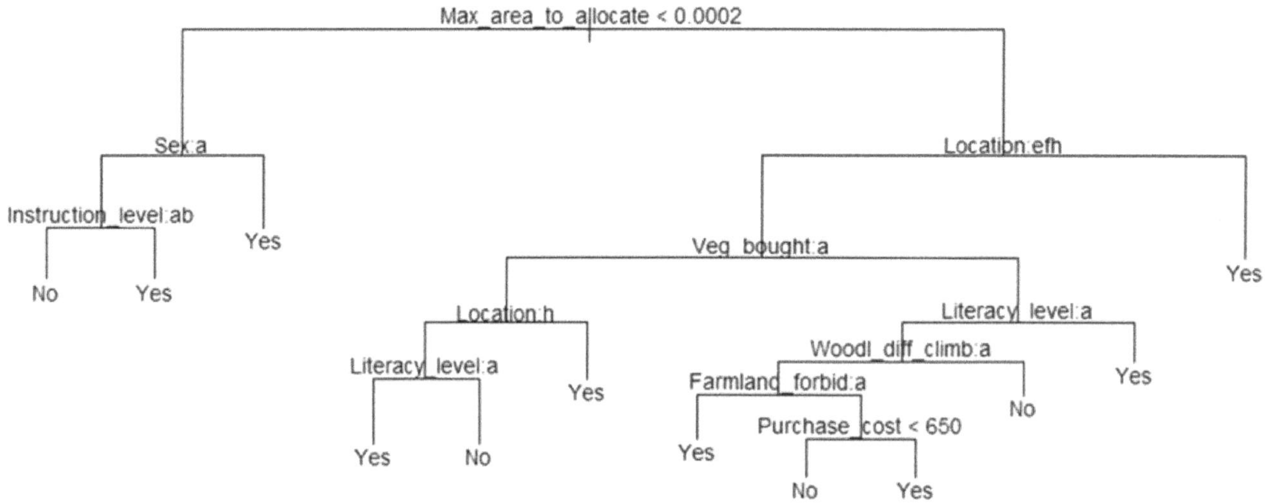

Figure 4: Conditional inference tree showing factors determining the decision of local people to cultivate *Vitex doniana*. The response Yes=the informant is willing to cultivate; and No=informant not likely to cultivate. Sex (a=Female, b=Male); Instruction level, with reference to western education (a=No schooling, b=Primary, c=Secondary and higher education); Literacy level (a=Cannot read nor write in mother tongue "illiterate", b=Can write and read in mother tongue "literate"); Location (a=Agbohoutogon, b=Dan, c=Djessi, d=Djidja centre, e=Kakatehou, f=Kome, g=Lakpo, h=Oungbega, i=Sekpongon); Veg_bought, i.e. Vegetable bought or purchase of *Vitex doniana* vegetable (a=No, b=Yes); Woodl_diff_climb, i.e. Difficult to climb *Vitex doniana* trees in the woodland (a=No, b=Yes); Famland_forbid, i.e. Harvesting is forbidden in farmland (a=No, b=Yes).

Among those farmers ready to allocate more than 0.0002 ha, people from Agbohoutogon, Dan, Djessi, Djidja centre, Lakpo, and Sekpongon villages were the most engaged to cultivate. In the other locations such as Kakatehou, Kome, and Oungbega, the type of access to resources (e.g. wild harvesting or purchase) was key in the decision making process. Among people who collect *V. doniana* in the wild, those from Kakatehou and Kome were ready to cultivate, without condition. Among this category, literacy level plays a role in the decision-making process in Oungbega and illiterate people (who cannot read and write their mother tongue) were most likely to cultivate the species; the literate were not. Among those who can only access the resource through purchasing, either for trading or for household consumption, in Kakatehou, Kome, and Oungbega; literacy level was also a determinant, and literate people were most willing to cultivate. Among the illiterate, perception of the hardship endured to climb trees to collect the leaves was crucial in the decision making process, and surprisingly those who find it difficult to climb trees in woodlands were not willing to cultivate the species. Among those who did not perceive it difficult to climb trees in woodlands, the existence or not of protection mechanisms of trees in private lands (farmlands) plays a role in the decision to cultivate. In fact, in farmlands where there is no ban on trees, people were more likely to cultivate. In communities where trees occurring on farmlands were private resources, the maximum amount the informant is ready to pay to acquire a seedling of the species was the best indicator of the willingness to cultivate with highest levels among those proposing over XOF 650 (USD 1.44) per seedling. This is well in line with the results presented in Table 6.

Determinants for the conservation of wild trees of Vitex doniana in-situ, using the conditional inference tree approach

The conditional inference tree (Figure 5) indicated that the maximum area people were ready to allocate was a main factor in respondents' willingness to conserve the species *in-situ*. For people allocating less than 0.18 ha of land to *V. doniana*, the total land available was key to the decision, and people with less than 1.1 ha of total available land were the most likely to conserve the species. Among those with more than 1.1 ha of land, availability of *V. doniana* trees (measured through abundance of mature trees for harvest) was determinant to the decision making process, and people who have easy access to trees were not likely to conserve wild populations. Among those with limited access to the resource (assessed as high pressure on immature trees), tenure has a bearing on the decision and only landowners were likely to conserve, land tenants were not.

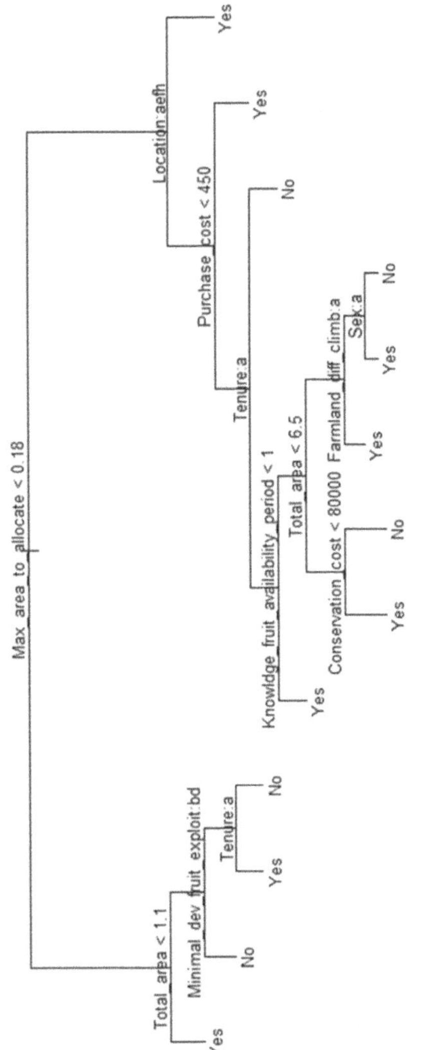

Figure 5: Conditional inference tree showing factors determining the decision to conserve wild trees of *Vitex doniana in-situ*.The response Yes=the informant is willing to conserve the species *in-situ*; and No=informant not likely to conserve the species *in-situ*. Minimal_dev_fruit_exploit, e.i. The minimal plant developmental stage required for exploitation of fruits (a=Juvenile stage, b=Fruits not used, c=Shrub-like stage, d=Mature tree stage); Tenure (a=Land owner, b=Tenant of land); Location (a=Agbohoutogon, b=Dan, c=Djessi, d=Djidja centre, e=Kakatehou, f=Kome, g=Lakpo, h=Oungbega, i=Sekpongon); Farmland_diff_climb, i.e. Difficult to climb *Vitex doniana* trees in the farmland (a=No, b=Yes); Sex (a=Female, b=Male).

For people who were ready to allocate more than 0.18 ha to the species, the decision to conserve depended mainly on location, with people from Dan, Djessi, Djidja centre, Lakpo, and Sekpongon villages more engaged into for its conservation. For other respondents from Agbohoutogon, Kakatehou, Kome, and Oungbega, the perceived cost of one seedling was key to the decision. People who were ready to pay more than XOF 450 (1 USD) were most engaged to conserve the species *in-situ*, supporting the results in Table 7. Among informants who proposed less than 1 USD, tenure system has to be taken into account, with tenants not likely to conserve. Among landowners of this category, main variables used in the decision making process included knowledge of plant phenology (fruiting period), the total land available, the trade-off value (TOV) or conservation cost, the perceived difficulties in climbing trees to harvest products and gender. In fact, among those farmers with less than 6.5 ha of total available land, people who requested a TOV below XOF 80,000 (USD 177) were likely to conserve, while those who requested more than USD 177 were not. Among the farmers with more than 6.5 ha of total land, people who perceived it less difficult to harvest the trees occurring within farmlands were more likely to conserve. Among those who find it difficult to harvest trees on their farm, only women were ready to conserve the species *in-situ*, men were not.

DISCUSSION

Valuation of tree resources in the surveyed area

Findings indicated that wild species represented the major part of the edible tree resources diversity in the twelve surveyed communities, meaning that local people still rely to a great extent on nature for their livelihoods. Out of the top 15 most valued woody plants, eleven have a high market value and species such as*A. occidentale, V. doniana*, and *P. biglobosa* combined high market and nutritional values. *Vitex doniana* stood out as the most important wild tree species followed by *P. biglobosa*, with both falling just behind *A. occidentale* which is an important cash crop in the surveyed communities. This result is consistent with findings by several authors who investigated the use of plant diversity in rural livelihoods in the West African sub-region [16,36-38]. *Parkia biglobosa* and *V. doniana* were already reported among the most valued NWFP resources in southern Benin [12], in Burkina Faso and Cameroon [39]. The two species are almost always used together in local cuisine where the local mustard "*afintin*" processed from seeds of *P. biglobosa* is used as seasoning for cooking leaves of *V. doniana* into a much appreciated popular vegetable dish in the region, known as "*fonman*". Therefore, the current results call for a close attention for safeguarding and valuing these species.

Cultivation appears to be the best management option to sustain both species and livelihoods when the value of a non-wood forest product (NWFP) and quantities harvested are high [1]. This opinion was also backed by Kupzow [5] who supported that plant cultivation is most stimulated by economic demand and highlighted that many cultivation experiments resulted in nothing because they failed to take economic aspects into consideration. Studies have revealed that black plum has an important market [13], that keeps expanding and this represents an incentive for its cultivation and conservation [40]. The species was also mentioned as present in at least 20% of traditional agroforests, from a countrywide survey in Benin [21]. In these conditions, cultivation of black plum is likely to be successful. In fact, *V. doniana* has the highest use value among the wild tree resources of importance to local livelihoods, especially for women and illiterate people, who tend to value it more than any other tree species. We also found that people with higher educational level seemed to value cash tree crops more highly than others, namely *E. guineensis, A. occidentale,* and *T. grandis* by decreasing order of importance. This could be an indication of the erosion of local value among the more "westernized" households. The educational elite classify resources almost solely based on market value of species, completely ignoring the social and cultural values associated with them. Another consideration is that they have less contact with wild resources in nature. Migrants also seemed to value *V. doniana* and *P. biglobosa*more highly than autochthons to whom the cash tree crop *A. occidentale* is the top most important. In fact, in the local land tenure systems migrants who are not owners of land have no right to plant perennial species. However, they have free access to wild resources in the nature, a fact which could explain this high value they attributed to the wild species. Overall, these variables are significantly important and should be carefully taken into account when selecting stakeholders in wild resource conservation programmes.

Uses, management and conservation strategies

Results clearly indicated that black plum is a multipurpose species, a characteristic of most NWFPs [41]. The inventoried uses confirmed the current state of knowledge on the species as reported by previous studies [12,16,40,42]. The high IAR values confirmed that knowledge of these uses is well distributed among user communities and the species is important in rural livelihoods. Exploitation of the species is profitable for all age groups, but more so for women than men. This is explained by the fact that men are most involved at the primary stage of exploitation, as primary collectors; while women are more involved in the wholesale and retail of processed leaves and fruits. By the law of profit distribution along the market chains of NWFPs, wholesalers and retailers (involving mainly women) obtain the bigger profit [43]. Species of this type are classified under the "*women goods*" categories of NWFPs [44].

Overall, juvenile trees are more vulnerable to disruption caused by destructive harvesting practices and wild fires. Therefore, much of any regeneration dies out at the end of the dry season. Overharvesting of leaves reduces fruiting ability of the overharvested trees, and this coupled with fruit harvesting obviously reduces the possibility of regeneration from seed. All these represent important threats to survival of wild populations. Nonetheless, the current level of roots and bark harvesting for medicinal purposes is low and does not add significantly to existing threats.

The analysis of access to resources of different landscapes indicated an incipient privatization of *V. doniana*trees occurring in farmlands and fallows. In fact, trees that are spared during land clearance for agriculture become the property of the farm owner. This tenurial system enhances conservation, as harvesting is better scheduled and competition with other harvesters is limited. On common lands (woodlands and savannas) trees are often cut down to harvest the leaves when they are too difficult to climb. The trunk and branches left out are then used to make charcoal or fuel wood. In farms, when the owner grants a usufruct right to a person to harvest trees on his or her fields, that person is not allowed to cut *V. doniana* trees. These kinds of measures protect the species.

From our data, local communities did not report any variants of *V. doniana* that can be subject to artificial selection in the sense of Blancas *et al.* [3]. However, referring to Blancas *et al.* [3]'s classification of the different management types, we distinguished the following categories which are considered as *in-situ*actions:

1. "*Gathering without recognition of variants*". Respondents did not report any selective harvesting of*V. doniana* resources. No variants were reported within the species.
2. The species is also "*tolerated without recognition of variants*", whereby stands were spared in farms but without any special tendering. This tolerance did not consider variants.
3. For the very few cases of regeneration reported by local communities (5.61%), the species is "*enhanced without recognition of variants*". People tried to regenerate the species in order to increase its abundance and subsequent availability, and this was not based on variants.
4. The species is "*protected without recognition of variants*". This protection mainly happens in farms where people spare *V. doniana* trees during land clearance and weeding, without any distinction of variants. No additional care is provided.

Other categories such as "*seed sowing*" and "*propagation of vegetative parts*" described by Blancas *et al.* [3] were observed at a very low proportion (3.93% used seeds and 1.68% used stem cuttings). These two categories were classified as *ex-situ* management by Blancas *et al.* [3]. However, based on the definitions of*in-situ* and *ex-situ conservation* by the CBD [34], we rather identified them as *in-situ* management, as these activities entirely happened in farms in the case of *V. doniana*.

Furthermore, future horticultural investigations should tackle the issue of keeping trees at a size and height that facilitate harvesting. This was further emphasized in the conditional inference model where it clearly appeared that some farmers put this as a condition before they engage into the cultivation of the species. Those farmers would prefer a shorter *V. doniana* tree which can be easily harvested.

Men have better knowledge of tree phenology (especially the fruiting period) than women, although women were the most involved in *V. doniana* exploitation. This is explained by the fact that men are the most involved in the primary collection of the resources and thus are more in contact with the species in its natural habitat. In addition, the autochthons have better knowledge of the biology of the species than migrants. Explanation to this might be that the autochthons have been in contact with the species for long time and knowledge is transmitted through generations. This indicates the importance of indigenous people's ecological knowledge on NWFPs. It was also found that local knowledge of people is eroded with schooling. This negative effect was also reported by Chukwuone [10] where schooling had negative effect on involvement of local people in NWFPs cultivation in Nigeria. At the current state, conservation initiatives are very poor in the study area, with very limited on-farm sparing of trees during land clearance for agriculture. This has been reported by N'Danikou *et al.* [12] and Dadjo *et al.* [40]. With regard to this and taking into account the pace with which the resource is harvested, wild populations of *V. doniana* are undoubtedly under threat. Based on the current level of local conservation initiatives, *V. doniana* falls under steps 1 and 2 of the domestication process [6,35]. Step 1 is characterized by wild plants maintained on farms during land preparation (e.g. clearance, burning and weeding) due to their proven utility and regular need, and their scarcity within household's compound limits. These spared plants are regularly observed to learn about their reproductive biology. In Step 2, farmers start giving more care to the protected plants (e.g. weeding, protection against animal damage) for their survival and their normal growth; a sort of ownership on the plants starts [35]. It is important to note that up to Step 2, no modification of plants has occurred due to selection; but plants just incidentally co-evolved to adapt to the disturbance imposed by human activities in the environment [6].

Motivations for cultivation and conservation of wild populations of *V. doniana* by user communities

Based on the current demand for *V. doniana* products and the intensity of the harvest, cultivation would be the best option to sustain the species and also the livelihoods of the different user groups. However, *V. doniana* is still at the earlier stage of the domestication process as indicated above and is still intensively collected from the wild. Thus, it is vital to understand what could motivate resource users to engage in cultivation of the species. From our study and based on the conditional inference tree statistical approach, four main variables have important effects on the decision-making process to cultivate *V. doniana* in the surveyed communities. This is partly consistent with previous studies in this domain [12,21]. First, an important proportion of users have expressed their willingness to cultivate the species for its importance in household food consumption and income. The second important parameter to be considered is schooling and literacy. Findings indicated that the use value of *V. doniana* to people declines with schooling. In this respect, awareness needs to be raised to sensitize elite classes. A third variable weighting in the cultivation of the species is the land tenure. Tenants were less likely to accept cultivation. In fact, the land tenure system in southern Benin and in any other parts of the country does not grant the tenants the right to plant trees [25]. Perennial species and most often trees are used to mark ownership of land. Thus, individuals with no property right on land may be willing but not ready to invest into cultivation of the species. At fourth and the most important, tenure size is key to the decision to cultivate and to maintain the species *in-situ*. People with larger parcels of land were eager to allocate bigger plots for the agroforestry of the species and were more likely to consider cultivation. Where land is available gender is an important factor in the likelihood for *V. doniana* cultivation. Although women's use value of the species was higher than men's, the men were more likely to plant the species. This could in fact be explained by the tenure system in which women have less access to land. There are also other variables such as the importance of wild-harvesting, the location and the cost of seedlings that were considered by people in their decision making. The threshold of XOF 650 (USD 1.44) found is well reflected in the positive relationship between the cultivation opportunity ratio (COR) and the likelihood to cultivate. All those who proposed a seedling acquisition cost below this threshold were not willing to cultivate.

Implications for management of overharvested wild resources

Several wild resource conservation programmes have failed around the world due to the lack of clear understanding of local drivers for success [10]. Gepts *et al.* [4] indicated that the success of conservation through use and restoration of biodiversity requires continued and increased research, and policies. Our findings pinpoint the main socioeconomic factors that trigger the success of conservation and management of threatened wild harvested plant resources with high economic value. Motivated local communities should be at the heart of conservation management, and we demonstrate that there is considerable willingness to engage in conservation of a locally threatened economically significant wild species. The availability of land was the most important element in our study with gender and lack of awareness of the importance of local foods among the elite class. The determined affordable cost of seedlings for farmers is a good entry point for nursery development. The negative values found for the cultivation opportunity ratio is also a good indication of the willingness of people to grow *V. doniana* in their farmlands and conserve wild populations. This means that less effort would be needed to get this category of participants involved.

The present study also encompasses the principle of complementarity between biological diversity conservation strategies. In fact, each of the elements treated in this paper (i.e. cultivation and *in-situ*conservation) have both positive and negative outcomes for biodiversity conservation, which need to be discussed. Plant cultivation can result in important setbacks if not well managed, despite its important advantages [45]. A plant cultivation programme that is not sustainably managed could lead to genetic drift and important biodiversity loss caused by high selection and breeding of superior lines. In the case of wild-harvested species such as *V. doniana* which have not been evaluated genetically, it is necessary to conserve wild populations in their natural habitats and in *ex-situ* genebanks (if possible) while they are undergoing domestication; in order to safeguard the gene pools in the species for future breeding purposes. Besides, there are examples of successful domestication of plant species whereby this undertaking has increased diversity. For instance in the case of maize which, despite the long history of selection and genetic improvement it has undergone, developed a huge diversity of cultivars. This was possible thanks to the ecological diversity of agricultural landscapes, which promoted the spread of cultivars and their adaptation to a diversity of agroecological niches [46], but also due to conservation of ancestors in the wild and *ex-situ* in genebanks. In the other hand, despite its positive effect in safeguarding diversity, *in-situ* conservation can also have adverse effects. In fact, populations that are conserved *in-situ* undergo natural selection. Although this favours coevolution and adaption to the changing environment, a fact which is sought in breeding for adaptive traits, a great diversity of genes is lost during the process. The weakest individuals which cannot survive changes that occur in the natural environment, though they may possess other useful traits for future breeding purposes, disappear and new genetic material of surviving individuals is established. Taking all the above into consideration, Article 9 of the Convention on Biological Diversity (CBD) advocates that *in-situ*conservation, as far as possible and as appropriate, be complemented by *ex-situ* measures [34].

Implications for research and development

From our findings, there is an avenue for domestication of *V. doniana* and conservation managers and horticulturalists should consider means to alleviate agronomic constraints such as keeping the plants at a shape and height to facilitate harvesting. Based on intrinsic characteristics and behavior of plants in the agricultural environment, domestication of *V. doniana* would just require adaptation to the agricultural environment [5]. Early results of horticultural trials indicated that the species responds very well in agroforestry systems (authors' unpublished data). In the debate on where and when domestication of wild species takes place it is often argued that the key determinant is the ecological but also the intrinsic characteristics of the wild plant or animal species which rendered them suitable for domestication which downplays the cultural, social and technological prowess of the hunter-gatherers themselves [4]. The authors, however, argue that in addition to biological and ecological traits, there are socio-cultural and economic factors influencing the genetic make-up of farmer bred varieties. In this regard, our findings make an important contribution to practical NWFP conservation. They are particularly relevant for the future of ethnobotanical studies aimed at providing advice to conservation management. The decision making process in the co-management or participatory natural resource management frameworks should use holistic approaches that integrate the social, cultural and economic dimensions of the local people in relation to wild plants use. Selection of participants in conservation and domestication programmes should be based on criteria that help to recruit members that are most likely to take actions to good ends. The approach used in the current research is simple and can be applied to any wild harvested resource similar to black plum. Nonetheless, future investigations of this type should dig further into the characteristics of locations that might make their inhabitants more active in wild resource cultivation. Chukwuone [10] identified the distance to reach harvesting sites and level of resource depletion. The effect of market access should also be further investigated. In our study access to market was not investigated as a variable, as all villages have relatively easy access to a market. Another important area of investigation is the analysis of the value chain of the species and the role of the stakeholders in a scenario of an organized commercial production of the species. Final, it is worthwhile noting the limitation of testing the relationship between presumed single explanatory variables and a response variable which is under the control of several factors. In our study the conditional inference tree model seemed most robust to unveil the rationale behind the decision making process about cultivation and *in-situ* conservation of species. This calls for caution when drawing general conclusions from inferential statistics on a phenomenon which is under the control of complex social, biological and environmental factors [47].

CONCLUSION

This study highlighted factors that trigger cultivation and *in-situ* conservation of wild harvested resources, with *V. doniana* as a case study. The species is thought to be declining at a high rate and all informants agreed that there is a need for action. Overall, we demonstrated that cultivation of the species is possible and eight main drivers were identified. These included land tenure, size of land holding, gender, cultivation opportunity ratio, schooling and literacy, resource availability measured through the importance of wild-harvesting, affordability of seedlings, and location. We also demonstrated that the Cultivation Opportunity Ratio is an important indicator for quick assessment of the likelihood of farmers to engage into cultivation and conservation of the species. The findings come in support to the agronomic investigations that are being developed for cultivation of the species. Only the combination of on-farm management and *in-situ*conservation would help take pressure off wild populations to safeguard genetic diversity in the species and enhance livelihoods of the various user groups.

ACKNOWLEDGEMENT

Authors are very grateful to New Alliance Trust for the financial support. We thank local communities of Djidja municipality for their kind participation in the study and for sharing their knowledge. We thank Jenny Wong for language editing. We also thank Christelle Komlan for her contribution to field data collection and Hounnankpon Yedomonhan of the National Herbarium of Benin, University of Abomey Calavi for the identification and maintenance of plant voucher specimens. The first author is also grateful to Hermane Avohou for his advices during manuscript preparation.

AUTHOR'S CONTRIBUTIONS

SN designed the study, analyzed data and finalized the manuscript. EGAD designed the study, gave conceptual advice, read and improved the manuscript. DAT supervised the field work, contributed to data analysis and manuscript preparation. COAA and CAH contributed to field data gathering, read and improved the manuscript. RSV and AA supervised the work and improved the manuscript. All authors read and approved the final manuscript.

REFERENCES

1. Clark L. Non-Timber Forest Products Economics and Conservation Potential. In Congo Basin information series: taking action to manage and conserve forest resources in the Congo Basin: results and lessons learned from the first phase (1996–2000). Volume Issue Brief #10. Edited by Byers B, Hakizumwami E, Hart T, Imboden C, Ritchie D, Swartzendruber F, Musters C: Central African Regional Program for the Environment, Biodiversity Support Program; 2001
2. Anderson KM, Wohlgemuth E. California Idian proto-agriculture: Its characterization and legacy. In: Gepts P, Bettinger R, Brush S, Damania A, Famula T, McGuire P, Qualset C, editors. Biodiversity in Agriculture: Domestication, Evolution and Sustainability. New York, USA: Cambridge University Press; 2012. p. 190–224.
3. Blancas J, Casas A, Pérez-Salicrup D, Caballero J, Vega E. Ecological and socio-cultural factors influencing plant management in Náhuatl communities of the Tehuacán Valley, Mexico. J Ethnobiol Ethnomed. 2013;9:39.
4. Gepts P, Famula TR, Bettinger RL, Brush SB, Damania AB, McGuire PE, et al. Biodiversity in Agriculture: Domestication, Evolution and Sustainability. New York, USA: Cambridge University Press; 2012.
5. Kupzow A. Theoretical basis of the plant domestication. Theor Appl Genet. 1980;57:65–74.
6. Clement CR. 1492 and the loss of Amazonian crop genetic resources. I. The relation between domestication and human population decline. Econ Bot. 1999;53:188–202.
7. Garrity D. Foreword. In: Akinnifesi FK, Leakey RRB, Ajayi OC, Sileshi G, Tchoundjeu Z, Matakala P, Kwesiga F, editors. Indigenous Fruit Trees in the Tropics: Domestication, Utilization and Commercialization. Oxfordshire: CABI; 2008. xvii–xviii.
8. Clutton-Brock J. The process of domestication. Mamm Rev. 1992;22:79–85.
9. Harris DR. Agriculture, cultivation and domestication: exploring the conceptual framework of early food production. In: Denham TP, Iriarte J, Vrydaghs L, editors. Rethinking Agriculture: Archaeological and Ethnoarchaeological Perspectives. Walnut Creek, California, USA: Left Coast Press, Inc; 2007. p. 16–35.
10. Chukwuone N. Socioeconomic determinants of cultivation of non-wood forest products in southern Nigeria. Biodivers Conserv. 2009;18:339–53.
11. Avohou HT, Vodouhe RS, Dansi A, Bellon M, Kpeki B. Ethnobotanical factors influencing the Use and management of wild edible plants in agricultural environments in benin. Ethnobotany Res Appl. 2012;10:571–92.
12. N'Danikou S, Achigan-Dako EG, Wong JL. Eliciting local values of wild edible plants in southern bénin to identify priority species for conservation. Econ Bot. 2011;65:381–95.
13. Achigan-Dako E, Pasquini M, Assogba Komlan F, N'Danikou S, Yédomonhan H, Dansi A, et al. Traditional Vegetables in Benin. Cotonou, Benin: Institut National des Recherches Agricoles du Bénin, Imprimeries du CENAP; 2010.
14. Burkill HM. Vitex. In: Burkill HM, editor. The Useful Plants of West Tropical Africa Families S–Z, Addenda, vol. 5. 2nd ed. Kew, Richmond, United Kingdom: Royal Botanic Gardens; 2000. p. 272–5.
15. Maundu P, Achigan-Dako E, Morimoto Y. Biodiversity of African vegetables. In: Shackleton C, Pasquini M, Drescher A, editors. African Indigenous Vegetables. London, UK: Earthscan; 2009. p. 65–104.
16. Achigan-Dako EG, N'Danikou S, Assogba-Komlan F, Ambrose-Oji B, Ahanchede A, Pasquini MW. Diversity, geographical, and consumption patterns of traditional vegetables in sociolinguistic communities in Benin: Implications for domestication and utilization. Econ Bot. 2011;65:129–45.
17. Kristensen M, Lykke AM. Informant-based valuation of use and conservation preferences of savanna trees in Burkina Faso. Econ Bot. 2003;57:203–17.
18. Ajenifujah-Solebo S, Aina J. Physico-chemical properties and sensory evaluation of jam made from black-plum fruit (*Vitex doniana*). Afr J Food Nutr Sci Dev. 2011;11:4772–84.
19. Okigbo RN. Fermentation of black plum (*Vitex doniana* Sweet) juice for production of wine. Fruits. 2003;58:363–9.
20. Ky K. *Vitex Doniana* Sweet. [Internet] Record from Protabase. In: Louppe D, Oteng-Amoako AA, Brink M, editors. Timbers 7(1). Wageningen, Netherlands: PROTA (Plant Resources of Tropical Africa); 2008. p. 578–81.
21. Assogbadjo AE, Glele Kakaï R, Vodouhê FG, Djagoun CAMS, Codjia JTC, Sinsin B. Biodiversity and socioeconomic factors supporting farmers' choice of wild edible trees in the agroforestry systems of Benin (West Africa). Forest Pol Econ. 2012;14:41–9.
22. Agossou OAC. Pressions anthropiques et stratégies locales de domestication de *Vitex doniana* Sweet dans la commune de Djidja au sud Bénin. Mémoire de Licence. Université d'Abomey-Calavi, Faculté des Sciences Agronomiques; 2011
23. N›Danikou S. Diversity, management and conservation of wild edible plants in the Fon community of Agbohoutogon, South Benin. MSc Thesis, University of Wales, Bangor, School of the Environment and Natural Resources; 2009
24. Oumorou M, Sinadouwirou T, Kiki M, Kakaï Glele R, Mensah GA, Sinsin B. Disturbance and population structure of Vitex doniana Sw. in northern Benin, West Africa. Int J Biol Chem Sci. 2010;4:624–32.
25. Biaou G. Régime foncier et système de culture sur le plateau Adja au Bénin. Bulletin de la Recherche Agronomique. 1994;9:11–8.
26. Fortmann L, Antinori C, Nabane N. Fruits of their labors: gender, property rights, and tree planting in two Zimbabwe villages. Rural Sociol. 1997;62:295–314.
27. DED-INSAE. Cahier des villages et quartiers de villes du Département du Zou. pp. 33. Cotonou, République

du Bénin: INSAE, Ministère Chargé du Plan, de la Prospective et du Développement; 2005:33.

28. Ethnobiology ISo. ISE Code of Ethics (with 2008 additions). http://ethnobiology.net/code-of-ethics/code-in-english/. 2006
29. Lawrence A, Phillips OL, Ismodes AR, Lopez M, Rose S, Wood D, et al. Local values for harvested forest plants in Madre de Dios, Peru: towards a more contextualised interpretation of quantitative ethnobotanical data. Biodiversity Conservat. 2005;14:45–79.
30. Phillips O, Gentry A, Reynel C, Wilkin P, Galvez-Durand B. Quantitative ethnobotany and Amazonian conservation. Conserv Biol. 1994;8:225–48.
31. Bakwaye FN, Termote C, Kibungu K, Van Damme P. Identification and local importance of medicinal plants used in the Mbanza-Ngungu region, Democratic Republic of Congo. Bois et Forêts des Tropiques 2013:63–77
32. Crawley MJ. Tree Models. In: The R Book. Secondth ed. Imperial College London at Silwood Park, UK: John Wiley & Sons, Ltd; 2013. p. 768–83.
33. Hothorn T, Hornik K, Zeileis A. Unbiased recursive partitioning: a conditional inference framework. J Comput Graph Stat. 2006;15:651–74.
34. Multilateral. Convention on Biological Diversity (with annexes). Concluded at Rio de Janeiro on 5 June 1992, registered ex officio on 29 December 1993. In *No 30619*. pp. 83: CBD; 1993:83.
35. Vodouhe R, Dansi A, Avohou HT, Kpèki B, Azihou F. Plant domestication and its contributions to in situ conservation of genetic resources in Benin. IJBC. 2011;3:40–56.
36. Lange D, Schippmann U. Trade Survey of Medicinal Plants in Germany. A Contribution to International Plant Species Conservation. Bonn, Germany: Bundesamt fur Naturschutz; 1997.
37. Uniyal RC, Uniyal MR, Jain P. Cultivation of Medicinal Plants in India. A Reference Book. New Delhi, India: TRAFFIC India and WWF India; 2000.
38. Ju Y, Zhuo J, Liu B, Long C. Eating from the wild: diversity of wild edible plants used by Tibetans in Shangri-la region, Yunnan. China J Ethnobiol Ethnomedicine. 2013;9:28.
39. Mapongmetsem P, Kapchie V, Tefempa B. Diversity of local fruit trees and their contribution in sustaining the rural livelihood in the Northern Cameroon. Ethiop J Environ Stud Manag. 2012;5:32–6.
40. Dadjo C, Assogbadjo AE, Fandohan B, Glèlè Kakaï R, Chakeredza S, Houehanou TD, et al. Uses and management of black plum (*Vitex doniana* Sweet) in Southern Benin. Fruits. 2012;67:239–48.
41. Quang DV, Anh TN. Commercial collection of NTFPs and households living near or in forests: case study in Que, Con Cuong and Ma, Tuong Duong, Nghe An, Vietnam. Ecol Econ. 2006;60:65–74.
42. Heubach K, Wittig R, Nuppenau E-A, Hahn K. Local values, social differentiation and conservation efforts: The impact of ethnic affiliation on the valuation of NTFP-species in Northern Benin, West Africa. Hum Ecol. 2013;41:513–33.
43. Neumann RP, Hirsch E. Commercialisation of Non-Timber Forest Products: Re and Analysis of Research. Bogor, Indonesia: SMT Grafika Desa Putera; 2000. CIFOR.
44. Shackleton S. Livelihood benefits from the local level commercialization of savanna resources: a case study of the new and expanding trade in marula (*Sclerocarya birrea*) beer in Bushbuckridge, South Africa. S Afr J Sci. 2004;100:651–7.
45. Jones TH, Steane DA, Jones RC, Pilbeam D, Vaillancourt RE, Potts BM. Effects of domestication on genetic diversity in*Eucalyptus globulus*. For Ecol Manage. 2006;234:78–84.
46. Hauser G, Little M, Roberts DF. Man, Culture and Biodiversity: Understanding interdependences. In: Biology International. Paris: International Union of Biological Sciences; 1994.
47. Guthery FS. Statistical ritual versus knowledge accrual in wildlife science. J Wildl Manag. 2008;72:1872–5.

CHAPTER 11

AXILLARY BUD PROLIFERATION APPROACH FOR PLANT BIODIVERSITY CONSERVATION AND RESTORATION

F. Ngezahayo[1] and B. Liu[2]

[1]Section Biologie-Chimie, Département des Sciences Naturelles, École Normale Supérieure, BP 6983, Bujumbura, Burundi

[2]Key Laboratory of Molecular Epigenetics of the Ministry of Education (MOE), Northeast Normal University, Changchun 130024, China

ABSTRACT

Due to mainly human population pressure and activities, global biodiversity is getting reduced and particularly plant biodiversity is becoming at high risk of extinction. Consequently, many efforts have been deployed to develop conservation methods. Because it does not involve cell dedifferentiation of differentiated cells but rather the development and growth of new shoots from preexisting meristems, the axillary bud proliferation approach is the method offering least risk of genetic instability. Indeed, meristems are more resistant to genetic changes than disorganized tissues. The present review explored through the scientific literature the axillary bud proliferation approach and the possible somaclonal variation that could arise from it. Almost genetic stability or low level of genetic variation is often reported. On the contrary, in a few cases studied to date, DNA methylation alterations often appeared in the progenies, showing epigenetic variations in the regenerated plants from axillary bud culture. Fortunately, epigenetic changes are often temporary and plants may revert to the normal phenotype. Thus, in the absence of genetic variations and the existence of reverting epigenetic changes over time, axillary bud culture can be adopted as an alternative nonconventional way of conserving and restoring of plant biodiversity.

INTRODUCTION

Global biodiversity is defined as the variation of all life on earth and the ecological complexes in which it occurs [1]. Biodiversity refers to genetic diversity, species diversity, and ecosystem diversity [2, 3] and includes the forest and agricultural ecosystems and the wild animals [4].

Among the above components, plants represent a vital part of biodiversity and healthy ecosystems. They provide multiple ecosystem services including production of oxygen for the rest of living organisms [5, 6], removal of atmospheric carbon dioxide emissions in the photosynthesis process, creation and stabilization of soil, protection of watersheds, and provision of natural resources including food, fibre, fuel, shelter, and medicine [7]. They also play an important role in the water cycle and constitute habitat for a wide range of other living organisms. Thus, plants are the basis for life on earth and humans are quite dependent on them [8–10] given that they are fundamental structural and nutrient-sequestering components of most ecosystems.

Due to dependency on biodiversity, the number of threatened plant species has gradually increased during the last decade, the maximum being observed in 2011 [11].

Citation: F. Ngezahayo and B. Liu, "Axillary Bud Proliferation Approach for Plant Biodiversity Conservation and Restoration," International Journal of Biodiversity, vol. 2014, Article ID 727025, 9 pages, 2014. doi:10.1155/2014/727025.

The key factor in threats to earth's biodiversity is often cited as the human population size, density, and growth [12–15]. As reported by the United Nations Population Division, the world had 2.5 billion people in 1950. This number was almost tripled in 2005 by reaching 6.5 billion people, while it is projected to rise to more than 9 billion people by 2050 [16]. The less developed world is showed to have the highest rate of human population growth [17]. Thus, the needs for this growing population must have much impact on biodiversity on which it depends for its survival.

The world's human population exponential growth and spatial expansion have been accompanied by changes in land-use, pollution, and overexploitation of natural resources [18], which in turn engender loss of species [19] and endangerment of ecosystem functioning [20–35] in three main ways.

First, human-driven land-use is among the greatest threats to terrestrial and aquatic biodiversity [36, 37] by causing habitat destruction in protected and nonprotected areas, which is motivated mostly by agricultural expansion [38–41].

Second, urbanization, another consequence of human population massive growth, is considered to be another major threat to biodiversity by a wide range of scientific literature [38, 42–53]. For instance, due to human population growth and migration, there will be nearly 2 billion new urban residents by 2030 [54], which means further habitat destruction accompanied by various sorts of pollution.

Third and corollary to the above impacts, human pressure is also responsible for different types of pollution, leading mainly to climate change, such as global warming [55]. The main consequences of climate change or global warming are (i) the extinction of plant pollinators [56, 57] which causes loss of genetic diversity and (ii) the habitat fragmentation resulting in loss of genetic diversity in local and global plant populations and bottleneck events in these populations [58].

These anthropogenic activities place global biodiversity and particularly plant species at risk of extinction even in the biodiversity hotspots [59, 60].

In order to reduce the growing extinction risk of plant biodiversity, its conservation and restoration was revealed to be more than a priority. Two conventional methods of conservation are used, in situ and ex situconservation, which are complementary to each other. In situ methods allow conservation to occur with ongoing natural evolutionary processes [61, 62]. Ex situ strategies maintain the biological materials outside their natural habitats [63] and often use plant biotechnology such as plant cryopreservation or micropropagation. The most important technique in micropropagation is meristem proliferation in which apical buds or nodal segments harbouring an axillary bud are cultured to regenerate multiple shoots without intervention of callus phase [64]. Meristem culture is also an efficient tool for regeneration, elimination of viruses from infected plants, and then production of virus-free seed material of different plant species [65–67].

In the present review, we briefly present the axillary bud proliferation approach as an alternative way that can be used for plant biodiversity conservation and restoration.

AXILLARY BUD CULTURE AND MASS PROPAGATION OF PLANT SPECIES

Among the methods developed for plant micropropagation, the axillary bud proliferation is the most used and is also considered the most suitable to guarantee genetic stability of the regenerated plants obtained. For rapidin vitro clonal propagation of plants, normally dormant axillary buds are induced to grow into multiple shoots by judicious use of growth regulators, cytokinins for activation and sprouting of dormant axillary bud and multiplication or cytokinin and auxin synergistic combinations for shoot multiplication.

In this way and because of its simplicity and reliability for clonal propagation, axillary bud culture was the method adopted for the mass micropropagation of various plants including almond [68], apple [69], Acacia mangium Willd [70], Cedrus [71], Eucalyptus [72], Anoectochilus formosanus Hayata [73], Swertia chirayita(Roxb. ex Fleming) H. Karst. [74], tea [75], Curcuma longa L. [76], Rosa rugosa Thunb. [77], hazelnut [78], marula tree [79], Fagopyrum dibotrys Hara mutant [80], Curcuma amada Roxb. [81], Alpinia galangal Linn [82], apple rootstock [83], apple rootstock Merton 793 [84], Cannabis sativa L. [85], hops [86], Allium ampeloprasum L. [87], squill [88], olive [89], Pinus thunbergii Parl. [90], Piper longum L. [91], Prosopis chilensis(Mol.) Stuntz [92], barley [93], rice [94], almond [95], Dendrobium longicornu Lindl. [96], and Mahonia leschenaultii Nutt. [97].

It is the stimulation of axillary buds to develop into a shoot [98] and this technique is comprised of the meristem and shoot tip culture and the bud culture [98]. It is a method exploiting the normal ontogenetic route for plant development by lateral meristems.

Since this technology does not involve cell dedifferentiation of differentiated cells but rather the development and growth of new shoots from preexisting meristems, it has been usually pointed out as the most faithful way of propagating plants in vitro. Indeed, it is thought to induce recovery of genetically stable and true-to-type progenies [99, 100], very little or no genetic variation [101], and epigenetic stability [102].

AXILLARY BUD CULTURE AND SOMACLONAL VARIATION

Axillary bud culture is one of the multiple techniques of plant in vitro culture. A major problem associated within vitro culture systems is the occurrence of somaclonal variation amongst subclones of one parental line [103,104]. Somaclonal variation is manifested as cytological abnormalities, frequent qualitative and quantitative phenotypic mutations, sequence change, gene activation and silencing [105, 106], and transposon and retrotransposon activation [107–110]. Thus, somaclonal variation is considered at both genetic and epigenetic levels.

Although somaclonal variation provides a valuable source of genetic variation for the improvement of crops through the selection of novel variants, which may show resistance to disease, improved quality, or higher yield [111–114], it may result in off-types that reduce the commercial value of resultant plants [115] and then shall be an important obstacle for plant biodiversity conservation. Plant in vitro culture, being comprised of sequential dedifferentiation (formation of callus) and redifferentiation (regeneration into plants) stages [116, 117], represents traumatic stress to plant cells and organs and often engenders an array of genetic and epigenetic alterations [118].

It is generally assumed that axillary bud culture is the method offering least risk of genetic instability since meristems are more resistant to genetic changes than disorganized tissues [119, 120]. Though it is sometimes species-specific [71, 75], genetic fidelity or no significant genetic variation is indeed often observed in the regenerants from axillary bud culture systems according to known scientific literature. In this concern, various DNA molecular markers were used such as cytophotometry, ISSR (intersimple sequence repeat), RAPD, RFLP (restriction fragment length polymorphism), and REMAP (retrotransposon microsatellite amplified polymorphism) markers and did not show genetic differences between regenerated progenies and their parental lines (Table 1).

Table 1: Clonally propagated plant species through axillary proliferation approach and the state of subsequent genetic somaclonal variation

Species	Used marker	Genetic variation	Reference
Cedrus atlantica L.	RAPD	Not significant	[71]
Cedrus libani L.	RAPD	Not significant	[71]
Eucalyptus tereticornis Smith	RFLP, RAPD	No	[72]
Eucalyptus camaldulensis Dehn.	RFLP, RAPD	No	[72]
Anoectochilus formosanus H.	ISSR	Low genetic instability	[73]
Swertia chirayita (Roxb. *ex* Fleming) H. Karst.	ISSR	No	[74]
Curcuma longa L.	Cytophotometry and RAPD	No	[76]
Prunus dulcis [Mill.] D. A. Webb	RAPD and ISSR	No	[95]
Alpinia galanga L.	RAPD and ISSR	No	[82]
Malus pumila Mill.	RAPD	Not significant	[83]
Apple rootstock Merton 1973	ISSR	No	[84]
Cannabis sativa L.	ISSR	No	[85]
Humulus lupulus L.	RAPD and REMAP	No	[86]
Allium ampeloprasum L.	ISSR	No	[87]
Vitis spp.	RAPD and ISSR	No	[136]
Hybrid hazelnut	RAPD	No	[78]
Sclerocarya birrea subsp. *caffra*	RAPD	No	[79]
Ochreinauclea missionis (Wall. ex G. Don) Ridsd.	ISSR	No	[108]
Fragaria x ananassa	RAPD-PCR	No	[137]
Doritaenopsis glenyle "Labios"	RAPD	No	[127]
Ocimum kilimandscharicum Guerke	RAPD	No	[138]

Contrary to a wide range of genetic investigations in regenerated plants from axillary bud culture, there is no much work in the epigenetic way. Nonetheless, the axillary bud culture system may cause epigenetic alteration according to few papers regarding the epigenetic aspect of somaclonal variation (Table 2). On the investigated cases, epigenetic stability was only observed in three plants including Bambusa balcooa Roxb. [121], Myrtus communis L. [122], and Solanum tuberosum L. [102]. The remaining show DNA methylation alterations inCedrus atlantica L. and C. libani L. [71], Corylus avellana L. [123], Malusxdomestica cv. "Gala" [124], Pisum sativum L. [125], Vitis vinifera L. [126], Doritaenopsis glenyle "Labios" [127], and Humulus lupulus L. [86].

Table 2: Clonally propagated plant species through axillary proliferation approach and the state of subsequent epigenetic somaclonal variation

Species	Used marker	Epigenetic variation	Reference
Bambusa balcooa Roxb.	MSAP	No	[121]
Cedrus atlantica L. and *C. libani* L.	HPLC	Yes	[71]
Corylus avellana L.	Isoschizomer restriction analysis	Yes	[123]
Malusxdomestica	MS-AFLP	Yes	[124]
Myrtus communis L.	HPLC	No	[122]
Pisum sativum L.	MSAP, HPCE	Yes	[125]
Solanum tuberosum L.	MS-AFLP	No	[102]
Vitis vinifera L.	MSAP	Yes	[126]
Doritaenopsis glenyle "Labios"	MSAP	Yes	[127]
Humulus lupulus L.	MSAP	Yes	[86]

Interestingly in the same studied plant species, genetic variation is quite absent while epigenetic alterations occur [86, 127], showing that only epigenetics will impact on the phenotype of the progenies from axillary bud culture in such plants and will then alone govern the consequent variant phenotype.

AXILLARY BUD CULTURE IN PLANT BIODIVERSITY CONSERVATION

Biotechnology such as axillary bud culture is supposed to help in conservation and restoration of plants without affecting main features of the plants. It is known that phenotypic variations in living organisms are strongly governed by both genetic and epigenetic fluctuations [128–130]. Thus, conservation and restoration methods that could induce genetic and/or epigenetic variation such as somaclonal variations shall produce mutations in plant progenies which may affect their stability and viability.

By using axillary bud culture for mass propagation of various plant species and particularly the endangered ones, it is evident through the present review that genetic variations are almost absent (Table 1). This confirms a clonal fidelity and then true-to-type progenies. The genetic stability will then remain through propagation in the field via seeds of the genetically stable progenies (Figure 1(a)) unless genetic variability is induced by sexual and natural selection processes.

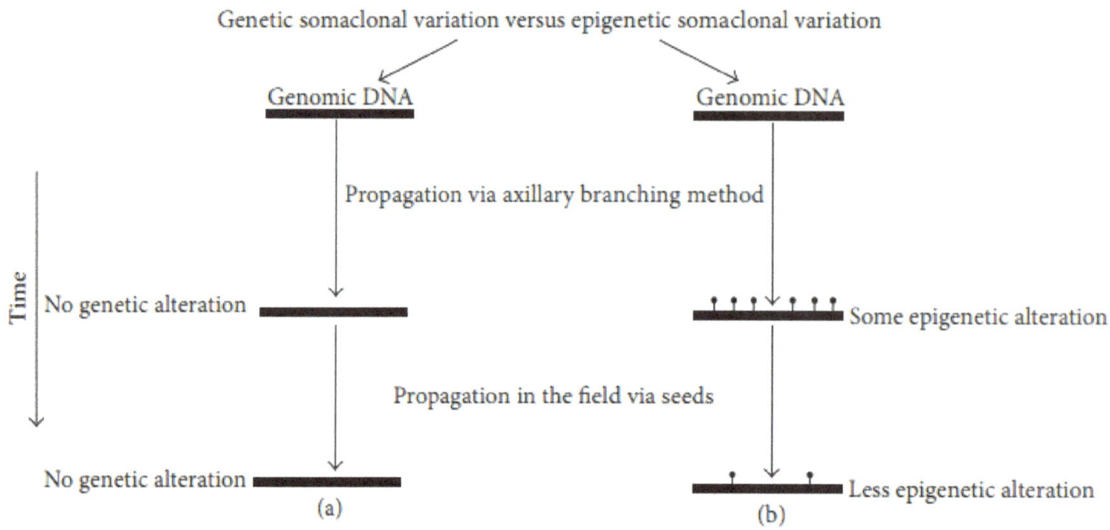

Figure 1: Schematic representation of the genetic (a) and epigenetic (b) variations arising from the axillary bud proliferation approach and their status over time, that is, after plant propagation in the field via the seeds (adapted from [135]).

The genetic stability explored throughout this review was depicted only by one DNA molecular marker in some cases (Table 1). Nonetheless, the use of one type of molecular marker to assess the stability of in vitropropagated plants may be insufficient. This is why different authors have recently exploited more than one molecular marker for studying of somaclonal variation in regenerants of several plant species. For instance, a relatively low level of polymorphism was detected with RAPD markers in Actinidia deliciosa A. Chev. cultures, whereas the polymorphism was higher with SSR markers [131].

On the other hand, epigenetic changes are often observed even in the absence of genetic variation (Table 2). This may then produce off-to-types progenies at the epigenetic level. Epigenetics such as cytosine methylation has been proposed to have diverse cellular functions in eukaryotes, but its primary role was believed to serve as a genome surveillance and defense system such as taming of transposable elements [132, 133]. This should then provoke their mobility throughout the genome and genetic mutations. Thus, assessment of epigenetic alterations arising from axillary bud culture process should be accompanied by the analysis of the possible consequent mobility of transposable elements. Fortunately, epigenetic changes are often temporary and plants may revert to the normal phenotype relatively easily though some can be long-lasting and may even be transferred during sexual propagation [134, 135]. By propagating the resulting progenies in the field via their seeds, there will be less epigenetic alteration in offspring (Figure 1(b)).

Thus, in the absence of genetic variations and the existence of reverting epigenetic changes over time, axillary bud culture can be adopted as an alternative way of conserving and restoring of endangered plant species. However, for a better planning of plant conservation and restoration by means of axillary bud proliferation approach, more than one genetic and epigenetic molecular marker should be used to rule out biased genetic and epigenetic results. This should be followed by the assessment of the status of transposable elements and the gene expression to assure the stability of the progenies.

CONCLUSION

The present review presented the benefits of the axillary bud culture as an alternative nonconventional way of conserving plant species which are facing extinction mainly due to human pressure. Evidence is increasing that axillary bud culture generates clonal fidelity and true-to-parental type progenies. It should however be synergistically used with other confirmed biotechnological methods of plant conservation such as cryopreservation in addition to conventional techniques.

Moreover, close attention should be paid to the genetic and epigenetic variation as well as karyological and morphological stability brought by axillary bud proliferation approach before any conservation plan.

ACKNOWLEDGMENT

The authors thank General Director of Ecole Normale Supérieure for facilities provided.

REFERENCES

1. P. Leadley, H. M. Pereira, R. Alkemade et al., Biodiversity Scenarios: Projections of 21st Century Change in Biodiversity and Associated Ecosystem Services, Convention on Biological Diversity, Montreal, Canada, 2010.
2. S. I. Dodson, T. F. H. Allen, S. R. Carpenter et al., Ecology, Oxford University Press, New York, NY, USA, 1998.
3. K. J. Gaston and J. I. Spicer, Biodiversity: An Introduction, Blackwell Science, Oxford, UK, 1998.
4. World Bank, "Ensuring the Future: The World Bank and Biodiversity, 1998–2004," Tech. Rep., World Bank, Environment Department, Washington, DC, USA.
5. M. A. Huston, Biological Diversity: The Coexistence of Species on Changing Landscapes, Cambridge University Press, Cambridge, Mass, USA, 1994.
6. R. B. Primack and R. T. Corlett, Tropical Rain Forests: An Ecological and Biogeographical Comparison, Blackwell Publishing, Oxford, UK, 2005.
7. E. O. Wilson, The Diversity of Life, Penguin, London, UK, 1992.
8. G. C. Daily, "Challenges in valuation," in Nature's Services: Societal Dependence on Natural ecosystems, G. C. Daily, Ed., pp. 365–374, Island Press, Washington, DC, USA, 1997.
9. A. Balmford, A. Bruner, P. Cooper et al., "Ecology: economic reasons for conserving wild nature,"Science, vol. 297, no. 5583, pp. 950–953, 2002.
10. A. Hamilton and P. Hamilton, Plant Conservation: An Ecosystem Approach, Earthscan, London, UK, 2006.
11. IUCN Red List version 2011.2, https://www.bolgermany.de/dateien/How_many_species_IUCN_Data.pdf.
12. G. L. Kirkland Jr. and R. S. Ostfeld, "Factors influencing variation among states in the number of federally listed mammals in the United States," Journal of Mammalogy, vol. 80, no. 3, pp. 711–719, 1999. ·
13. K. Thompson and A. Jones, "Human population density and prediction of local plant extinction in Britain," Conservation Biology, vol. 13, no. 1, pp. 185–189, 1999.
14. M. Cardillo, A. Purvis, W. Sechrest, J. L. Gittleman, J. Bielby, and G. M. Mace, "Human population density and extinction risk in the world's carnivores," PLoS Biology, vol. 2, no. 7, 2004. · ·
15. E. C. Ellis, E. C. Antill, and H. Kreft, "All is not loss: plant biodiversity in the anthropocene," PLoS ONE, vol. 7, no. 1, Article ID e30535, 2012.
16. United Nations Population Division, World Population Prospects, 2008.
17. United Nations Population Division, Briefing Packet, 1998 Revision of World Population Prospects, and World Population Prospects, the 2006 Revision.
18. A. K. Diraiappah and S. Naeem, Millennium Ecosystem Assessment: Ecosystems and Human Well-Being: Biodiversity Synthesis, World Resources Institute, Washington, DC, USA, 2005.
19. S. H. M. Butchart, M. Walpole, B. Collen et al., "Global biodiversity: indicators of recent declines,"Science, vol. 328, no. 5982, pp. 1164–1168, 2010.
20. F. S. Chapin III, E. S. Zavaleta, V. T. Eviner et al., "Consequences of changing biodiversity," Nature, vol. 405, no. 6783, pp. 234–242, 2000.
21. M. Loreau, S. Naeem, P. Inchausti et al., "Ecology: biodiversity and ecosystem functioning: current knowledge and future challenges," Science, vol. 294, no. 5543, pp. 804–808, 2001.
22. B. J. Cardinale, D. S. Srivastava, J. E. Duffy et al., "Effects of biodiversity on the functioning of trophic groups and ecosystems," Nature, vol. 443, no. 7114, pp. 989–992, 2006.
23. D. U. Hooper, F. S. Chapin III, J. J. Ewel et al., "Effects of biodiversity on ecosystem functioning: a consensus of current knowledge," Ecological Monographs, vol. 75, no. 1, pp. 3–35, 2005. ·
24. P. Balvanera, A. B. Pfisterer, N. Buchmann et al., "Quantifying the evidence for biodiversity effects on ecosystem functioning and services," Ecology Letters, vol. 9, no. 10, pp. 1146–1156, 2006.
25. D. S. Srivastava and M. Vellend, "Biodiversity-ecosystem function research: is it relevant to conservation?" Annual Review of Ecology, Evolution, and Systematics, vol. 36, pp. 267–294, 2005.
26. H. Hillebrand and B. Matthiessen, "Biodiversity in a complex world: consolidation and progress in functional biodiversity research," Ecology Letters, vol. 12, no. 12, pp. 1405–1419, 2009. · ·
27. J. Emmett Duffy, "Why biodiversity is important to the functioning of real-world ecosystems," Frontiers in Ecology and the Environment, vol. 7, no. 8, pp. 437–444, 2009.
28. J. T. Kerr and D. J. Currie, "Effects of human activity on global extinction risk," Conservation Biology, vol. 9, no. 6, pp. 1528–1538, 1995.
29. D. J. Forester and G. E. Machlis, "Modeling human factors that affect the loss of biodiversity,"Conservation Biology, vol. 10, no. 4, pp. 1253–1263, 1996. · ·
30. K. Thompson and A. Jones, "Human population density and prediction of local plant extinction in Britain," Conservation Biology, vol. 13, no. 1, pp. 185–189, 1999.
31. R. P. Cincotta, J. Wisnewski, and R. Engelman, "Human population in the biodiversity hotspots," Nature, vol. 404, no. 6781, pp. 990–992, 2000.

32. M. L. McKinney, "Role of human population size in raising bird and mammal threat among nations,"Animal Conservation, vol. 4, no. 1, pp. 45–57, 2001. ·

33. A. H. Harcourt and S. A. Parks, "Threatened primates experience high human densities: adding an index of threat to the IUCN Red List criteria," Biological Conservation, vol. 109, no. 1, pp. 137–149, 2002.

34. A. H. Harcourt, S. A. Parks, and R. Woodroffe, "Human density as an influence on species/area relationships: double jeopardy for small African reserves?" Biodiversity and Conservation, vol. 10, no. 6, pp. 1011–1026, 2001.

35. G. Ceballos and P. R. Ehrlich, "Mammal population losses and the extinction crisis," Science, vol. 296, no. 5569, pp. 904–907, 2002.

36. M. Parkes, "Personal commentaries on 'ecosystems and human well-being: Health synthesis—a report of the millennium ecosystem assessment'," EcoHealth, vol. 3, no. 3, pp. 136–140, 2006. · ·

37. Y. A. Izrael, S. M. Semenov, O. A. Anisimov et al., "The fourth assessment report of the intergovernmental panel on climate change: working group II contribution," Russian Meteorology and Hydrology, vol. 32, no. 9, pp. 551–556, 2007.

38. O. E. Sala, F. S. Chapin III, J. J. Armesto et al., "Global biodiversity scenarios for the year 2100," Science, vol. 287, no. 5459, pp. 1770–1774, 2000.

39. R. E. Green, S. J. Cornell, J. P. W. Scharlemann, and A. Balmford, "Farming and the fate of wild nature,"Science, vol. 307, no. 5709, pp. 550–555, 2005.

40. J. A. Foley, N. Ramankutty, K. A. Brauman et al., "Solutions for a cultivated planet," Nature, vol. 478, no. 7369, pp. 337–342, 2011.

41. http://www.fao.org/fishery/topic/3541/en.

42. B. Czech, P. R. Krausman, and P. K. Devers, "Economic associations among causes of species endangerment in the United States," BioScience, vol. 50, no. 7, pp. 593–601, 2000. ·

43. M. L. McKinney, "Urbanization, biodiversity, and conservation," BioScience, vol. 52, no. 10, pp. 883–890, 2002. ·

44. M. L. McKinney, "Measuring floristic homogenization by non-native plants in North America," Global Ecology and Biogeography, vol. 13, no. 1, pp. 47–53, 2004. · ·

45. P. M. Vitousek, H. A. Mooney, J. Lubchenco, and J. M. Melillo, "Human domination of Earth's ecosystems," Science, vol. 277, no. 5325, pp. 494–499, 1997. · ·

46. R. I. Mcdonald, P. Kareiva, and R. T. T. Forman, "The implications of current and future urbanization for global protected areas and biodiversity conservation," Biological Conservation, vol. 141, no. 6, pp. 1695–1703, 2008.

47. G. Buczkowski and D. S. Richmond, "The effect of urbanization on ant abundance and diversity: a temporal examination of factors affecting biodiversity," PLoS ONE, vol. 7, no. 8, 2012. ·

48. P. Bolund and S. Hunhammar, "Ecosystem services in urban areas," Ecological Economics, vol. 29, no. 2, pp. 293–301, 1999.

49. G. McGranahan, P. Marcotullio, X. Bai et al., "Urban systems," in Ecosystems and Human Well-Being: Current State and Trends, R. Hassan, R. Scholes, and N. Ash, Eds., Island Press, Washington, DC, USA, 2006.

50. R. Forman, Urban Regions: Ecology and Planning Beyond the City, Cambridge University Press, New York, NY, USA, 2008.

51. G. McGranahan and D. Satterthwaite, "Urban centers: an assessment of sustainability," Annual Review of Environment and Resources, vol. 28, pp. 243–274, 2003.

52. P. Clergeau, J.-P. L. Savard, G. Mennechez, and G. Falardeau, "Bird abundance and diversity along an urban-rural gradient: a comparative study between two cities on different continents," Condor, vol. 100, no. 3, pp. 413–425, 1998. ·

53. S. P. D. Riley, R. M. Sauvajot, T. K. Fuller et al., "Effects of urbanization and habitat fragmentation on bobcats and coyotes in southern California," Conservation Biology, vol. 17, no. 2, pp. 566–576, 2003.

54. UNPD (United Nations Population Division), World Urbanization Prospects: The 2005 Revision, New York, NY, USA, 2005.

55. Science for Environment Policy, DG Environment News Alert Servce, 24, 2010.

56. W. Thuiller, S. Lavorel, M. B. Araújo, M. T. Sykes, and I. C. Prentice, "Climate change threats to plant diversity in Europe," Proceedings of the National Academy of Sciences of the United States of America, vol. 102, no. 23, pp. 8245–8250, 2005.

57. J. Memmott, P. G. Craze, N. M. Waser, and M. V. Price, "Global warming and the disruption of plant-pollinator interactions," Ecology Letters, vol. 10, no. 8, pp. 710–717, 2007.

58. S. Dubey and R. Shine, "Restricted dispersal and genetic diversity in populations of an endangered montane lizard (Eulamprus leuraensis, Scincidae)," Molecular Ecology, vol. 19, no. 5, pp. 886–897, 2010.

59. T. J. Davies, G. F. Smith, D. U. Bellstedt et al., "Extinction risk and diversification are linked in a plant biodiversity hotspot," PLoS Biology, vol. 9, no. 5, Article ID e1000620, 2011.

60. K. J. Feeley and M. R. Silman, "Land-use and climate change effects on population size and extinction risk of Andean plants," Global Change Biology, vol. 16, no. 12, pp. 3215–3222, 2010. · ·

61. B. S. P. Wang, P. J. Charest, and B. Downie, "Ex situ storage of seeds, pollen, and in vitro cultures of perennial woody plant species," FAO Forestry Paper, vol. 113, p. 83, 1993.

62. L. Glowka, F. Burhene-Guilmann, H. Synge, J. A. McNeely, and L. Gündling, A Guide To the Convention on Biological Diversity (Environmental Policy and Law Paper No. 30), IUCN, Switzerland, 1994.

63. United Nations Conference on Environment and Development (UNCED), Convention on Biological Diversity, Geneva, Switzerland, 1992.

64. P. K. Pati, S. P. Rath, M. Sharma, A. Sood, and P. S. Ahuja, "In vitro propagation of rose—a review,"Biotechnology

Advances, vol. 24, no. 1, pp. 94–114, 2006.

65. W. Grout and W. Brian, "Meristem-tip culture for propagation and virus elimination," in Methods in Molecular Biology, R. D. Hall, Ed., pp. 115–125, Plant Cell Culture Protocol, Humana Press, Totowa, NJ, USA, 1999.

66. G. Faccioli, "Control of potato viruses using meristem and stem-cutting cultures, thermotherapy and chemotherapy," in Virus and Virus-Like Diseases of Potatoes and Production of Seed Potatoes, G. Loebenstein, H. P. Berger, A. A. Brunt, and G. R. Lawsan, Eds., pp. 365–390, Kluwer Academic Publisher, Dordrecht, The Netheralands, 2001.

67. G. R. Rout, A. Mohapatra, and S. M. Jain, "Tissue culture of ornamental pot plant: a critical review on present scenario and future prospects," Biotechnology Advances, vol. 24, no. 6, pp. 531–560, 2006.

68. C. M. Miguel, Adventitious regeneration and genetic transformation of almond (Prunus dulcis Mill.) [Ph.D. thesis], Faculty of Sciences, University of Lisbon, Lisbon, Portugal, 1998.

69. O. McMeans, R. M. Skirvin, A. Otterbacher, and G. Mitiku, "Assessment of tissue culture-derived 'Gala' and 'Royal Gala' apples (Malus x domestica Borkh.) for somaclonal variation," Euphytica, vol. 103, no. 2, pp. 251–257, 1998.

70. O. Monteuuis, F. C. Baurens, D. K. S. Gogh, M. Quimado, S. Doulbeau, and J. L. Verdeil, "DNA methylation in acacia mangium in vitro and ex-vitro buds, in relation to their within-shoot position, age and leaf morphology of the shoots," Silvae Genetica, vol. 58, no. 5-6, pp. 287–292, 2009.

71. B. Renau-Morata, S. G. Nebauer, I. Arrillaga, and J. Segura, "Assessments of somaclonal variation in micropropagated shoots of Cedrus: consequences of axillary bud breaking," Tree Genetics and Genomes, vol. 1, no. 1, pp. 3–10, 2005.

72. V. Rani and S. N. Raina, "Genetic analysis of enhanced-axillary-branching-derived Eucalyptus tereticornisSmith and E. camaldulensis Dehn. plants," Plant Cell Reports, vol. 17, no. 3, pp. 236–242, 1998.

73. F. Zhang, Y. Lv, H. Dong, and S. Guo, "Analysis of genetic stability through intersimple sequence repeats molecular markers in micropropagated plantlets of Anoectochilus formosanus Hayata, a medicinal plant,"Biological and Pharmaceutical Bulletin, vol. 33, no. 3, pp. 384–388, 2010.

74. P. Joshi and V. Dhawan, "Assessment of genetic fidelity of micropropagated Swertia chirayita plantlets by ISSR marker assay," Biologia Plantarum, vol. 51, no. 1, pp. 22–26, 2007.

75. R. M. Devarumath, S. Nandy, V. Rani, S. Marimuthu, N. Muraleedharan, and S. N. Raina, "RAPD, ISSR and RFLP fingerprints as useful markers to evaluate genetic integrity of micropropagated plants of three diploid and triploid elite tea clones representing Camellia sinensis (China type) and C. assamica ssp. assamica (Assam-India type)," Plant Cell Reports, vol. 21, no. 2, pp. 166–173, 2002.

76. M. K. Panda, S. Mohanty, E. Subudi, L. Acharya, and S. Nayak, "Assessment of genetic stability of micropropagated plants of Curcuma longa L. by cytophotometry and RAPD analyses," International Journal of Integrative Biology, vol. 1, no. 3, pp. 189–195, 2007.

77. W. Xing, M. Bao, H. Qin, and G. Ning, "Micropropagation of Rosa rugosa through axillary shoot proliferation," Acta Biologica Cracoviensia Series Botanica, vol. 52, no. 2, pp. 69–75, 2010.

78. M. N. Nas, N. Mutlu, and P. E. Read, "Random amplified polymorphic DNA (RAPD) analysis of long-term cultured hybrid hazelnut," HortScience, vol. 39, no. 5, pp. 1079–1082, 2004.

79. M. H. N. Mollel and E. M. A. Goyvaerts, "Micropropagation of marula, Sclerocarya birrea subsp. caffra(Anarcadiaceae) by axillary bud proliferation and random amplified polymorphic DNA (RAPD) analysis of plantlets," African Journal of Biotechnology, vol. 11, no. 93, pp. 16003–16012, 2012.

80. C. Chen, J. Lan, S. Xie, S. Cui, and A. Li, "In vitro propagation and quality evaluation of long-term micro-propagated and conventionally grown Fagopyrum dibotrys Hara mutant, an important medicinal plant," Journal of Medicinal Plants Research, vol. 6, no. 15, pp. 3003–3012, 2012.

81. S. Mohanty, R. K. Joshi, E. Subudi, S. Sahoo, and S. Nayak, "Genetic stability assessment of micropropagated mango Ginger (Curcuma amada Roxb.) through RAPD and ISSR markers," Research Journal of Medicinal Plant, vol. 6, no. 7, pp. 529–536, 2012.

82. R. Parida, S. Mohanty, and S. Nayak, "Evaluation of genetic fidelity of in vitro propagated greater galangal (Alpinia galangal L.) using DNA based markers," International Journal of Plant, Animal and Environmental Sciences, vol. 1, no. 3, pp. 123–133, 2011.

83. R. Gupta, M. Modgil, and S. K. Chakrabarti, "Assessment of genetic fidelity of micropropagated apple rootstock plants, EMLA 111, using RAPD markers," Indian Journal of Experimental Biology, vol. 47, no. 11, pp. 925–928, 2009.

84. H. Pathak and V. Dhawan, "ISSR assay for ascertaining genetic fidelity of micropropagated plants of apple rootstock Merton 793," In Vitro Cellular and Developmental Biology, vol. 48, no. 1, pp. 137–143, 2012.

85. H. Lata, S. Chandra, N. Techen, I. A. Khan, and M. A. Elsohly, "Assessment of the genetic stability of micropropagated plants of cannabis sativa by ISSR markers," Planta Medica, vol. 76, no. 1, pp. 97–100, 2010.

86. E. L. Peredo, R. Arroyo-García, and M. Á. Revilla, "Epigenetic changes detected in micropropagated hop plants," Journal of Plant Physiology, vol. 166, no. 10, pp. 1101–1111, 2009.

87. S. Gantait, N. Mandal, and P. K. Das, "Field evaluation of micropropagated vs. conventionally propagated elephant garlic," Journal of Agricultural Technology, vol. 7, no. 1,

pp. 97–103, 2010.

88. A. Kongbangkerd, A. Köpf, P. Allacher, C. Wawrosch, and B. Kopp, “Micropropagation of squill (Charybdis numidica) through nodule culture,” Plant Cell Reports, vol. 23, no. 10-11, pp. 673–677, 2005.

89. L. Annarita, “Morphological evaluation of olive plants propagated in vitro culture through axillary buds and somatic embryogenesis methods,” African Journal of Plant Science, vol. 3, no. 3, pp. 037–043, 2009.

90. S. Goto, R. C. Thakur, and K. Ishii, “Determination of genetic stability in long-term micropropagated shoots of Pinus thunbergii Parl. using RADP markers,” Plant Cell Reports, vol. 18, no. 3-4, pp. 193–197, 1998.

91. E. V. Soniya and M. R. Das, “In vitro micropropagation of Piper longum—an important medicinal plant,”Plant Cell, Tissue and Organ Culture, vol. 70, no. 3, pp. 325–327, 2002.

92. L. A. Caro, P. A. Polci, L. I. Lindström, C. V. Echenique, and L. F. Hernández, “Micropropagation ofProsopis chilensis (Mol.) Stuntz from young and mature plants,” Biocell, vol. 26, no. 1, pp. 25–33, 2002. ·

93. P. Bregitzer, S. Zhang, M.-J. Chob, and P. G. Lemaux, “Reduced somaclonal variation in barley is associated with culturing highly differentiated, meristematic tissues,” Crop Science, vol. 42, no. 4, pp. 1303–1308, 2002. ·

94. R. Medina, M. Faloci, M. A. Marassi, and L. A. Mroginski, “Genetic stability in rice micropropagation,”Biocell, vol. 28, no. 1, pp. 13–20, 2004. ·

95. M. Martins, D. Sarmento, and M. M. Oliveira, “Genetic stability of micropropagated almond plantlets, as assessed by RAPD and ISSR markers,” Plant Cell Reports, vol. 23, no. 7, pp. 492–496, 2004.

96. S. Dohling, S. Kumaria, and P. Tandon, “Multiple shoot induction from axillary bud cultures of the medicinal orchid, Dendrobium longicornu,” AoB PLANTS, vol. 2012, 2012. ·

97. R. K. Radha, A. M. Varghese, and S. Seeni, “Conservation through in vitro propagation and restoration ofMahonia leschenaultii, an endemic tree of the Western Ghats,” ScienceAsia, vol. 39, pp. 219–229, 2013.·

98. H. S. Chawla, Introduction To Plant Biotechnology, Science press, 2nd edition, 2002.

99. T. Murashige, “Plant propagation through tissue culture,” Annual Review of Plant Physiology, vol. 25, pp. 135–166, 1974. ·

100. C. Y. Hu and P. J. Wang, “Handbook of plant cell culture,” in Meristem Shoot Tip and Bud Culture, D. A. Evans, W. R. Sharp, P. V. Ammirato, and Y. Yamada, Eds., pp. 177–227, Macmillan, New York, NY, USA, 1983.

101. G. J. De Klerk, “How to measure somaclonal variation,” Acta Botanica Neerlandica, vol. 39, pp. 129–144, 1990.

102. S. K. Sharma, G. J. Bryan, M. O. Winfield, and S. Millam, “Stability of potato (Solanum tuberosum L.) plants regenerated via somatic embryos, axillary bud proliferated shoots, microtubers and true potato seeds: a comparative phenotypic, cytogenetic and molecular assessment,” Planta, vol. 226, no. 6, pp. 1449–1458, 2007.

103. P. J. Larkin and W. R. Scowcroft, “Somaclonal variation—a novel source of variability from cell cultures for plant improvement,” Theoretical and Applied Genetics, vol. 60, no. 4, pp. 197–214, 1981.

104. A. R. Gould, “Factors controlling generations of variability in vitro,” in Cell Culture and Somatic Cell Genetics in Plants. Plant Regeneration and Genetic Variability, I. K. Vasil, Ed., pp. 549–567, Academic, Orlando, Fla, USA, 1986.

105. S. M. Kaeppler, H. F. Kaeppler, and Y. Rhee, “Epigenetic aspects of somaclonal variation in plants,” Plant Molecular Biology, vol. 43, no. 2-3, pp. 179–188, 2000. ·

106. F. Ngezahayo, Y. Dong, and B. Liu, “Somaclonal variation at the nucleotide sequence level in rice (Oryza sativa L.) as revealed by RAPD and ISSR markers, and by pairwise sequence analysis,” Journal of Applied Genetics, vol. 48, no. 4, pp. 329–336, 2007. ·

107. H. Hirochika, K. Sugimoto, Y. Otsuki, H. Tsugawa, and M. Kanda, “Retrotransposons of rice involved in mutations induced by tissue culture,” Proceedings of the National Academy of Sciences of the United States of America, vol. 93, no. 15, pp. 7783–7788, 1996.

108. Z. L. Liu, F. P. Han, M. Tan et al., “Activation of a rice endogenous retrotransposon Tos17 in tissue culture is accompanied by cytosine demethylation and causes heritable alteration in methylation pattern of flanking genomic regions,” Theoretical and Applied Genetics, vol. 109, no. 1, pp. 200–209, 2004.

109. F. Ngezahayo, C. Xu, H. Wang, L. Jiang, J. Pang, and B. Liu, “Tissue culture-induced transpositional activity of mPing is correlated with cytosine methylation in rice,” BMC Plant Biology, vol. 9, article 91, 2009.

110. X. Wang, R. Wu, X. Lin et al., “Tissue culture-induced genetic and epigenetic alterations in rice pure-lines, F1 hybrids and polyploids,” BMC Plant Biology, vol. 13, p. 77, 2013. ·

111. Y. R. Mehta and D. C. Angra, “Somaclonal variation for disease resistance in wheat and production of dihaploids through wheat x maize hybrids,” Genetics and Molecular Biology, vol. 23, no. 3, pp. 617–622, 2000. ·

112. S. Predieri, “Mutation induction and tissue culture in improving fruits,” Plant Cell, Tissue and Organ Culture, vol. 64, no. 2-3, pp. 185–210, 2001.

113. A. Karp, “Somaclonal variation as a tool for crop improvement,” Euphytica, vol. 85, no. 1–3, pp. 295–302, 1995.

114. E. Unai, T. Iselen, and E. de Garcia, “Comparison of characteristics of bananas (Musa sp.) from the somaclone CIEN BTA-03 and its parental clone Williams,” Fruit, vol. 59, pp. 257–263, 2004. ·

115. T. J. Oh, M. A. Cullis, K. Kunert, I. Engelborghs, R. Swennen, and C. A. Cullis, “Genomic changes associated with somaclonal variation in banana (Musa spp.),” Physiologia Plantarum, vol. 129, no. 4, pp. 766–774, 2007.

116. G. Grafi and Y. Avivi, “Stem cells: a lesson from

dedifferentiation," Trends in Biotechnology, vol. 22, no. 8, pp. 388–389, 2004.

117. G. Grafi, A. Florentin, V. Ransbotyn, and Y. Morgenstern, "The stem cell state in plant development and in response to stress," Frontiers in Plant Science, vol. 2, no. 53, pp. 1–10, 2011.

118. B. McClintock, "The significance of responses of the genome to challenge," Science, vol. 226, no. 4676, pp. 792–801, 1984. ·

119. M. R. Ahuja, "Somaclonal genetics of forest trees," in Somaclonal Variation and Induced Mutations in Crop Improvement, S. M. Jain, D. S. Brar, and B. S. Ahloowalia, Eds., pp. 105–121, Kluwer Academic, Dordrecht, The Netherlands, 1998.

120. V. Rani and S. N. Raina, "Genetic fidelity of organized meristem-derived micropropagated plants: a critical reappraisal," In Vitro Cellular and Developmental Biology, vol. 36, no. 5, pp. 319–330, 2000. ·

121. K. Gillis, J. Gielis, H. Peeters, E. Dhooghe, and J. Oprins, "Somatic embryogenesis from mature Bambusa balcooa Roxburgh as basis for mass production of elite forestry bamboos," Plant Cell, Tissue and Organ Culture, vol. 91, no. 2, pp. 115–123, 2007.

122. R. Parra, M. T. Pastor, E. Pérez-Payá, and J. B. Amo-Marco, "Effect of in vitro shoot multiplication and somatic embryogenesis on 5-methylcytosine content in DNA of Myrtus communis L.," Plant Growth Regulation, vol. 33, no. 2, pp. 131–136, 2001.

123. C. Diaz-Sala, M. Rey, A. Boronat, R. Besford, and R. Rodriguez, "Variations in the DNA methylation and polypeptide patterns of adult hazel (Corylus avellana L.) associated with sequential in vitro subcultures,"Plant Cell Reports, vol. 15, no. 3-4, pp. 218–221, 1995. ·

124. X. Li, M. Xu, and S. S. Korban, "DNA methylation profiles differ between field- and in vitro-grown leaves of apple," Journal of Plant Physiology, vol. 159, no. 11, pp. 1229–1234, 2002. ·

125. P. Smýkal, L. Valledor, R. Rodríguez, and M. Griga, "Assessment of genetic and epigenetic stability in long-term in vitro shoot culture of pea (Pisum sativum L.)," Plant Cell Reports, vol. 26, no. 11, pp. 1985–1998, 2007.

126. M. Baránek, B. Křižan, E. Ondrušíková, and M. Pidra, "DNA-methylation changes in grapevine somaclones following in vitro culture and thermotherapy," Plant Cell, Tissue and Organ Culture, vol. 101, no. 1, pp. 11–22, 2010.

127. S. Y. Park, H. N. Murthy, D. Chakrabarthy, and K. Y. Paek, "Detection of epigenetic variation in tissue-culture-derived plants of Doritaenopsis by methylation-sensitive amplification polymorphism (MSAP) analysis," In Vitro Cellular and Developmental Biology, vol. 45, no. 1, pp. 104–108, 2009. View at Publisher

128. A. H. C. Wong, I. I. Gottesman, and A. Petronis, "Phenotypic differences in genetically identical organisms: the epigenetic perspective," Human Molecular Genetics, vol. 14, no. 1, pp. R11–R18, 2005.

129. E. A. Smith, S. B. Collette, T. A. Boynton et al., "Developmental contributions to phenotypic variation in functional leaf traits within quaking aspen clones," Tree Physiology, vol. 31, no. 1, pp. 68–77, 2011.

130. S. Hirsch, R. Baumerger, and U. Grossniklaus, "Epigenetic variation,inheritance, and selection in plant populations," Cold Spring Harbor Symposia on Quantitative Biology, vol. 77, pp. 97–104, 2012. ·

131. M. A. Palombi and C. Damiano, "Comparison between RAPD and SSR molecular markers in detecting genetic variation in kiwifruit (Actinidia deliciosa A. Chev)," Plant Cell Reports, vol. 20, no. 11, pp. 1061–1066, 2002.

132. J. A. Yoder, C. P. Walsh, and T. H. Bestor, "Cytosine methylation and the ecology of intragenomic parasites," Trends in Genetics, vol. 13, no. 8, pp. 335–340, 1997.

133. R. A. Martienssen and V. Colot, "DNA methylation and epigenetic inheritance in plants and filamentous fungi," Science, vol. 293, no. 5532, pp. 1070–1074, 2001. · ·

134. R. I. S. Brettell and E. S. Dennis, "Reactivation of a silent Ac following tissue culture is associated with heritable alterations in its methylation pattern," Molecular and General Genetics, vol. 229, no. 3, pp. 365–372, 1991. ·

135. M. J. M. Smulders and G. J. de Klerk, "Epigenetics in plant tissue culture," Plant Growth Regulation, vol. 63, no. 2, pp. 137–146, 2011. ·

136. M. Alizadeh and S. K. Singh, "Molecular assessment of clonal fidelity in micropropagated grape (Vitisspp.) rootstock genotypes using RAPD and ISSR markers," Iranian Journal of Biotechnology, vol. 7, no. 1, pp. 37–44, 2009. ·

137. A. E. Mohamed, "Somaclonal variation in micropropagated strawberry detected at the molecular level,"International Journal of Agricultural and Biological Engineering, vol. 9, no. 5, pp. 721–725, 2007.

138. S. Saha, T. Dey, and P. Ghosh, "Micropropagation of Ocimum kilimandscharicum guerke (Labiatae)," Acta Biologica Cracoviensia Series Botanica, vol. 52, no. 2, pp. 50–58, 2010.

CHAPTER 12

CONSERVATION AND USE OF GENETIC RESOURCES OF UNDERUTILIZED CROPS IN THE AMERICAS—A CONTINENTAL ANALYSIS

Gea Galluzzi [1], and Isabel López Noriega [2]

[1]Bioversity International, Regional Office for the Americas, c/o CIAT, km 17 Recta Cali-Palmira, Cali, Colombia

[2]Bioversity International, Via dei Tre Denari, 472/a Maccarese, Rome 00057, Italy

ABSTRACT

Latin America is home to dramatically diverse agroecological regions which harbor a high concentration of underutilized plant species, whose genetic resources hold the potential to address challenges such as sustainable agricultural development, food security and sovereignty, and climate change. This paper examines the status of an expert-informed list of underutilized crops in Latin America and analyses how the most common features of underuse apply to these. The analysis pays special attention to if and how existing international policy and legal frameworks on biodiversity and plant genetic resources effectively support or not the conservation and sustainable use of underutilized crops. Results show that not all minor crops are affected by the same degree of neglect, and that the aspects under which any crop is underutilized vary greatly, calling for specific analyses and interventions. We also show that current international policy and legal instruments have so far provided limited stimulus and funding for the conservation and sustainable use of the genetic resources of these crops. Finally, the paper proposes an analytical framework for identifying and evaluating a crop's underutilization, in order to define the most appropriate type and levels of intervention (international, national, local) for improving its status.

INTRODUCTION

In an increasingly globalized and interdependent world, eradicating hunger and poverty is not only an ethical imperative but an essential prerequisite for peace and world security. The challenge of feeding the expected 9 billion world population by 2050 in a sustainable manner can be met, among other measures, by rescuing and using more diversity in agricultural and food production systems, both in terms of crop as in terms of varieties within any given crop [1,2]. The human population today derives most of its calories from a very narrow set of crops, with only about 30 species providing 95% of the global food energy [3,4]. On the contrary, over 7,000 species are known as edible and are either partly or fully domesticated, suggesting that a large share of potential food sources is underutilized [5,6].

Latin America is home to a number of dramatically diverse agroecological regions which harbor a high concentration of crop diversity. This is particularly the case for the broadly defined diversity hotspots of Mesoamerica, the Andean region and the Amazon Basin, identified by Vavilov as key centers of plant domestication [7,8] and where genetic resources in traditional farming systems initially developed by indigenous

Citation: Gea Galluzzi and Isabel López Noriega, Conservation and Use of Genetic Resources of Underutilized Crops in the Americas—A Continental Analysis, doi:10.3390/su6020980.

people have undergone centuries of cultural and biological evolution. Some of the crops from those centers of origin and diversification have acquired global relevance (e.g., maize, potatoes, cassava), whereas others have retained a more local distribution. Early marginalization of some native crops may be traced back to the time of European colonization of Latin America; local species which were part of Pre-Columbian food systems were gradually substituted by crops from the Old World or local crops which attracted newcomers' attention [9]. More recent agricultural developments, particularly since the Green Revolution, deepened this marginalization by focusing on a package of staples (mostly wheat, maize and rice) on which a large percentage of the world population was already dependent for food security and to which priority investments were dedicated [10]. For these and other reasons, a number of Latin American edible plant species are nowadays considered minor, underutilized or neglected and, have joined a category of useful plant species often referred to as "NUS" (neglected and underutilized species). We will use the acronym as well as the adjectives listed above interchangeably throughout the study to refer to these crops.

Although a standard definition of NUS does not exist, a number of studies have described the typical features of NUS and the overriding issues affecting the conservation and use of their genetic resources [11,12,13,14,15]. In this paper, we aim to contribute to the discussion by critically analyzing the status of an expert-informed list of underutilized food crops in Latin America in light of the most commonly described features of underutilized crops. We also discuss the extent to which existing international policy and legal frameworks dealing with wild and cultivated biological diversity effectively support or not the conservation and sustainable use of underutilized crops' genetic resources.

METHODS

In order to obtain an expert-informed list of crops which could be considered underutilized in Latin America, we prepared and circulated a survey to 120 experts involved in the conservation and use of plant genetic resources for food and agriculture (PGRFA) in 15 Latin American countries. In the survey, we asked them to list crops they considered underutilized in their country and to define their potential relevance, the limiting factors affecting their conservation in *ex situ* facilities, in the wild and on farm, and the effect of international biodiversity conventions on their conservation and sustainable use (text of the survey is available in Annex 1). The list of crops emerging from the survey was used to test the features commonly attributed to an underutilized crop and to analyze the influence of selected international policy frameworks on the conservation and use of their genetic resources. The selection of the common features was guided by Padulosi *et al.* 2004 [15]: the authors identified a series of characteristics which are attributed to NUS and their work is often cited as a reference in this subject area. Table 1 shows the criteria and the relative sources of information, in addition to the surveyees anweres, we used to evaluate the status of the crops in relation to each of the commonly attributed features considered.

Table 1: Criteria and relative sources of information for the evaluation of crops' status

Feature Commonly Attributed to Underutilized Crops	Criteria and Sources of Information for Testing the Feature against the Listed Crops
Limited research efforts devoted to the species	Number of hits in Google Scholar [16], using the common name and the scientific name of each of the species [1]. Google Scholar searches publications in scholarly literature of all disciplines.
Limited representation of the species in globally available *ex situ* collections	Number of accessions of the species in Genesys (the database of the collections held by the consortium of international agricultural research centers of the CGIAR, the Department of Agriculture of the United States of America, USDA and public research institutes of the European countries) [17], in the catalogue of CATIE (Centro Agronómico Tropical de Investigación y Enseñanza) [18] and the database of the genebank at the World Vegetable Center (AVDRC) [19].
Limited representation of the species in national *ex situ* collections	Number of accessions maintained in selected national genebanks (Bolivia, Brazil, Ecuador, Mexico, Peru) [2] according to the information provided by the World Information and Early Warning System (WIEWS) [20] of the Food and Agriculture Organization of the United Nations (FAO) and the countries' reports [21] for the preparation of the *Second State of the World's Plant Genetic Resources for Food and Agriculture* [22].
Limited efforts in germplasm characterization	Availability of internationally developed descriptors, according to Bioversity International's list [23]
Lack of plant breeding efforts and commercial varieties of the crop species	Number of long-term research and improvement programmes in selected countries (Bolivia, Brazil, Ecuador, Mexico, Peru) from the countries' reports for the preparation of the *Second State of the World's Plant Genetic Resources for Food and Agriculture*. Number of varieties in the Plant Variety Database (PLUTO) [26] of the Union for Protection of Plant Varieties (UPOV), which includes varieties registered in the national listings of varieties admitted for commercialization, varieties subject to plant breeders' rights and varieties subject to plant *sui generis* patents.
Limited knowledge of the species' distribution and production levels	Number of hits in the Global Biodiversity Information Facility (GBIF) [24] which contains species' occurrence data associated to samples deposited in a number of national and international institutions working with plants (genebanks, herbaria, botanical gardens, *etc.*). [3] Presence/absence of each crop in FAOSTAT [25], the database maintained by the Statistics Division of FAO, which provides data relating to food and agriculture production for 200 countries.

[1] We used English, Spanish and Portuguese common names for each of the species. The search was limited to the title of the publications included in Google Scholar from 1970 to 2013. Patents and citations were excluded.

[2] These countries were selected because of the predominance of answers we received from experts working there, and as an attempt to cover albeit partially some of the main hotspots of the continent's diversity.

[3] The search was restricted to Latin America and the Caribbean. Duplicates were removed and only georeferenced observations were retained, given that these are the only ones capable of providing direct, meaningful information on a species' distribution.

Table 2 shows the sources of information that we used to assess the impact of three selected international policy instruments on the conservation and use of the genetic resources of minor crops.

Table 2: Sources of information for the assessment of the influence of international policy instruments on the conservation and use of the genetic resources of minor crops

International Instrument	Source of the Information and Methodology for Analysis
Convention on Biological Diversity	In the project data base of the Global Environmental Facility (GEF) [27] we performed a search of national and global projects dealing with agricultural biodiversity between 1991 and 2012, measured against the total funds disbursed in the biodiversity focal area.
International Treaty on Plant Genetic Resources for Food and Agriculture	We used the information on the Benefit Sharing Fund's webpage [28] to identify projects that deal with minor crops listed by the experts consulted in the survey. The share of efforts devoted to underutilized crops by the Global Crop Diversity Trust (GCDT) was measured by checking the list of the technical support projects to *ex situ* collections which have been carried out [29].
Global Plan of Action for the Conservation and Sustainable Utilization of Plant Genetic Resources for Food and Agriculture (GPA)	We filtered information about NUS projects from the GPA project database [30] and the national reports submitted to FAO for the preparation of the *Second Report on the State of the World's Plant Genetic Resources for Food and Agriculture*. This analysis focused on Bolivia, Brazil, Ecuador, Mexico and Peru [1]. The information obtained from these sources was complemented with data available on the websites of the largest public research organizations of these countries.

RESULTS

An Expert-Informed List of Underutilized Crops in Latin America

We received 40 complete answers from researchers in 12 countries, resulting in a response rate of 30 per cent. Table 3shows the respondents' field of expertise.

Although the response rate was relatively low, the number of crops listed as underutilized totalled 84 between genera and species (listed in Annex 2, Table A1). Species were more frequently reported, as requested in the survey, but in a few cases entire genera considered predominantly underutilized were also referred to (such as *Cucurbita*, *Carica*, *Xanthosoma* and*Passiflora*). We removed the three medicinal species which were reported, since the focus of the present article is on food crops. It is worth noting that 70% of the crops in our list are included in the list of Crops for the Future [31], an international organization with the mandate to document and promote research and awareness on underutilized species globally.

Table 3: Respondents' field of expertise. The total exceeds 100% because some respondents are involved in more than one field

Respondents Field of Expertise	Percentages of Answers
Ex situ conservation	81.82
In situ conservation	27.27
Crop improvement	50
Policies	18.18
Other	34.09

Acknowledging this list in no way represents a neither exhaustive nor definitive picture, we considered it as an informative guide to orient our analyses and discussions. As shown in Figure 1, fruits were by far the category hosting the largest share of underutilized crops with 38 highlighted species, followed by roots and tubers with 15. All of the 38 fruit species are native to the continent, as well as all the 15 roots and tubers. Most of the species from other categories are also natives (88 per cent of the total), originating either in the Andean (e.g., grains and tubers), Mesoamerican (e.g., beans, sapotes) or Amazonian centers of diversity (cassava, peach palm, copuazú, among others).

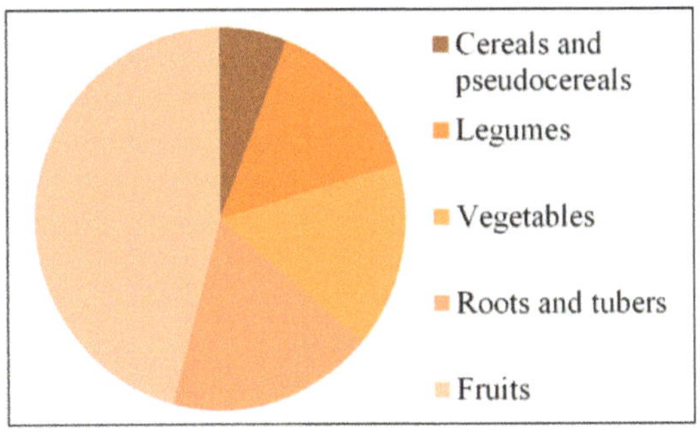

Figure 1: Crop categories represented in the list of underutilized species resulting from the survey among Latin American experts.

When asked to describe the potential use and relevance of each listed species, respondents almost invariably (in 95% of the cases) selected more than one option, suggesting that underutilized crops have a multi-purpose potential. Nevertheless, the most frequent answer was dietary diversification, equaled by diversification of agricultural systems and followed by food security.

Testing NUS Features against the List of Latin American Underutilized Crops

This section presents the results of our testing of the common features attributed to neglected and underutilized crops against the list of species obtained from the survey, combined with the respondents' perceptions. Some of the tested features have to do with research and conservation (Section 3.2.1 and Section 3.2.2) whereas others are more related to use (Section 3.2.3 and Section 3.2.4).

Limited Research Efforts Dedicated to the Species

Our survey revealed that the lack of research efforts only came third in the ranking of the main limiting factors affecting conservation and use of minor crops' genetic resources. When complementing this perception with an analysis of the degree of widely available scientific literature using Google Scholar searches performed in November 2013, we found that the average number of hits for our species was 1,954. Out of the 84 crops constituting our list, 67 species (*i.e.*, 79 per cent) have a lower than average number of hits, whereas the remaining 17 are above average. For an idea of scale, the score in literature hits for major cereals (rice, wheat and maize) with the same search criteria was just below 99,000, while for major vegetables (average between cabbage and tomato) it was 42,000. For chili pepper, common bean and potato (considered the major spice, legume and root/tuber of reference) we retrieved respectively 63,300, 31,600 and 67,400 hits. Average number of hits for three major fruit crops (pineapple, papaya and apple) was 25,540. All these values, as expected, largely exceed the average of hits retrieved for the crops treated here (Table 4).

Table 4: Comparison between literature hits in Google Scholar between the crops in the Latin American list of underutilized species and selected reference major crops

NUS Categories	Average N. Hits		Average N. Hits	Major Crop of Reference
Cereals, pseudocereals	3,494	↔	98,967	Maize, wheat, rice
Fruits	1,056	↔	25,540	Apple, pineapple, pawpaw
Legumes	2,013	↔	31,600	Common bean
Roots and tubers	2,697	↔	67,400	Potato
Vegetables	1,768	↔	31,345	Cabbage, tomato

Species which feature in Google Scholar with a number of hits below the dataset's average (*i.e.*, less than 1,954 hits) are mostly crops natives to the continent (with only four exceptions, being exotic vegetables like verdolaga, winter melon, sponge gourd and lablab bean). Species at the top end of the range, with over 5,000 literature hits, are mostly crops of broad worldwide cultivation and commercial distribution (asparagus, sweet potato, cowpea, chilies and cassava).

Limited Representation in ex situ Collections

International Collections with Facilitated Access

It is often argued that a common feature of NUS is that they are conserved mostly on-farm and scarcely represented in *ex situ* collections [15], particularly in those collections which make their samples and the related documentation freely available for research and development purposes. Most PGRFA-based agricultural research in developing countries has been possible thanks to the germplasm conservation, documentation, characterisation and distribution services of the network of genebanks of the CGIAR [32]. Other institutions with vast, well documented and freely available collections which have been and still are instrumental in providing the genetic basis for breeding and improvement programs of a number of crops are the European genebanks, and those of the USDA-ARS (Agricultural Research Service of the US Department of Agriculture), CATIE (Centro Agronómico Tropical de Investigación y Enseñanza) and AVRDC (The World Vegetable Center).

Of all the crops in the list, 17 (*i.e.*, 20%) have no accession in these genebanks and another 26 have less than 10 accessions, confirming that many of the species reported to be neglected are indeed poorly represented in international and regional collections. On the other hand, 22 of the crops listed as underutilized in the survey have between 100 and 5,000 accessions, countering the above mentioned negative picture. Crops which are absent or poorly represented are mostly fruit species (together with a root and two vegetables, all native); this is partly explained by the fact that most of the checked genebanks' mandate focuses on annual crops and partly by the difficultly and high cost of storing and maintaining perennials in field collections or *in vitro*. Even in the case of CATIE, with its extensive field collection in Costa Rica, fruit trees still occupy the lower end of the range in terms of numbers of accessions, with the notable exception of peach palm (*Bactris gasipaes* with over 600 samples). At the opposite end of the range of representation in the surveyed genebanks are crops such as sweet potato, cassava, cowpea, pigeon pea and squashes, which are the only ones in our list with collections exceeding 5,000 accessions.

National **ex situ** *Collections*

In our survey, regional experts from Latin America did not rank *ex situ* conservation high among the critical areas of neglect for the species they listed. Our analyses show that, for many underutilized species in selected countries, there is a good level of representation in national genebanks (Annex 3). Collections of national agricultural research organizations in Bolivia, Ecuador and Peru (Table A2 of Annex 3) contain on average 38 of the underutilized species reported in the survey with an average 270 accessions each. Brazilian collections (Table A3 of Annex 3) store a wealth of tropical fruit trees and horticultural plants native to the continent, containing 27 of the species reported here (represented by an average 647 accessions); in Central America, the Mexican network of germplasm collections includes 53 of the crops reported in the survey, with average accession number of 365 (Table A4 of Annex 3).

However, when measuring the size of each species' collection against the average size of all underutilized crops' collections in each country, it appears that most underutilized species in national genebanks are conserved by a lower than average number of accessions (Figure 2), with a few champions contributing to the relatively high average number recorded. Underutilized species whose collections are well above average in the Andean region are mostly native grains and tubers. Quinoa collections reach 4,000 and 1,980 accessions in Bolivian and Peruvian collections, together with 487 and 2,217 accessions of oca (*Oxalis tuberosa*); over 2,000 accessions of South American lupin are conserved in Peru; native potatoes (a different set of species than the common potato) are represented by 1,760 accessions in Bolivia. In Brazil, the species whose collections greatly exceed average are legumes (*Phaseolus lunatus* with 2,673 accessions, *Vigna* with 1,787 accessions) and squashes (5,675 entries) and the large Passifloraceas fruit family. Only three crops in the Mexican genebanks have collections with more than average accessions, namely those of amaranth, chilli peppers and squashes.

A commonly observed trend in all countries is the tendency of most fruits species to occupy the lowest end of the collection size ranges, for likely the same reasons described for international collections.

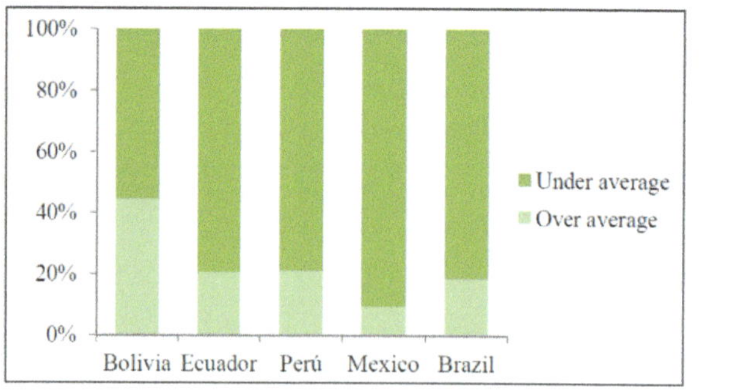

Figure 2: Percentage of underutilized species conserved in national gene banks whose collections are represented by a lower than average number of accessions (light green), and percentage of those which are above average (dark green).

However, number of accessions per se is not a sufficient indicator of a species' status of *ex situ* conservation. Quality of the material in terms of viability and availability of accurate passport and/or characterization and evaluation data is an additional, essential measure of a collection´s conservation status, since availability of this data is a precondition for using the genetic material in breeding or research. Survey respondents pointed to the technical and financial limitations to maintaining collections in good shape. Crop descriptors, which provide unambiguous guidelines for the documentation of accessions, have been developed at international level and made publicly available for only 22 of the 87 crops listed here. For one (a palm species from southern Brazil, *Butia odorata*) descriptors are currently under preparation thanks to a collaboration between Bioversity International and Brazilian experts. This of course does not exclude the likelihood of scientists using locally developed descriptors in their daily work, but this approach may limit broader sharing and comparison of their results.

Limited Knowledge on the Species' Distribution and Production

The lack of efforts in on-farm conservation was considered to be the most serious limiting factor to the enhancement of minor crops and, in turn, their use within farming systems by 75 per cent of the surveyed experts in the Americas. An important step towards improving the on-farm conservation of any species is having data about its current (and potential) distribution and cultivation. In addition to individuals' knowledge on these aspects, publicly available georeferenced data is a powerful starting point for studies on a crop's agroecological potential and adaptive capacity which in turn are useful for planning *in situ* conservation initiatives [33,34,35]. Searches performed in the data portal of the GBIF resulted in retrieval of georeferenced data for almost all the species. There is great variation in the number of observations across our dataset, ranging from no points (for only three species, all fruits) to over 2,000 for bean species (*Phaseolus lunatus* and *Phaseolus coccineus*), sweet potato and chili peppers (the latter considered as whole genus); cassava still totaled almost 5,000 observation points. Observation points for each species were spread across an average of nine countries within Latin America. Half of the species are represented by a lower than average number of observations, most often less than 500 (Figure 3). It is interesting to observe that half of the species with less than 50 observations are fruits and roots and tubers.

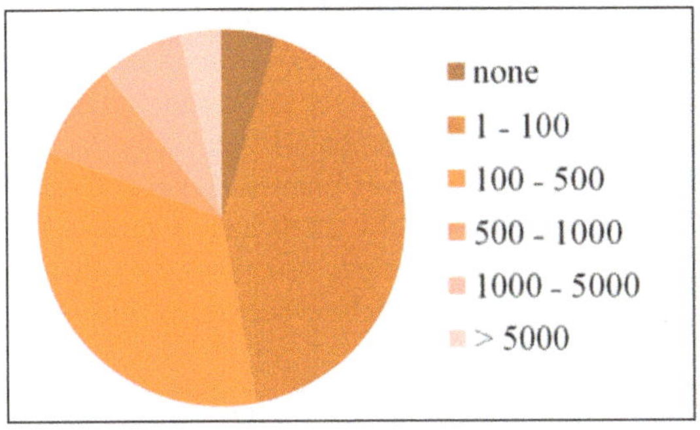

Figure 3: Number of observations recorded in GBIF for the species listed in our survey. Most species are recorded with less than 500 georeferenced observation points.

In terms of publicly available information on the production levels of our crops, the FAOSTAT database, which gathers and elaborates national statistics from FAO member countries, covers only 11 of our underutilized species, plus 7 generic categories which are likely to include them (such as "tropical fruits" and "roots and tubers"), over the 156 crop or crop categories represented. FAOSTAT's section on methods and standards acknowledges that generally, production data relates to plantation crops or orchard crops grown mainly for sale, excluding many minor crops which are conserved for self-consumption or sale through local, informal channels. In particular, it is flagged that statistics on some species, particularly tropical fruits, are unavailable in many countries, and where reported often lack uniformity. We were able to access public agricultural statistics data only for Brazil [36], Mexico [37] and Peru [38]. While Brazilian statistics only mentioned major crops already covered in FAOSTAT, the Mexican and Peruvian databases allowed consulting information on some of the underutilized crops listed by our respondents (but only an additional seven in each country).

Lack of Plant Breeding Efforts and Commercial Varieties

A varying number of underutilized crops among those highlighted in the survey are object of institutional, longer term research and improvement programmes (see Table 5 and Table 6).

These lists are likely not to be complete, but highlight significant attention to minor crops in Andean countries and to a certain extent in Mexico and Brazil (this may also depend on the smaller number of answers received from these countries and thus the smaller representation of their NUS in our list). It is important to recall that species which have been somewhat neglected by breeding are likely to harbor significant levels of genetic diversity, which not only makes them of great conservation value may also be allowing their continued adaptation to marginal environments and small-scale, low-input farming systems [39]. Therefore, breeding approaches which strive for maintaining a broad genetic basis while achieving the needed improvements in agronomic performance, product uniformity or other important aspects may be particularly relevant. The approaches commonly used in participatory plant breeding frequently aim at striking such a balance and encouraging the maintenance of diversity [40]. Curiously, in participatory breeding (PPB) programs in the Americas, underutilized crops are rarely represented: we are aware only of experiences of participatory variety selection only in the Andes on native potatoes [41,42] and of an interest and willingness to expand the maize and bean-based activities carried out in a number of Central American countries under the umbrella of the Mesoamerican Programme on Participatory Breeding to minor crops [43,44].

Table 5: Underutilized crops for which a public breeding or evaluation program exists in selected Andean countries

Crop		Country		
		Peru	Ecuador	Bolivia
Amaranth	*Amaranthus caudatus*	✓	✓	
Chirimoya	*Annona cherimola*	✓	✓	
Soursop	*Annona muricata*		✓	
Borojó	*Borojoa patinoi*		✓	
Pigeon pea	*Cajanus cajan*	✓		
Chili pepper	*Capsicum* spp.	✓		
Papayas	*Carica/Vasconcellea* spp.	✓	✓	
Cañihua	*Chenopodium pallidicaule*	✓		✓
Quinoa	*Chenopodium quinoa*	✓	✓	✓
Squashes	*Cucurbita* spp.	✓		
Tree tomato	*Cyphomandra betacea*		✓	
Arazá	*Eugenia stipitata*	✓	✓	
Sweet potato	*Ipomoea batatas*	✓		
Andean lupin	*Lupinus mutabilis*		✓	
Cassava	*Manihot esculenta*	✓		
Camu camu	*Myrciaria dubia*	✓		
Maracuyá	*Passiflora edulis*	✓	✓	
Granadilla	*Passiflora ligularis*	✓		
Avocado	*Persea americana*	✓	✓	
Lima bean	*Phaseolus lunatus*	✓		
Inca berry	*Physalis peruviana*	✓	✓	
Lúcuma	*Pouteria lucuma*	✓		
Guava	*Psidium guajava*		✓	
Sapote	*Puteria sapota*	✓	✓	
Andean berries	*Rubus* spp.		✓	
Lulo	*Solanum quitoense*		✓	
Native potatoes	*Solanum tuberosum* ssp. *andigenum*	✓	✓	✓
Pitaya	*Stenocereus* spp.	✓	✓	
Copuazú	*Theobroma grandiflorum*		✓	
Cowpea	*Vigna unguiculata*	✓	✓	

Commercialization of seed of varieties resulting from breeding efforts is often preceded by a series of procedures aimed at officially evaluating and releasing the plant variety in the country where its commercialization is sought. Officially released plant varieties are commonly listed in national registers of commercial varieties. The presence of minor species in such registers can therefore be taken as an indication of, first, public and private organizations' breeding work on such species and, second, their interest in the commercialization of seed of improved varieties of the species, which suggests, in turn, that there is a demand for such species among farmers. The same can be said about species included in the national and international catalogues of varieties subject to plant breeders' rights. The Union for the Protection of Plant Varieties maintains a database of plant varieties included in the registers of commercial varieties and in the catalogues of varieties protected by plant breeders' rights or plant patents in UPOV member countries. According to the information in this database, out of the 84 crops in our list, 37 have at least one variety registered in national listings or protected by plant breeder's rights or plant patents, representing around 43% of the total. The number of registered varieties however is highly variable, ranging from one as in the case of Andean fruit *Physalis peruviana* to over 500 as for asparagus. The species with no registered varieties under UPOV's database are in the great majority fruits and roots and tubers. Only 13 of the fruits species out of 38 reported by experts have commercially registered varieties, the average being five varieties; of the 15 roots and tubers searched in UPOV's database, only five have commercial varieties, the average being nine (including cassava which alone has over 17 varieties). For comparison, major staples such as wheat and maize, have 39,000 and 89,000 registered commercial varieties, whereas major fruit species such as apple and avocado have over 3,000 and over 100 respectively.

Table 6: Underutilized crops for which a public breeding or evaluation program exists in Brazil (left) and Mexico (right)

Brazil		Mexico	
Crop		Crop	
Copuazu	*Theobroma grandiflorum*	Chili peppers	*Capsicum* spp.
Cassava	*Manihot esculenta*	Amaranth	*Amaranthus caudatus*
Agraz	*Vaccinium* spp.	Squashes	*Cucurbita* spp.
Peach palm	*Bactris gasipaes*	Papayas	*Carica/Vasconcellea* spp.
Chili pepper	*Capsicum* spp.	Prickly pear	*Opuntia ficus-indica*
Squashes	*Cucurbita* spp.	Avocado	*Persea americana*
Quinoa	*Chenopodium quinoa*	Chirimoya	*Annona cherimola*
Passifloras	*Passiflora* spp.	Chayote	*Sechium edule*
Surinam Cherry	*Eugenia uniflora*	Pitaya	*Stenocereus* spp.
Feijoa	*Acca sellowiana*	Chili pepper	*Capsicum* spp.
Jelly Palm	*Butia capitata*	Amaranth	*Amaranthus caudatus*
Rio Grande Cherry	*Eugenia involucrata*	Squashes	*Cucurbita* spp.

The Influence of International Biodiversity Conventions and Programs on the Conservation and Use of Genetic Resources of Underutilized Crops in the Americas

In the last decade, the attention paid to minor crops by international policy making has increased considerably. As international conventions, programs, initiatives and projects dealing with the conservation and use of wild and domesticated biodiversity and food security have proliferated, the interest in underutilized species as key examples of the genetic diversity under threat and as potential contributors to food security has grown. However, to what extent do these international instruments have an influence on countries' management of NUS' genetic resources? Have they been able to effectively improve their conservation and sustainable use? In this section, based on data from Latin America, we critically analyze the commitments related to minor crops assumed by parties to two international conventions and one international plan of action that represent the most relevant international instruments in the area of plant diversity conservation and sustainable use.

The Convention on Biological Diversity

The Convention on Biological Diversity (CBD) was the first international instrument to address conservation, sustainable use and equitable sharing of benefits derived from the utilization of all biological diversity. All Latin American countries are members of the CBD. Since its entry into force in 1993, the parties to the CBD have agreed on an innumerable amount of strategies and programmes that translate the general obligations stated in the convention's text into concrete measures to be adopted at the local, national and regional levels. Some of these are more relevant than others for the conservation and use of domesticated species. In 2002, the member countries of the CBD agreed to put in place a Global Strategy for Plant Conservation (Annex to decision VI/9) with the ultimate goal to halt the continuing loss of plant diversity. It includes 16 outcome targets to be met by 2010, target 9 being the most relevant for cultivated species: "70 per cent of the genetic diversity of crops and other socioeconomically valuable is plant species conserved, and associated indigenous and local knowledge maintained". The CBD's program on agricultural biodiversity (established with Decision VI/5 of the Conference of the Parties in 2000) includes an initiative on biodiversity for food and nutrition which explicitly calls for "identification and promotion of species currently underutilized or of potential value to human food and nutrition, including those important in times of crisis, and their conservation and sustainable use" [45].

Around 20 per cent of the respondents to the survey acknowledged that increased attention had been devoted to *in situ*conservation of biological diversity, after the entry into force of the CBD in 1993. However, 60 per cent of respondents pointed that due to the convention's focus on wild biodiversity conservation and the fact that its implementation resides in the hands of Ministries of the Environment, the consideration of crop and agricultural issues is relatively weak within this framework. Interventions in the field of agricultural genetic resources directly supported by the CBD's financial mechanism are much less far-reaching than those in the field of wild biodiversity conservation. The Global Environment Facility (GEF) has been functioning since 1996 as the institutional structure to operate the financial mechanism of the Convention, providing countries and institutions with funds for carrying out conservation and sustainable use initiatives. From 1992 to 2012, agricultural biodiversity projects in the Americas received only 1.8% of the overall GEF funds disbursed at national level in the biodiversity focal area, spread out over seven projects out of 211. These seven projects (in Argentina, Bolivia, Chile, Colombia, Cuba, Ecuador and Peru) include at least some minor crops among their target portfolio, the greatest emphasis being observed in the Andean region with a focus on grains, roots and tubers and a few fruit species. In terms of GEF funds for regional or global projects including a Latin American country, agricultural biodiversity projects received a greater share of funds, reaching 9.5 per cent of the total biodiversity budget, spread over four projects out of 43 of the biodiversity focal area. Three of these include at least a partial focus on underutilized species. Figure 4 shows the small share of agro-biodiversity projects within the overall funding of the GEF. The CBD recognized countries' sovereign rights over the genetic resources within their territories, and introduces an obligation for users of such resources to share the benefits arising from their utilization with the countries where the resources were obtained. The international recognition of countries' fundamental rights and obligations in relation to the access to, and use of genetic resources (commonly referred to as access and benefit-sharing or ABS) was brought up to avoid appropriation of plant genetic resources from developing countries, by commercial actors usually operating in developed countries, with no benefit or involvement of the provider [46,47]. In 2010, the Conference of the Parties to the CBD adopted the Nagoya Protocol on Access to Genetic Resources and the Fair and Equitable Sharing of Benefits Arising from their Utilization, which elaborates the rights and obligations of the parties to the Convention in relation to ABS. The Nagoya Protocol has not entered into force yet. The complexity of the ABS procedures developed in many countries and the limited capacity to implement them at national level are often cited as major limitations to research on any country's biodiversity, including agricultural biodiversity and native crops [48,49,50]. The results of our survey point to the same direction: a strong point raised by 40 per cent of respondents when surveyed on the effect of the CBD on conservation and use of neglected crops' germplasm, resides in access and benefit sharing rules. This area of influence of the CBD was the only one in which a specific (negative) effect was detected in particular on minor species, conserved both in and *ex situ*. Perhaps the most loudly voiced concerns about access, emerging from our survey and confirmed in literature, have to do with the effects of the Andean Decision 391, directly or through derived legislation, on the Andean Community of Nations (Bolivia, Colombia, Ecuador, and Peru). Difficulties with the Brazilian and Mexican laws have also been reported [51]. Since the main scope of these legislations is *in situ* material [52], those genetic resources of minor crops which are mostly conserved on-farm rather than in genebanks fall under these usually complex and lengthy legal arrangements. Although some countries developed somewhat simplified procedures for access to material from genebanks (mostly through the design of institutional material transfer agreements, MTAs), the unclear legal scenario created around genetic resources is perceived to have affected the ease with which genebank curators grant access to their collections, for fear of infringing national law and the country's sovereign rights, particularly when the object are collections of their native crops. Data from Peru exemplify this uncertainty: the national genebank sent out genetic material under 35 MTAs for research purposes over the 2001–2009 time period but rejected applications which upfront mentioned a commercial development [53]. Respondents to the survey from other countries confirmed this tendency, in which fears of not complying with the national ABS legislation and thus exposing the nation's genetic heritage to misappropriation, combined with limited negotiation capacities of the competent authorities, make rejection a common reaction to a number of germplasm requests.

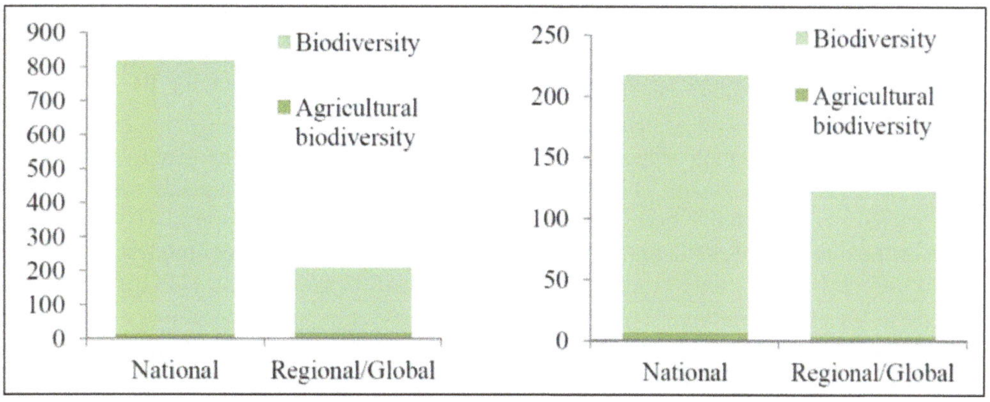

Figure 4: Share of Global Environmental Facility (GEF) funds (in million US dollars) and of GEF projects (numbers) dedicated to biodiversity and agricultural biodiversity within the overall funding strategy.

The International Treaty on Plant Genetic Resources for Food and Agriculture

The International Treaty (ITPGRFA) operates in harmony with the CBD, sharing the objectives of conservation, sustainable use and equitable benefit-sharing but with a narrower focus on plant genetic resources for food and agriculture, in recognition of their "special nature […], […] distinctive characteristics and problems, which require specific solutions" (Decision II-15 of the International Undertaking on PGRFA, Jakarta 1995). After seven years of negotiations under the auspices of FAO, the ITPGRFA was adopted in 2001, and entered into force in 2004. At the time we write this article (November 2013), 16 out of the 26 countries in Latin America are parties to the ITPGRFA.

The ITPGRFA creates a multilateral system of access and benefit sharing in which parties provide facilitated access to plant genetic resources for research, breeding, conservation and training, and commit to contribute part of the monetary benefits which may arise from the use of such resources to a common fund managed by the Governing Body of the Treaty (the Benefit Sharing Fund). The genetic resources subject to the multilateral system of the Treaty are those of species listed in the Annex 1 of the Treaty (35 globally relevant crops and 29 forages [54]), which are under the management and control of the Treaty parties and in the public domain; hosted in international *ex situ* collections like the ones of the CGIAR centers and institutions like CATIE in Latin America (according to the Treaty's Article 15 which deals with the agreements established between the Treaty's Governing Body and International Agricultural Research Centers, with regard to their *ex situ*collections); and those voluntarily included by other national and international organizations or individuals, for example companies or private collectors. The ITPGRFA's multilateral system offers an alternative to bilateral ABS agreements inspired by the CBD, for an important subset of the world's biodiversity. The text of the Treaty explicitly mentions the importance of conserving "those PGR that are under threat or are of potential use" [54] (Article 5b) and promoting the use of "local and locally adapted crops, varieties and underutilized species" [54] (Article 6.2e). Through an entire section on farmers' rights [54] (Article 9), the Treaty also calls for protection of the rights of small scale farmers and for recognition of their traditional production systems, which provides additional space for initiatives enhancing conservation and use of minor crops and traditional varieties.

Almost all respondents to the survey (average 66 per cent) feel that the ITPGRFA has not determined significant changes in the status of conservation and use of underutilized crops; this perception of a lack of influence on minor species is particularly felt in terms of their on-farm conservation (80 per cent of respondents) *versus ex situ* conservation. The fewer respondents which did feel the ITPGRFA had a somewhat differentiated effect depending on the species, reported that germplasm distribution (30 per cent), breeding (29 per cent) and international funding (36 per cent) had improved only for major, globally distributed staples.

The correlation between countries' obligations under the ITPGRFA and their efforts to conserve and promote the sustainable utilization of minor crops in Latin American countries is not easy to measure. The number of projects targeting NUS which have been funded by the two financial mechanisms of the ITPGRFA may be used as a tentative measure of the Treaty's influence. One of these mechanisms is the Global Crop Diversity Trust (GCDT), which makes funds available for improving and upgrading the status of *ex situ* collections included in the multilateral system, and the second one is the benefit sharing fund, which supports actions covering all aspects of PGRFA management. Underutilized species in the Americas have only partially benefitted from the GCDT's funds. Over the three rounds of funding released to date (2008 to 2010) only few of the minor crops listed by our experts, namely Peruvian and Argentinian sweet potato collections and Mexican collections of common bean, lima bean and cowpea, have been targeted by evaluation projects. No collection project has been funded in Latin America in the 2010 round of funding. In terms of regeneration, funds from the Trust have allowed securing vulnerable accessions of 22 priority crops, including cassava, cowpea, taro, pigeon pea, sweet potato and yam among those listed here. The Trust has also launched a project in partnership with the Millennium Seed Bank of the Royal Botanic Gardens in Kew which will safeguard and use the wild relatives of 29 crops of major importance to food security, of which only four coincide with species in our list (cowpea, lima bean, pigeon pea, sweet potato). One of the explanations for this low coverage of crops from our list in the GCDT's portfolio is that its operational strategy, which is independent from the Treaty's Funding Strategy, explicitly prioritizes *ex situ* collections of species listed in the Annex 1 of the Treaty, *i.e.*, major species of global importance and distribution. Furthermore, even if exceptions are made to this Annex 1 preference, the Trust funding mechanism also requires that the target collections be effectively exchanged according to the rules of the multilateral system of access and benefit-sharing, *i.e.*, be placed within the MLS. These conditions are seen by some respondents as a limitation to the possibility of obtaining funds for the conservation of those underutilized crops which parties are not necessarily willing to put in the multilateral system and prefer keeping under the CBD-inspired access and benefit sharing mechanism in place. It is anyhow important to acknowledge that the Trust has also supported broader initiatives such as the drafting of the Hemispheric Strategy for the Americas [55], which aimed at assessing the most important crops and collections in the hemisphere and providing recommendations for their continued *ex situ* conservation. Forty-nine of the total 85 species in the Hemispheric Strategy's list overlap with those highlighted in the present survey.

The benefit sharing fund of the Treaty, whose priorities are informed by the Treaty's Funding Strategy (adopted by the Governing Body of the Treaty at its first session, Resolution 1/2006) and its Annexes on eligibility criteria and funding priorities (adopted by the Governing Body at its Second Session in 2007), does not have a formally declared intention to focus on specific crop categories. Admittedly, a requirement to focus on Annex 1 crops was introduced in one of the two windows which made up the second call for proposals (released in 2010), making roughly half of the funding portfolio unavailable to most NUS in that cycle. Overall, out of the nine projects awarded funding in Latin America, only two explicitly target underutilized crops from our list. Another three include underutilized species among the portfolio of target crops (see Table 7 for details).

The creation of the Treaty's multilateral system of access and benefit sharing has not until now addressed the shortcomings of CBD-inspired ABS rules in Latin America due to a great extent to the very modest level of implementation of the system in most of the ITPGRFA member countries [56]. The only Latin American country that has officially notified the national *ex situ* collections included in the multilateral system is Brazil with rice, maize, bean and cassava collections held by Embrapa. Except for Embrapa's cassava collection, no national collection of any minor crop among those listed here has been included in the multilateral system. However, even under the best-case scenario of a well implemented multilateral system, many underutilized crops are likely not to be covered by its scope: only 14 (16%) of the species reported as underutilized here are in the Annex 1 of the ITPGRFA (although parties are free to place non-Annex 1 collections in the system), and most of the genetic resources of those species are found on-farm, for which the application of the rights and obligations of the Treaty's multilateral system is unclear. Only three of the 22 respondents who expressed and motivated their opinion on the possible benefits of the Treaty's implementation to conservation and use of NUS germplasm, mentioned they did not see any benefits; the other experts feel there is scope for ITPGRFA-driven improvements in minor crops' status.

Table 7: Projects supported by the Treaty's Benefit Sharing Fund in its current two rounds of funding and coverage of underutilized crops

	Title	Targeted country or countries	Focus on underutilized crops
First round projects	Identification of useful potato germplasm adapted to biotic and abiotic stresses caused by global climate change	Costa Rica	no
	Rescue, conservation and sustainable management of teocintle in Nicaragua (*Zea nicaraguensis* Iltis & Benz) in the Apacunca Genetic Reserve	Nicaragua	no
	Contribution of traditional methods for the *in situ* conservation and management of maize (*Zea mays* L.) and bean (*Phaseolus* spp.) to the food security of farming families in Cuba	Cuba	no
	Conservation and Sustainable Use of Native Potato Diversity in the "Potato Park"; Cusco, Peru	Peru	yes
	Broadening of potato (*Solanum tuberosum*) genetic basis through introgression of local wild species, *Solanum commersonii*	Uruguay	no
Second round projects	Participatory and science-based formulation of a Strategic Action Plan to strengthen the conservation of plant genetic resources and their enhanced use in adapting to climate change in Mesoamerica	Guatemala, Belize, El Salvador, Honduras, Nicaragua, Costa Rica, Panama, Mexico	partial
	Shared management and use of (agro)biodiversity by indigenous and the traditional communities from the semi-arid region of Minas Gerais State as a strategy for food security and to reduce climate risks	Brazil with partner applicants in: Costa Rica, Nicaragua, Guatemala, Cuba, Haiti, Mozambique	partial
	Conservación y manejo sostenible del germoplasma de papas nativas en las comunidades campesinas de la Provincia de Andahuaylas	Peru	yes
	Establecimiento de una red preliminar de bancos comunitarios de semillas, en regiones vulnerables del país, para disponer de semillas en caso de desastres naturales	Guatemala	partial

The Global Plan of Action for the Conservation and Sustainable Use of Plant Genetic Resources for Food and Agriculture

The Global Plan of Action for the Conservation and Sustainable Utilization of Plant Genetic Resources for Food and Agriculture (GPA) is a set of recommendations developed by the Commission on Genetic Resources for Food and Agriculture of FAO, and based on the Report on the State of the World's Plant Genetic Resources for Food and Agriculture. The GPA was adopted by 150 countries in 1996, and renewed in 2011, and is intended as a framework, guide and catalyst for action at community, national, regional and international levels. Unlike the CBD and the ITPGRFA, the GPA does not spell out member countries' rights and obligations but rather 20 priority activity areas grouped into four thematic sections: *in situ*conservation and development; *ex situ* conservation; utilization of plant genetic resources and institutions and capacity building. Although most activity areas of the plan are relevant for the conservation and sustainable use of minor and underutilized plant species, activity area 12 deals explicitly with minor species, encouraging actions to "identify underused species, develop sustainable management practices, develop post-harvest and marketing methods, and promote policies for the development and use of under-utilized species", based on their "potential for wider contributions to food security, agricultural diversification and income generation".

According to their reports to FAO under the GPA, Andean countries have directed 39% of their total *ex situ* projects to underutilized species as defined in the survey. Peru has the highest ratio, with 51% of projects being on minor crop collections (covering 28 of the species resulting from the survey). Bolivia has been giving high priority to quinoa, Ecuador to Andean fruit species such as cherimoya (*Annona cherimola*) and naranjilla (*Solanum quitoense*), while Peru has largely focused on Andean tubers [57,58,59], as exemplified in Figure 5. Respondents from the Andean region noted that although there had been some degree of national research and development efforts on NUS, they tended to focus on a narrow set of crops, especially those with increasing demand on national and international markets; this is the case of Andean grains, spearheaded by quinoa. Brazil's report to FAO (no information is available through the GPA portal) highlights significant investments in the upgrading and improvement of decentralized *ex situ* collections of cucurbits in the North East and fruit tree collections elsewhere. Among these, most work focuses on Passifloraceae, peach palm (*Bactris gasipaes*), Surinam cherry (*Eugenia uniflora*), feijoa (*Acca sellowiana*), guabiju (*Myrcianthes pungens*), jelly palm (*Butia capitata*) and Rio Grande cherry (*Eugenia involucrata*) [60].

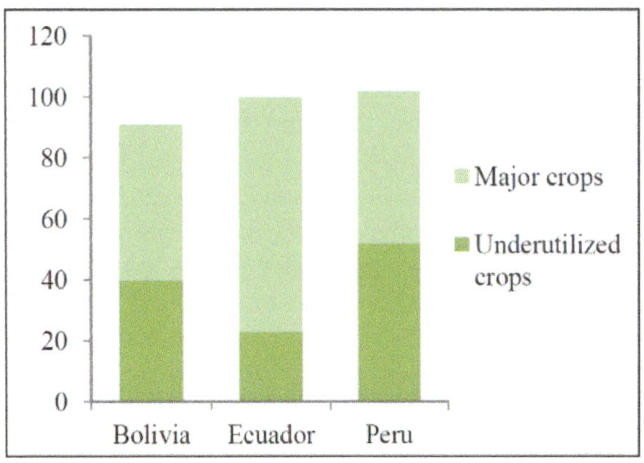

Figure 5: Balance between *ex situ* projects dedicated to underutilized (darker green)*versus* major crops (light green) in Andean countries.

Notwithstanding admirable progress, countries' reports to FAO still highlight a worrying limitation in the funds available, both in amount and regularity. All countries report greater financial shortcomings for species whose conservation is not through seed but *in vitro* or, even more, as field collections, which is the case for most fruit species and is clearly reflected in the numbers presented above.

Within the projects reported under the GPA *in situ* priority activities, an average 28% of conservation projects in selected countries are dedicated to underutilized species figuring in our list. Top percentages were reached in Peru (43%), covering 22 underutilized crops as defined in our survey (Figure 6). Crops in which most effort has been placed are Andean grains (particularly quinoa in Bolivia) and tubers (highest effort being in Peru), followed by specific fruit families (Passifloraceae, Caricaceae and Annonceae).

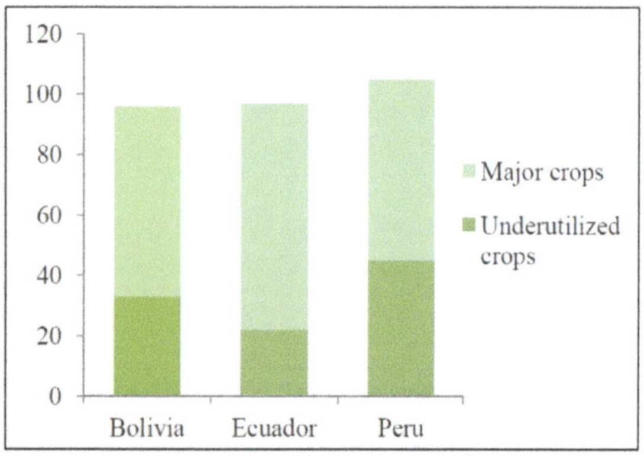

Figure 6: Balance between *in situ* projects dedicated to underutilized (darker green)*versus* major crops (light green) in Andean countries.

In these countries, many of the projects have focused on registering and documenting on-farm crop diversity and its associated traditional knowledge at local levels. In Peru these efforts have led to the identification of specially managed areas of agricultural biodiversity such as the well-known Potato Park [61]. In Bolivia, ten micro-centers of crop diversity have been identified across six departments and special initiatives such as inventories of on-farm agricultural biodiversity and traditional knowledge, seed fairs and repatriation of native varieties from genebanks have been carried out there [57]. In Ecuador, initiatives ranged from the local establishment of conservation gardens for Andean roots and tubers in farmers' fields to characterization and on-farm conservation projects on fruit species, particularly papayas and cherimoyas [58]. The Brazilian Ministry of Environment, in collaboration with research institutions (such as national research organization Embrapa and the Amazon National Research Institute, INPA) led the identification and the mapping of landraces and wild relatives of some of the main crops grown in Brazil. Seven subprojects involve two of the species here listed as underutilized (cucurbits and peach palm). A number of public institutions often in collaboration with NGOs and the civil society have been active in recovering and documenting local crop landraces, and reintroducing them in cultivation among small scale farmers. However, most initiatives have focused on major staples, namely maize and beans [60]. In Mexico, national surveys of the flora were conducted to establish a baseline of the conservation status for future monitoring: of the 62 plant families prioritized, 19 included food crops, only few of which correspond to those listed here, namely species of the *Opuntia* genus, squashes, beans and chilli (*Capsicum chinense*) [62].

DISCUSSION AND CONCLUSIONS

The first part of the study confirmed the complexity of defining neglect and underutilization and the great variation which exists among NUS species: not all of the common assumptions hold true for different species, each of which may suffer more under some aspects than others. This makes a univocal definition of NUS a challenging endeavor, which, in the words of regional experts, also affects the possibility of well-grounded priority setting exercises for improving the conservation and sustainable use of the genetic resources of minor crops at regional level. A way to address this issue could be to adopt a framework for the analysis of any crop's status based on the number of the typical features of neglect and underuse which affect it, and the degree to which each such features limit its development. We chose the following approach to somehow validate and quantify our crops' status by checking if they met at least three among the following thresholds, which were established on the basis of the results of our analyses:

- be relatively neglected by scientific research (here, we chose the average number of Google Scholar hits across our sample set of species as a threshold, *i.e.*, species with less than 1,954 hits are neglected in this aspect)
- be under-represented in the publicly available dataset of the GBIF (here, less than 126 georeferenced observations in Latin America, the average across our sample set, is the threshold)
- be not represented in FAOSTAT
- have a low number of registered varieties, a proxy of scarce plant breeding and seed commercialization (less than 20 registered commercial varieties, with this threshold corresponding to the average number of varieties registered in national catalogues of commercial plant varieties and in the national lists of plant varieties protected by plant breeders' rights and plant patents for the species from our list)
- be under-represented in globally accessible germplasm collections (by less than 320 accessions, the threshold corresponding to the average accession number of the species considered here in such collections)

The application of the above filtering criteria and thresholds (calculations were performed in R [63]) retained 74 of the crops (*i.e.*, those for which more than three typical features of neglect and underuse held true), while the ten species listed inTable 8 did not qualify as (fully) neglected or underutilized.

The top three crops in Table 8 are globally distributed and consumed, likely explaining why they retain none of the features commonly related to neglect and/or underuse. Other crops have a more intermediate status: cassava (*Manihot esculenta*), albeit not seriously neglected by research and conservation efforts, is still relatively underused partially because of the limited availability of commercial varieties for larger scale cultivation. The presence of quinoa (*Chenopodium quinoa*), often referred to as a typical NUS, in Table 8, suggests it is not as seriously overlooked as other Andean native crops, particularly in terms of research, but that it still suffers from limitations in terms of accessibility of germplasm and of conservation-relevant data. On the other hand, the absence of fruit species and most roots and tubers from Table 8 confirms that the majority (often all) of NUS features hold true for the two most numerous categories of NUS reported in the survey. Most of these species appear to be both neglected by research and conservation efforts as well as underused in commercial production, because of the limited availability of agricultural statistics and of commercial varieties. The non-staple, non-commodity nature of fruit species, the generally limited long-term storability of their products and the inherent complexities of conserving and breeding species with vegetative propagation are probably the main reasons for their status. For promoting these species, efforts are required all along the conservation-to-use realm. In particular, a focus on on-farm conservation may be particularly appropriate given the difficulties of conserving fruit trees, roots and tubers *ex situ*; engagement of farmers in collaborative research projects for improving agronomic and processing aspects would provide incentives for the continued on-farm conservation of these species' genetic resources, enhancing their use potential; awareness raising efforts to highlight the role of these crops in diets and health, will be crucial to foster consumer demand and in turn, enhance commercial use.

Table 8: Crops which were not validated as underutilized according to the pilot framework proposed here

Crop	Common Name	Origin	Aspects in which Underutilized (<3)
Capsicum spp.	Chili peppers	native	None
Ipomoea batatas	Sweet potato	native	None
Phaseolus vulgaris	Common bean	native	None
Cajanus cajan	Pigeon pea	introduced	Lack of widely available distribution data (GBIF)
Fagopyrum esculentum	Buckwheat	introduced	Lack of widely available distribution data (GBIF)
Manihot esculenta	Cassava	native	Limited commercial varieties available
Vigna unguiculata	Cowpea	introduced	Lack of widely available distribution data (GBIF)
Chenopodium quinoa	Quinoa	native	Low number of accessions in international genebanks and lack of widely available distribution data (GBIF)
Cucurbita spp.	Squashes	native	Not represented in FAOSTAT and no international descriptors available
Phaseolus coccineus	Runner bean	native	Limited representation in literature and not represented in FAOSTAT

While local, context-dependent expert knowledge and perceptions remain essential to assessing the possibly underexploited potential of any species, the availability of a validation method may provide for a more systematic definition of the extent to which crops present features of neglect and/or underuse and highlight where the major limitations and opportunities for improvement lie (e.g., if in research or conservation efforts, or in options for increased commercial use). This may better support the identification and prioritization of species or categories affected by similar constraints and planning for specific interventions to strengthen different areas along the conservation-use realm. Naturally, different threshold values and a different number of minimum conditions to be fulfilled can be tested, adapting the approach flexibly in consultation with experts′ and depending on different contexts and time frames.

International frameworks aimed at conservation and use of plant genetic resources have had mixed, sometimes conflicting effects on the conservation and use of neglected crops. The GPA has probably helped to channel national and international funds to projects which meet its goals, allowing them to gain visibility, but the extent to which it has been the actual driver of these initiatives is not clear, among others because it lacks a financial scheme. While the Convention on Biological Diversity and the Treaty on Plant Genetic Resources for Food and Agriculture have contributed to overall international awareness of the importance of agricultural biodiversity, their translation into practical measures for the conservation and sustainable use of minor crops at national levels has been limited. Important areas of influence of the CBD and the Treaty on minor crops are perceived to be those of access and benefit sharing, and experts have highlighted the importance of addressing the widespread lack of harmonized implementation of ABS mechanisms in countries which are Parties to both conventions, particularly when they harbor native, underutilized crops. Such harmonization is crucial for promoting the mobilization of genetic diversity of these species and encouraging research and development while ensuring equitable participation of all relevant actors.

The financial mechanisms of these two international instruments have provided very modest funds to the conservation, characterization and use of genetic resources of NUS to date. The relative dilution of agricultural biodiversity and NUS projects within the GEF's funding on biodiversity is somewhat understandable considering the broad mandate of the CBD (which encompasses all biodiversity) and the quantity of proposals focused on wild biodiversity conservation submitted to the GEF by organizations working in a mega-diverse region such as Latin America. The relatively limited coverage of NUS by the projects supported by the Treaty's benefit sharing fund may be due to a combination of factors, among which the partial reduction in scope in the latest BSF's call, the limited funds yet available through this funding mechanisms, and possibly the lack of NUS-based proposals received from applicants. If the latter hypothesis were the case, it may in turn reflect limited capacity, interest or resources at national level for participating in the competitive tender with NUS-based proposals, or a perception that this framework is not entirely appropriate for conservation and use of minor crops. Results of our survey suggest that the latter hypothesis may be at least partly true; there appears to be a widespread perception that the mechanisms and provisions of the Treaty are tailored exclusively around the Annex 1 list of crops, *i.e.*, mostly major species. While an Annex 1 focus (which is a matter of prioritization rather than exclusivity) is correct for the functioning of the MLS and the Trust's disbursements, we have explained how the benefit sharing fund has only partially restricted, in one of its funding windows, its scope to Annex 1 crops and therefore still promises to be an important opportunity for NUS-centred projects to receive support. It would be important to confirm if this somewhat narrow interpretation of the Treaty is indeed in place, and in that case, address it through capacity building and awareness raising in developing countries. Efforts in this direction should cover other important sections of the Treaty, such as [54] Article 6 on the sustainable use of PGRFA, and [54] Article 9 on farmers' rights, which also constitute very appropriate umbrellas for non-project based initiatives targeting NUS. A certain lack of awareness around the potential for NUS of these provisions may be the result of their rather general nature, which leaves it up to countries to devise relevant national measures. In order for these Treaty provisions to translate into national-level measures favourable to conservation and use of minor species parties may need more guidance on what types of interventions would be relevant. Some advances in this direction have been made in relation to [54] Article 9 on farmers' rights: recommendations to the Treaty's Governing Body on means for recognizing and implementing such rights have included an emphasis on participatory plant breeding [64,65] and other approaches to conservation and use that rely on farmers' participation, such as community seed banks and local seed production [66]. Given the fact that a considerable amount of the genetic resources of minor crops and of the knowledge about their diversity resides in farmers' fields, adopting such community-based and participatory approaches may be particularly advantageous for meeting the combined goal of recognizing farmers' rights and halting genetic erosion.

It is important to acknowledge that other drivers are likely to be far more influential in defining countries' priority crops and research and development strategies. Market demand is likely one of the most powerful; increasing customer demand on national and international markets for specialty products from a number of minor crops may already be an important incentive for moving forward national level research and development (and to a certain extent conservation through use), becoming more powerful than any international instrument or funding scheme. Although markets have been quite extensively analyzed for their sometimes controversial effects leading to erosion of diversity and unequal participation [67,68], their potentially positive driving role in the conservation and sustainable use of PGRFA has also been recognized [69,70,71,72,73].

As expressed by the experts in Latin America, the contribution of international frameworks has so far remained mostly at the level of rhetorical discourse; however, these instruments do have a potential to better contribute to improving minor crops' status, by raising worldwide awareness and funds for the conservation and sustainable utilization of their genetic resources. At the same time, they cannot replace national efforts: initiatives at national and local levels are necessary to prioritize, define and implement practical measures that can effectively address the diversity of issues affecting each particular species. Our results allow a better appreciation of the complex nuances which characterize the status of crops commonly bulked under the NUS label; we show how different crops or crop categories are underutilized to very different degrees and in different aspects and suggest practical means to quantify these differences. We believe the application of an analytical approach such as the one presented here provides a basis for informing more targeted, case-specific measures in the field of minor crops' genetic resource management. Finally, in addition to better targeted interventions in the genetic resource conservation and use area, to truly leverage the role of these species, it will be important to link up with national and international initiatives in the fields of sustainable agriculture, food security and sovereignty [74], health [75,76], and adaptation to climate change [77], as minor crops can substantially contribute to the achievement of these objectives.

ACKNOWLEDGMENTS

The authors wish to thank all participants to the survey for taking time to provide the initial data for this article. They also wish to thank Toby Hodgkin and Marleni Ramirez for their insights and intellectual inputs, and Stefano Padulosi for facilitating the authors' participation in relevant meetings and providing contacts of relevant regional experts. Finally, they wish to thank the four anonymous reviewers for their useful comments. The financial support to publish this work open-access comes from the CGIAR Research Programme on Policies, Institutions and Markets. The authors are grateful for this support.

AUTHOR CONTRIBUTIONS

Galluzzi, G. designed and undertook data collection and analyses. López Noriega, I. provided substantial contributions to the survey design and the section on international policies, both in terms of analyses and writing.

APPENDIX

Annex 1. Regional survey on underutilized crops of the Americas and the influence of international policies on biodiversity and plant genetic resources

Background and Assumptions

Over the last 20 years, significant political developments have changed the regulatory landscape governing the access to, sharing and use of plant genetic resources, especially with the adoption of the Convention on Biological Diversity (CBD) and the International Treaty on Plant Genetic Resources for Food and Agriculture (ITPGRFA).

However, simplification of agricultural landscapes and practices is causing food production to increasingly rely on a small number of globally distributed crops, to the detriment of a variety of local or regional species. Many of these are threatened by socio-economic and cultural factors and neglected by research and development (hence their definition as "NUS, Neglected and Underutilized Species").

Some representatives of the international scientific community have expressed concern about the potential negative effect that policy frameworks could have on NUS, further undermining the recognized role these species can play in diversification, sustainability and resilience of agro-ecosystems as well as their contribution to income, health and nutrition of the poor.

One concern is related to the effects of the concept of national sovereignty over genetic resources, formalized by the CBD. It is perceived that the flow of germplasm may have suffered restrictions, especially in native crops—many of which are NUS—out of some countries' fear of misappropriation of the genetic resources on their territory by a third party.

Another concern lies in the perception that the ITPGRFA is failing to recognize the importance of NUS, by designing its mechanism of access, benefit sharing and funding around a limited number of crops (listed in the Annex 1 of the Treaty,http://www.planttreaty.org/training/annex1_en.htm) selected based on their global importance and countries' interdependence.

While acknowledging the difficulty of isolating effects of political and legal frameworks from economic, social and cultural issues affecting the conservation and use of NUS, we believe that gathering opinions and experiences from regional experts is extremely important for better understanding the relative influence of the different factors and developing a research agenda that will contribute to better conservation and use of NUS in the region.

Respondent Information

Q1

Please indicate your name and position/institute your work for. These data are useful for analyzing responses and necessary for following up and involving you in subsequent steps of the study on NUS.

Q2

Within the PGRFA theme, which are your areas of main expertise/interest? Please tick where relevant.

ex situ conservation	
in situ conservation	
breeding	
policies	
Other (please specify)	

NUS and Policies

Q3

Please list the crops in your collection/context of work/country which can be considered NUS (please provide scientific names).

Note: For general information on the definition of NUS, please consult http://www.underutilised-species.org/spotlight/what_are_underutilised_species.asp.

For the purpose of this survey, we invite you to refer to those crop species with under-exploited potential to contribute to food security, ecosystem and nutritional diversification.

Q4

In what area do you consider the NUS indicated above to have the greatest potential? Please insert the name of the species in the most relevant spaces.

For example, it could be "quinoa" in the box corresponding to "Food Security".

Climate change mitigation	
Climate change adaptation	
Breeding	
Food security	
Dietary Diversity	
High value products (potential for niche markets)	
Agro-ecosystem diversification	
Other	

Note: The contribution to mitigation derives from an increased sequestration (mostly perennials) or a decreased emission of carbon dioxide (less chemical inputs required). The contribution to adaptation derives from greater resistance to droughts, floods or other extreme events, or from constituting a food or income generating alternative for farmers faced with increasing climate risks.

Q5

Please indicate (by ticking the box) the effect of each of the following limiting factors on conservation, distribution and use of NUS.

	No effect	Limited effect only in a few NUS	Limited effect common to most NUS	Strong effect in only a few NUS	Strong effect common to most NUS
Low number of accessions conserved in genebanks					
Poor regeneration status of *ex situ* material					
Poor documentation on *ex situ* material					
Complex regulations for access and distribution of material					
Lack of funding for *ex situ* conservation/management					
Lack of funding for *in situ* conservation/management					
Decreasing on farm conservation/loss of traditional knowledge on use					
Lack of attention by research and development					

Please expand your comments and views on the above or other limiting factors.

Q6

Based on your knowledge of the Convention on Biological Diversity (CBD), please indicate its effects in your working context and—most importantly—if it has influenced NUS and other crops differently over the past 20 years.

	Positive, stronger in NUS	Positive, stronger in other crops	Negative, stronger in NUS	Negative, stronger in other crops	Positive trend in all crops	Negative trend in all crops	No significant effect
Germplasm acquisition (*ex situ*)							
Germplasm characterization and evaluation (*ex situ*)							
Germplasm distribution							
In situ/on farm conservation							
Breeding and crop improvement							
Availability of national funds (for conservation, use, research)							
Availability of international funds (for conservation, use, research)							

Note: Please consider the support granted by the CDB to conservation projects and initiatives through the Global Environmental Facility (GEF), which serves as the CBD's financial mechanism .

Please explain how this policy framework can have affected or may in the future affect the status of NUS in your working context.

Q7

Based on your knowledge of the International Treaty on Plant Genetic Resources for Food and Agriculture (ITPGRFA), please indicate its effects in your working context and—most importantly—if it has influenced NUS and other crops differently since its entry into force.

Note: Please keep in mind the support granted by the Treaty to *ex situ* conservation initiatives/projects through its financial mechanisms, the Global Crop Diversity Trust (GCDT) .

	Positive, stronger in NUS	Positive, stronger in other crops	Negative, stronger in NUS	Negative, stronger in other crops	Positive trend in all crops	Negative trend in all crops	No significant effect
Germplasm acquisition (*ex situ*)							
Germplasm characterization and evaluation (*ex situ*)							
Germplasm distribution							
In situ/on farm conservation							
Breeding and crop improvement							
Availability of national funds (for conservation, use, research)							
Availability of international funds (for conservation, use, research)							

Please explain how this policy framework can have affected or may in the future affect the status of NUS in your working context.

Q8

Do you believe that conservation, management and use of NUS can benefit from the participation of your country in the International Treaty on Plant Genetic Resources for Food and Agriculture? Please explain how. If not, please give reasons.

Annex 2. List of NUS emerging from the survey

Table A1: List of crops reported as underutilized by plant genetic resource experts in Latin America

Scientific name	Common name in Spanish or local language	English name	Origin
Acca sellowiana	Guayabo del país - feijoa	Feijoa	native
Amaranthus caudatus	Amaranto - kiwicha - achita	Amaranth	native
Amaranthus quitensis	Ataco - sangorache	Amaranth	native
Anacardium occidentale	Marañón	Cashew	native
Annona cherimola	Chirimoyo	Cherimoya	native
Annona muricata	Guanábana	Soursop	native
Aristotelia chilensis	Maqui	Chilean Wineberry	native
Arracacia xanthorrhiza	Arracacha	White carrot	native
Asparagus officinalis	Espárrago	Asparagus	introduced
Bactris gasipaes	Pejibaye - chontaduro	Peach palm	native
Benincasa hispida	Calabaza china	Winter melon	introduced
Berberis spp.	Berberis sp.	Barberry	native
Borojoa patinoi	Borojó	Borojó	native
Butia capitata	Butiá	Jelly palm	native
Cajanus cajan	Guandul	Pigeon pea	introduced
Calathea allouia	Dale dale	Guinea arrow root	native
Canavalia ensiformis	Canavalia	Jack bean	native
Canna edulis	Achira	Achira	native
Capsicum spp.	Ajíes	Chili	native
Chamaedorea tepejilote	Pacaya	Pacaya palm	native
Chenopodium pallidicaule	Cañihua	Cañihua	native

Chenopodium quinoa	Quinua	Quinoa	native
Chrysobalanus icaco	Icaco	Cocoplum	native
Cnidoscolus chayamansa	Chaya	Tree spinach	native
Colocasia esculenta	Pituca	Taro	native
Cucurbita spp.	Calabazas - zapallos	Pumpkins	native
Curcuma longa	Cúrcuma	Turmeric	introduced
Cyclanthera pedata	Achocha or cayhua	Stuffing cucumber	native
Cynara scolymus	Alcaucil	Artichoke	introduced
Cyphomandra betacea	Tomate de árbol	Tree tomato	native
Dioscorea trifida	Sacha papa	Indian yam	native
Eugenia aggregata	Cereza de Río grande	Rio Grande cherry	native
Eugenia stipitata	Arazá	Arazá	native
Eugenia uniflora	Grosella - pitanga - cereza de Cayenna	Suriname cherry	native
Fagopyrum esculentum	Trigo sarraceno	Buckwheat	introduced
Fernaldia pandurata	Loroco	Loroco	native
Fragaria chiloensis	Frutilla chilena - fresa chilena o frutilla blanca	Chiloé strawberry	native
Frantzia tacaco	Tacaco	Tacaco	native
Hexachlamis edulis	Ubajay	Ubajay	native
Inga feuillei	Pacay	Pacay or ice-cream bean	native
Ipomoea batatas	Batata - papa dulce - camote	Sweet potato	native
Lablab purpureus	Zarandaja, frijol trepador	Hyacinth bean	introduced
Luffa cylindrica	Estropajo	Sponge gourd	introduced
Lupinus mutabilis	Tarwi	South American Lupin	native
Mammea americana	Mamey	Mamey	native
Manihot esculenta	Yuca	Cassava	native
Maranta arundinacea	Jamachipeke	Arrowroot	native
Mirabilis expansa	Chago	Pepper weed	native
Myrcianthes pungens	Guaviyú	Guaviyú	native
Myrciaria dubia	Camu Camu	Camu Camu	native
Opuntia ficus indica	Tuna	Prickly pear	native
Oxalis tuberosa	Oca	Oca	native
Pachyrhizus ahipa	Ashipa - ajipa	Andean yam bean	native
Pachyrhizus erosus	Jicama	Yam bean	native
Passiflora ligularis	Granadilla	Sweet granadilla	native
Passiflora mollissima	Tumbo - curuba	Banana passionfruit	native
Paullinia cupana	Guaraná	Guaraná	native
Persea schiedeana	Aguacate silvestre	-	native
Phaseolus acutifolius	Frijoles "comunes"	-	native
Phaseolus coccineus	Ayocote	Runner bean	native
Phaseolus dumosus	Frijol cacha	-	native
Phaseolus lunatus	Pallar - frijol Lima	Lima bean	native
Phaseolus vulgaris	Frijol ñuña	Popping beans	native
Physalis peruviana	Aguaymanto - Uchuva o Uvilla	Inca berry	native
Plukenetia volubilis	Sacha inchi	Sacha inchi, Sacha peanut	native
Portulaca oleracea	Verdolaga	Pigweed	introduced
Pouteria obovata	Lúcuma	Lucuma	native
Pouteria sapota	Zapote	Sapote	native
Psidium guajava	Guayaba	Guava	native
Rubus glaucus	Mora de Castilla	Andes berry	native
Sechium edule	Chayote	Pear squash, vegetable pear	native
Smallanthus sonchifolius	Yacón	Yacón	native
Solanum muricatum	Pepino dulce	Pepino melon or melon pear	native
Solanum quitoense	Lulo	Lulo	native
Solanum andigenum	Papas nativas	Native potatoes	native
Spondias spp.	Spondiaceas	Mombins	native
Stenocereus spp.	Pitaya	Pitaya	native
Theobroma grandiflorum	Copoazú	Copoazú	native
Tropaeolum tuberosum	Mashua - Isaño	Mashua	native
Ullucus tuberosus	Olluco - Ulluco - Papalisa	Ulluco	native
Vaccinium meridionale	Agraz - mortiño	-	native
Vasconcella spp.	Papaya de altura	Highland papaya	native
Vigna unguiculata	Caupi	Cowpea	introduced
Xanthosoma spp.	Malanga - quequisque - Walusa	Tannia, yautia	native

Annex 3. The *ex situ* conservation status of NUS in selected countries

Table A2: Number of accessions of selected underutilized species, conserved in genebanks of Andean countries. Source: country report to FAO and WIEWS

Crop	No. Accessions	Holding Institutions
Ipomoea batatas	8	Vallecito Bolivia
Xanthosoma sagittifolium	11	INIAF Bolivia
Lupinus mutabilis	12	INIAF Bolivia
Canna edulis	17	INIAF Bolivia
Pachyrhizus ahipa	18	INIAF Bolivia
Amaranthus caudatus	19	CIBREF-FCAP-UTO Bolivia
Pachyrhizus tuberosus	22	INIAF Bolivia
Annona cherimola	29	INIAF Bolivia
Manihot esculenta	38	Vallecito Bolivia
Arracacia xanthorrhiza	41	INIAF Bolivia
Smallanthus sonchifolius	43	INIAF Bolivia
Passiflora spp.	49	INIAF Bolivia
Amaranthus caudatus	51	INIAF Bolivia
Tropaeolum tuberosum	79	INIAF Bolivia
Lupinus mutabilis	105	INIAF Bolivia
Amaranthus caudatus	134	INIAF Bolivia
Ullucus tuberosus	197	INIAF Bolivia
Phaseolus	326	INIAF Bolivia
Solanum sp.	350	CIBREF-FCAP-UTO Bolivia
Capsicum spp.	447	INIAF Bolivia
Chenopodium pallidicaule	448	CIBREF-FCAP-UTO Bolivia
Cucurbita spp.	450	INIAF Bolivia
Oxalis tuberosa	487	INIAF Bolivia
Chenopodium pallidicaule	801	INIAF Bolivia
Chenopodium quinoa	1,700	CIBREF-FCAP-UTO Bolivia
Solanum spp. *	1,760	INIAF Bolivia
Chenopodium quinoa	3,121	INIAF Bolivia
Dioscorea spp.	1	DENAREF-INIAP Ecuador
Opuntia spp.	1	DENAREF-INIAP Ecuador
Spondias dulcis	1	DENAREF-INIAP Ecuador
Myrciaria dubia	1	Station Napo Payamino - INIAP Ecuador
Annona cherimola	1	Station Napo Payamino - INIAP Ecuador
Theobroma grandiflorum	1	Station Napo Payamino - INIAP Ecuador
Annona muricata	1	Station Napo Payamino - INIAP Ecuador
Eugenia stipitata	1	Station Napo Payamino - INIAP Ecuador
Spondias purpurea	1	Station Napo Payamino - INIAP Ecuador
Annona cherimola	1	Station Portoviejo - INIAP Ecuador
Annona muricata	1	Station Portoviejo - INIAP Ecuador
Annona cherimola	2	DENAREF-INIAP Ecuador
Annona muricata	2	DENAREF-INIAP Ecuador
Vigna unguiculata	2	DENAREF-INIAP Ecuador
Chenopodium pallidicaule	3	DENAREF-INIAP Ecuador
Passiflora spp.	3	Station Napo Payamino - INIAP Ecuador
Amaranthus hybridus	5	DENAREF-INIAP Ecuador
Pachyrhizus erosus	13	DENAREF-INIAP Ecuador
Eugenia stipitata	15	Station Portoviejo - INIAP Ecuador
Pachyrhizus ahipa	17	DENAREF-INIAP Ecuador
Vaccinium spp.	29	DENAREF-INIAP Ecuador
Borojoa patinoi	30	Station Napo Payamino - INIAP Ecuador
Cyclanthera pedata	33	DENAREF-INIAP Ecuador
Ipomoea batata	33	DENAREF-INIAP Ecuador
Cyphomandra betacea	38	DENAREF-INIAP Ecuador
Pachyrhizus tuberosus	39	DENAREF-INIAP Ecuador
Vasconcella pentagona	47	UNL Ecuador
Amaranthus caudatus	54	DENAREF-INIAP Ecuador
Physalis peruviana	64	DENAREF-INIAP Ecuador
Rubus spp.	79	DENAREF-INIAP Ecuador
Carica	113	DENAREF-INIAP Ecuador
Bactris gasipaes	145	Station Napo Payamino - INIAP Ecuador

Cucurbita spp.	147	DENAREF-INIAP Ecuador
Solanum quitoense	163	UNL Ecuador
Solanum quitoense	168	DENAREF-INIAP Ecuador
Solanum tuberosum (incl. ssp. *andigena)*	237	DENAREF-INIAP Ecuador
Passiflora spp.	239	DENAREF-INIAP Ecuador
Carica sp.	336	UNL Ecuador
Capsicum annuum	355	UNL Ecuador
Capsicum spp.	370	DENAREF-INIAP Ecuador
Lupinus mutabilis	396	DENAREF-INIAP Ecuador
Chenopodium quinua	672	DENAREF-INIAP Ecuador
Annona cherimola	961	UNL Ecuador
Phaseolus spp.	3,100	DENAREF-INIAP Ecuador
Dioscorea alata	1	INIA SURDIGEB Peru
Eugenia uniflora	1	INIA SURDIGEB Peru
Passiflora edulis	1	INIA SURDIGEB Peru
Spondias spp.	1	INIA SURDIGEB Peru
Theobroma grandiflora	1	INIA SURDIGEB Peru
Carica pubescens	2	INIA SURDIGEB Peru
Annona muricata	3	INIA SURDIGEB Peru
Physalis peruviana	3	INIA SURDIGEB Peru
Pachyrrhyzus ahipa	4	INIA SURDIGEB Peru
Xanthosoma sagittifolium	4	INIA SURDIGEB Peru
Colocasia esculenta	6	INIA SURDIGEB Peru
Calathea allouia	8	INIA SURDIGEB Peru
Amaranthus hybridus	12	INIA SURDIGEB Peru
Anacardium occidentale	12	INIA SURDIGEB Peru
Cucurbita spp.	14	INIA SURDIGEB Peru
Dioscorea trifida	15	INIA SURDIGEB Peru
Mirabilis expansa	19	INIA SURDIGEB Peru
Lepidium meyenii	21	INIA SURDIGEB Peru
Canna indica	25	INIA SURDIGEB Peru
Plukenetia volubilis	42	INIA SURDIGEB Peru
Camu camu	43	INIA SURDIGEB Peru
Cyclanthera pedata	46	INIA SURDIGEB Peru
Phaseolus lunatus	47	INIA SURDIGEB Peru
Lucuma obovata	95	INIA SURDIGEB Peru
Phaseolus vulgaris (ñuña)	98	INIA SURDIGEB Peru
Capsicum spp.	105	INIA SURDIGEB Peru
Bactris gasipaes	113	INIA SURDIGEB Peru
Vigna unguiculata	114	INIA SURDIGEB Peru
Smallanthus sonchifolius	136	INIA SURDIGEB Peru
Passiflora spp.	158	INIA SURDIGEB Peru
Arracacia xanthorrhiza	174	INIA SURDIGEB Peru
Opuntia ficus indica	176	INIA SURDIGEB Peru
Cyphomandra betacaea	193	INIA SURDIGEB Peru
Capsicum pubescens	200	INIA SURDIGEB Peru
Chenopodium pallidicaule	267	INIA SURDIGEB Peru
Tropaeolum tuberosum	310	INIA SURDIGEB Peru
Annona cherimola	383	INIA SURDIGEB Peru
Amaranthus caudatus	486	INIA SURDIGEB Peru
Ullucus tuberosus	702	INIA SURDIGEB Peru
Manihot esculenta	740	INIA SURDIGEB Peru
Chenopodium quinoa	1,936	INIA SURDIGEB Peru
Lupinus mutabilis	2,103	INIA SURDIGEB Peru
Oxalis tuberosa	2,217	INIA SURDIGEB Peru

* = including ssp. andigena, native potatoes.

Table A3: Number of accessions of selected underutilized species, conserved in Brazilian genebanks. Source: country report to FAO and WIEWS

Crop	No. Accessions	Holding Institutions
Colocasia	3	various
Eugenia stipitata	6	various
Pachyrrhizus sp.	8	various
Dioscorea spp.	10	various
Spondias purpurea	11	various
Spondias spp.	21	various
Arracacia xanthorrhiza	22	various
Anacardium occidentale	35	various
Spondias spp.	42	various
Rubus spp.	60	various
Myrciaria dubia	70	various
Acca sellowiana	76	various
Annona muricata	105	various
Spondias tuberosa	123	various
Eugenia uniflora	132	various
Spondias mombin	136	various
Acca sellowiana	165	various
Acca sellowiana	193	various
Carica papaya	210	various
Psidium guayaba	323	various
Theobroma grandiflora	529	various
Anacardium occidentale	643	various
Passiflora spp.	1,292	various
Vigna unguiculata	1,787	various
Phaseolus lunatus	2,673	various
Capsicum spp.	3,137	various
Cucurbita spp.	5,675	various

Table A4: Number of accessions of selected underutilized species, conserved in genebanks of Mexico. Source: WIEWS

Crop (genus)	No. Accessions	Holding Institutions
Amaranthus spp.	700	University of Chapingo
Anacardium occidentale	14	University of Chapingo
Annona cherimola	2	INIFAP
Annona chirimola	12	University of Chapingo
Annona muricata	8	University of Chapingo
Annona squamosa	1	INIFAP
Bactris gasipaes	8	INIFAP
Bactris spp.	1	University of Chapingo
Capsicum annuum	248	University of Chapingo
Capsicum annuum	1,211	INIFAP
Capsicum chinense	25	University of Chapingo
Capsicum chinense	100	INIFAP
Capsicum spp.	3,350	INIFAP
Carica papaya	1	INIFAP
Carica spp.	60	INIFAP
Chrysobalanus icaco	1	University of Chapingo
Cnidoscolus chayamansa	23	University of Chapingo
Colocasia esculenta	8	University of Chapingo
Cucurbita spp.	1,580	INIFAP
Dioscorea alata	6	University of Chapingo
Dioscorea bulbifera	8	University of Chapingo
Eugenia spp.	5	University of Chapingo
Feijoa sellowiana	2	INIFAP

Hylocereus undatus	9	University of Chapingo
Ipomea batatas	15	University of Chapingo
Ipomoea batatas	178	INIFAP
Lagenaria siceraria	6	INIFAP
Luffa cylindrica	2	INIFAP
Mammea americana	3	University of Chapingo
Manihot esculenta	30	University of Chapingo
Manihot spp.	200	INIFAP
Manilkara zapota	9	University of Chapingo
Maranta arundinacaea	2	University of Chapingo
Opuntia spp.	133	INIFAP
Pachyrhizus erosus	49	INIFAP
Passiflora edulis	4	University of Chapingo
Passiflora edulis	2	INIFAP
Persea americana	14	University of Chapingo
Persea americana	58	INIFAP
Phaseolus acutifolius	88	University of Chapingo
Phaseolus acutifolius	40	INIFAP
Phaseolus coccineus	209	University of Chapingo
Phaseolus dumosus	104	University of Chapingo
Phaseolus lunatus	93	University of Chapingo
Pouteria sapota	5	University of Chapingo
Psidium guajaba	8	University of Chapingo
Psidium guajava	4	INIFAP
Psidium spp.	4	University of Chapingo
Rubus occidentalis	6	INIFAP
Sechium edule	93	University of Chapingo
Spondias mombim	3	University of Chapingo
Spondias purpurea	17	University of Chapingo

REFERENCES AND NOTES

1. PAR. *Biodiversity for Food and Agriculture: Contributing to food security and Sustainability in a Changing World*; Outcomes of an Expert Workshop held by FAO and the Platform on Agrobiodiversity Research: Rome, Italy; 14–16; April; 2010.
2. FAO. *Save and Grow: A Policymakers' Guide to the Sustainable Intensification of Smallholder Crop Production*; Food and Agriculture Organization of the United Nations: Rome, Italy, 2011. Available online: http://www.fao.org/docrep/014/i2215e/i2215e.pdf (accessed on 10 July 2013).
3. FAO. *Report on the State of the World's Plant Genetic Resources for Food and Agriculture*; Food and Agriculture Organization of the United Nations: Rome, Italy, 2010.
4. Prescott, A.R.; Prescott, A.C. How many plants feed the world? *Conserv. Biol.* 1990, *4*, 365–374.
5. Rehm, S.; Espig, G. *The Cultivated Plants of the Tropics*; Verlag Josef Margraf and CTA: Weikersheim, Germany, 1991.
6. Wilson, E.O. *The Diversity of Life*; Penguin: London, UK, 1992.
7. Vavilov, N.I. Studies on the origin of cultivated plants. *Russ. Bull. Appl. Bot. Plant Breed.* 1926, *14*, 245.
8. Vavilov, N.I. *Origin and Geography of Cultivated Plants*; Cambridge University Press: Cambridge, UK, 1992.
9. National Research Council. *Lost Crops of the Incas: Little-Known Plants of the Andes with Promise for Worldwide Cultivation*; National Academy Press: Washington, DC, USA, 1989.
10. Hazell, P. The Green Revolution. In *Oxford Encyclopedia of Economic History*; Mokyr, J., Ed.; Oxford University Press: Oxford, UK, 2003.
11. Padulosi, S.; Eyzaquirre, P; Hodgkin, T. Challenges and strategies in promoting conservation and use of neglected and underutilised crop species. In *Perspectives on New Crops and New Uses*; Janick, J., Ed.; ASHS Press: Alexandria, VA, USA, 1999.
12. Padulosi, S.; Hoeschle-Zeledon, I.; Bordoni, P. Minor crops and underutilised species: Lessons and prospects. In *Crop Wild Relative Conservation and Use*; Maxted, N., Ford-Lloyd, B.V., Kell, S.P., Iriondo, J.M., Dulloo, M.E., Turok, J., Eds.; CAB International: Wallingford, UK, 2008; pp. 605–624.
13. Thies, E. *Promising and Underutilised Species, Crops and Breeds*; Deutsche Gesellschaft F√r Technische Zusammenarbeit (GTZ): Eschborn, Germany, 2000. Available online: http://www.ilri.cgiar.org/InfoServ/Webpub/fulldocs/AnGenResCD/docs/PDFFORMATFILES/PromUnderutilizeSpecCropBreed.pdf (accessed on 15 August 2013).
14. Williams, J.T.; Haq, N. *Global Research on Underutilised Crops. An Assessment of Current Activities and Proposals for Enhanced Cooperation*; International Centre for Underutilised Crops (ICUC): Southampton, UK, 2002.
15. Padulosi, S.; Hoeschle-Zeledon, I. Underutilised plant species: What are they? Available online: http://www.mtnforum.org/sites/default/files/publication/files/3800.pdf (accessed on 15 August 2013).

16. Google Scholar. Available online: http://scholar.google.com.co/ (accessed on 03 July 2013).

17. Genesys. Available online: http://www.genesys-pgr.org/ (accessed on 03 July 2013).

18. Ebert, A. *Asegurando nuestro futuro: colecciones de germoplasma del CATIE. Securing our future: CATIE's germoplasm collections*; Boletín Técnico (CATIE): Turrialba, Costa Rica, 2007. (in Spanish)

19. AVGRIS. AVDRC Vegetable Genetic Resources Information System. Available online: http://203.64.245.173/search.asp(accessed on 03 July 2013).

20. World Information and Early Warning System on Plant Genetic Resources for Food and Agriculture. Available online:http://apps3.fao.org/wiews/wiews.jsp (accessed on 21 June 2013).

21. Country Reports. Available online: http://www.fao.org/agriculture/crops/core-themes/theme/seeds-pgr/sow/sow2/country-reports/en/ (accessed on 21 June 2013).

22. FAO. *Second Report on the State of the World's Plant Genetic Resources for Food and Agriculture*; Commission on Genetic Resources for Food and Agriculture, Food and Agriculture Organization of the United Nations: Rome, Italy, 2010.

23. List of International Crop Descriptors. Available online: http://www.bioversityinternational.org/fileadmin/user_upload/research/research_portfolio/Information_systems_for_crop/List_of_Descriptor_Titles_2013.pdf (accessed on 29 September 2013).

24. Global Biodiversity Information Facility. Available online: http://data.gbif.org/ (accessed on 20 June 2013).

25. FAOSTAT. Available online: http://faostat.fao.org/ (accessed on 20 June 2013).

26. Plant Variety Database of the Union for Protection of Plant Varieties (UPOV). Available online: https://www3.wipo.int/pluto/user/en/index.jsp (accessed on 29 September 2013).

27. Project database of the Global Environmental Facility. Available online: http://www.thegef.org/gef/gef_projects_funding (accessed on 1 June 2013).

28. Project database of the Benefit Sharing Fund of the International Treaty on Plant Genetic Resources for Food and Agriculture. Available online: http://www.planttreaty.org/content/projects-2009-2011 (accessed on 1 June 2013).

29. Project database of the Global Crop Diversity Trust. Available online: http://www.croptrust.org/content/evaluation(accessed on 17 July 2013).

30. Project database of the Global Plan of Action on Plant Genetic Resources for Food and Agriculture. Available online: http://www.globalplanofaction.org/servlet/CDSServlet?status=ND0xMDcyJjY9ZW4mMzM9KiYzNz1rb3M~ (accessed on 17 July 2013).

31. Crops for the Future. Available online: http://www.cropsforthefuture.org/ (accessed on 1 June 2013).

32. Renkow, M.; Byerlee, D. The impacts of CGIAR research: A review of recent evidence. *Food Policy* 2010, *35*, 391–402.

33. Padulosi, S.; Hodgkin, T.; Williams, J.T.; Haq, N. Underutilised Crops: Trends, challenges and opportunities in the 21st century. In *Managing Plant Genetic Resources*; Engels, J.M.M., Ramanta Rao, V., Brown, A.H.D., Jackson, M.T., Eds.; CABI: Wallingford, UK; IPGRI: Rome, Italy, 2002; pp. 323–338.

34. Van Zonneveld, M; Scheldeman, X.; Escribano, P.; Viruel, M.; van Damme, P.; Garcia, W.; Tapia, C.; Romero, J.; Sigueñas, M.; Hormaza, J.I. Mapping genetic diversity of cherimoya (*Annona cherimola* Mill.): Application of spatial analysis for conservation and use of plant genetic resources. *Plos One* 2012.

35. Thomas, E.; van Zonneveld, M.; Loo, J.; Hodgkin, T.; Galluzzi, G.; van Etten, J. Present spatial diversity patterns of*Theobroma cacao* L. in the neotropics reflect genetic differentiation in Pleistocene refugia followed by human-Influenced dispersal. *Plos One* 2012.

36. Instituto Brasileiro de Geografia e Estatistica. Available online: http://downloads.ibge.gov.br/downloads_estatisticas.htm (accessed on 20 October 2013). (In Portuguese).

37. Portal de la Secretaría de Desarrollo Agropecuario del Gobierno del Estado de México. Available online: http://www.edomex.gob.mx/portal/page/portal/sedagro/estadisticas-agropecuarias (accessed on 20 October 2013). (In Spanish).

38. Series históricas de producción agrícola, Ministerio de Agricultura del Perú (in Spanish). Available online: http://frenteweb.minag.gob.pe/sisca/?mod=consulta_cult (accessed on 20 October 2013).

39. Altieri, M.A.; Merrick, L.C. In situ conservation of crop genetic resources through maintenance of traditional farming systems. *Econ. Bot.* 1987, *41*, 86–96.

40. Ceccarelli, S. Evolution, plant breeding and biodiversity. *J.Agric. Environ. Int. Dev.* 2009, *103*, 131–145.

41. Gabriel, J.; Torrez, R.; Thiele, G. Participatory approaches in potato improvement: Experience of PROINPA in Bolivia. In *Encouraging Diversity: The Conservation and Development of Plant Genetic Resources*; Almekinders, C.J.M., de Boef, W.S., Eds.; Intermediate Technology Publications: London, UK, 2000.

42. Devaux, A.; Thiele, G. *Compendio de papa andina, logros y experiencias de la primera fase (1998–2002)*; (in Spanish). Compendio, International Potato Center (CIP): Lima, Peru, 2002.

43. Gomez, M. Fundación para la Investigación Participativa con Agricultores de Honduras (FIPAH), Yorito, Honduras. Personal communication, 2013.

44. Alonso, S. Asociación de Organizaciones de los Cuchumatanes, ASOCUCH, Huehuetenango, Guatemala. Personal communication, 2013.

45. CBD. *Annex of Decision V/5 of the Conference of the Parties to the Convention on Biological Diversity*; Convention on Biological Diversity: Montreal, Canada, 2000. Available online: https://www.cbd.int/decision/cop/default.shtml?id=7147 (accessed on 18 February 2014).

46. Fowler, C. Biological diversity in a north-south context. In *Green Globe Yearbook*; Oxford University Press: Oxford, UK, 1993; pp. 35–44.

47. Sedjo, R. Property rights, genetic resources, and biotechnological change. *J Law Econ.* 1992, *35*, 199–213.

48. Reji, K.J. International regime on access and benefit sharing: Where are we now? *Asian Biotechnol. Dev. Rev.* 2010, *12*, 77–94.

49. Dutfield, G. *Intellectual Property Rights, Trade and Biodiversity: The Case of Seeds and Plant Varieties*; Oxford University Press: Oxford, UK, 1999.

50. Petit, M.; Fowler, C.; Collins, W.; Correa, C.; Thornström, C.G. *Why Governments Can't Make Policy. The Case of Plant Genetic Resources in the International Arena*; International Potato Center (CIP): Lima, Peru, 2001.

51. Cabrera Medaglia, J.; Perron-Welch, F.; Rukundo, O. *Overview of National and Regional Measures on Access and Benefit Sharing. Challenges and Opportunities in Implementing the Nagoya Protocol*, 2nd ed.; Centre for International Sustainable Development Law (CISDL): Montreal, Canada, 2012.

52. Correa, C.M. Do national access regimes promote the use of genetic resources and benefit sharing? *Int. J. Environ. Sus. Dev.* 2005, *4*, 444–463.

53. Lapeña, I.; Sig✓eñas Saavedra, M. Incentives and disincentives for Peru to participate in the multilateral system of the International Treaty on Plant Genetic Resources for Food and Agriculture. *The Multilateral System of Access and Benefit-Sharing. Case Studies on Implementation in Kenya, Morocco, Philippines and Peru*; López Noriega, I., Halewood, M., Lapeña, I., Eds.; Bioversity International: Rome ,Italy, 2013. Available online: http://www.bioversityinternational.org/nc/publications/publication/issue/the_multilateral_system_of_access_and_benefit_sharing.html (accessed on 16 October 2013).

54. FAO. *International Treaty on Plant Genetic Resources for Food and Agriculture*; Food and Agriculture Organization of the United Nations: Rome, Italy, 2001. Available online: http://www.planttreaty.org/content/texts-treaty-official-versions(accessed on 19 October 2013).

55. GCDT. Towards a Rational Hemispheric Conservation Strategy for Plant Genetic Resources for Food and Agriculture in the Americas. The Global Crop Diversity Trust: Rome, Italy, 2008. Available online: http://www.croptrust.org/documents/regionalstrategies/Americas.pdf (accessed on 12 July 2013).

56. López Noriega, I.; Wambugu, P.; Mejías, A. Assessment of progress to make the multilateral system functional: Incentives and challenges at the country level. In *Crop Genetic Resources as a Global Commons. Challenges in International Law and Governance*; Halewood, M., López Noriega, I., Louafi, S., Eds.; Routledge: London, UK; New York, NY, USA, 2013; pp. 199–225.

57. Representación de la FAO en Bolivia; Oficial de Agricultura del Servicio de Semillas y Recursos Fitogenéticos; Vice ministerio de Medio Ambiente, Biodiversidad y Cambios Climáticos del Ministerio de Medio Ambiente y Aguas; Instituto Nacional de Innovación Agropecuaria y Forestal; y Ministerio de Desarrollo Rural y Tierras. *Bolivia. Informe Nacional sobre el Estado de los Recursos Fitogenéticos para la Agricultura y la Alimentación*; FAO: Rome, Italy, 2009. Available online: http://www.fao.org/docrep/013/i1500e/Bolivia.pdf (accessed on 20 October 2013). (In Spanish).

58. Tapia, C.; Zambrano, E.; Monteros, A. *Ecuador. Informe Nacional sobre el Estado de los Recursos Fitogenéticos para la Agricultura y la Alimentación*; INIAP and FAO, 2009. Available online: http://www.fao.org/docrep/013/i1500e/Ecuador.pdf (accessed on 20 October 2013). (In Spanish).

59. Becerra Gallardo, R.; Sig✓eñas Saavedra, M. *Perú: Segundo informe sobre el estado de los recursos fitogenéticos para la alimentación y la agricultura*; INIA and FAO: Lima, Peru and Rome, Italy, 2009; Available online: ://www.fao.org/docrep/013/i1500e/Peru.pdf (accessed on 20 October 2013). (In Spanish)

60. Da Silva Mariante, A., Amstalden Sampaio, M.J., Valadares Inglis, M.C., Eds.; *State of Brazil's Plant Genetic Resources. Second Report*; Embrapa and FAO: Brasilia, DF, Brazil and Rome, Italy, 2009. Available online: http://www.fao.org/docrep/013/i1500e/Brazil.pdf (accessed on 20 October 2013).

61. Ruiz, M. *Agrobiodiversity Zones and the Register of Native Crops in Peru: Learning from Ourselves*; Sociedad Peruana de Derecho Ambiental and Bioversity International: Lima, Perú and Rome, Italy, 2009.

62. Molina Moreno, J.C., Córdova Téllez, L., Eds.; *Recursos Fitogenéticos en México para la Alimentación y la Agricultura. Informe Nacional*; (in Spanish). SAGARPA, SOMEFI and FAO, 2006. Available online: http://www.fao.org/docrep/013/ i1500e/Mexico.pdf (accessed on 20 October 2013).

63. R Development Core Team. *R: A Language and Environment for Statistical Computing*; version 3.0.0. R Foundation for Statistical Computing: Vienna, Austria, 2013. Available online: http://www.R-project.org (accessed on 10 May 2013).

64. Chiarolla, C.; Jungcurt, S. *Outstanding Issues on Access and Benefit Sharing under the Multilateral System of the International Treaty of Plant Genetic Resources for Food and Agriculture*; Background study paper, The Bern Declaration and the Norwegian Development Fund: Lausanne, Switzerland and Lisaker, Norway, 2011. Available online: http://www.evb.ch/ cm_data/ITPGR_ABS_Study_1.pdf (accessed on 25 October 2013).

65. Halewood, M.; Deupmann, P.; Sthapit, B.; Vernooy, R.; Ceccarelli, S. *Participatory Plant Breeding to Promote Farmers' Rights*; Bioversity International: Rome, Italy, 2007. Available online: http://www.bioversityinternational.org/e-library/publications/detail/participatory-plant-breeding-to- promote-farmers-rights/ (accessed on 23 October 2013).

66. Andersen, R. *Realising Farmers' Rights under the*

International Treaty on Plant Genetic Resources for Food and Agriculture. Summary of Findings from the Farmers' Rights Project, Phase 1; The Fridtjof Nansen Institute: Lisaker, Norway, 2006. Available online: http://www.fni.no/doc&pdf/FNI-R1106.pdf (accessed on 23 October 2013).

67. Jacobsen, S.E. The situation for quinoa and its production in Southern Bolivia: From economic success to environmental disaster. *J. Agron. Crop Sci.* 2011, *197*, 390–399.

68. Winkel, T.; Bertero, H.D.; Bommel, P.; Bourliaud, J.; Chevarría Lazo, M.; Cortes, G.; Gasselin, P.; Geerts, S.; Joffre, R.; Léger, F.; *et al.* The sustainability of quinoa production in Southern Bolivia: From misrepresentations to questionable solutions. *J. Agron. Crop Sci.* 2012, *198*, 314–319.

69. Giuliani, A., Hintermann, F., Rojas, W., adulosi, S., Eds.; *Biodiversity of Andean Grains: Balancing Market Potential and Sustainable Livelihoods*; Bioversity International: Rome, Italy, 2012. Available online: http://www.bioversityinternational.org/uploads/tx_news/Biodiversity_and_Andean_grains__balancing_market_potential_and_sustainable_livelihoods_1635.pdf (accessed on 20 October 2013).

70. Kruijssen, F.; Giuliani, A.; Sudha, M. Marketing underutilised crops to sustain agrobiodiversity and improve livelihoods. *Acta Hort.* 2009, *806*, 415–422.

71. Gruere, G.L.; Nagarajan, O.; King, E.D.I. The role of collective action in the marketing of underutilized plant species: Lessons from a case study on minor millets in South India. *Food Policy* 2009, *34*, 39–45.

72. Will, M. *Promoting Value Chains of Neglected and Underutilized Species for Pro-Poor Growth and Biodiversity Conservation. Guidelines and Good Practices*; Bioversity International and Global Facilitation Unit for Underutilized Species (GFU): Rome, Italy, 2008. Available online: http://www.underutilized-species.org/documents/PUBLICATIONS/promoting_vc.pdf (accessed on 20 October 2013).

73. Hellin, J.; Higman, S. Crop diversity and livelihood security in the Andes. *Dev. Pract.* 2005, *15*, 165–174.

74. Altieri, M.A.; Nicholls, C.I. Scaling up agroecological approaches for food sovereignty in Latin America. *Development*2008, *51*, 472–480.

75. Johns, T.E.; Eyzaguirre, P.B. Linking biodiversity, diet and health in policy and practice. *P. Nutr. Soc.* 2006, *65*, 182–189.

76. Toledo, A.; Burlingame, B. Biodiversity and nutrition: A common path toward global food security and sustainable development. *J. food compo. Anal.* 2007, *19*, 477–483.

77. Padulosi, S.; Heywood, V.; Hunter, D.; Jarvis, A. Underutilised species and climate change: Current status and outlook. In *Crop Adaptation to Climate Change*; Yadav, S.S., Redden, R., Hatfield, J.L., Hall, A., Eds.; Wiley-Blackwell: Hoboken, NJ, USA, 2011; pp. 507–521.

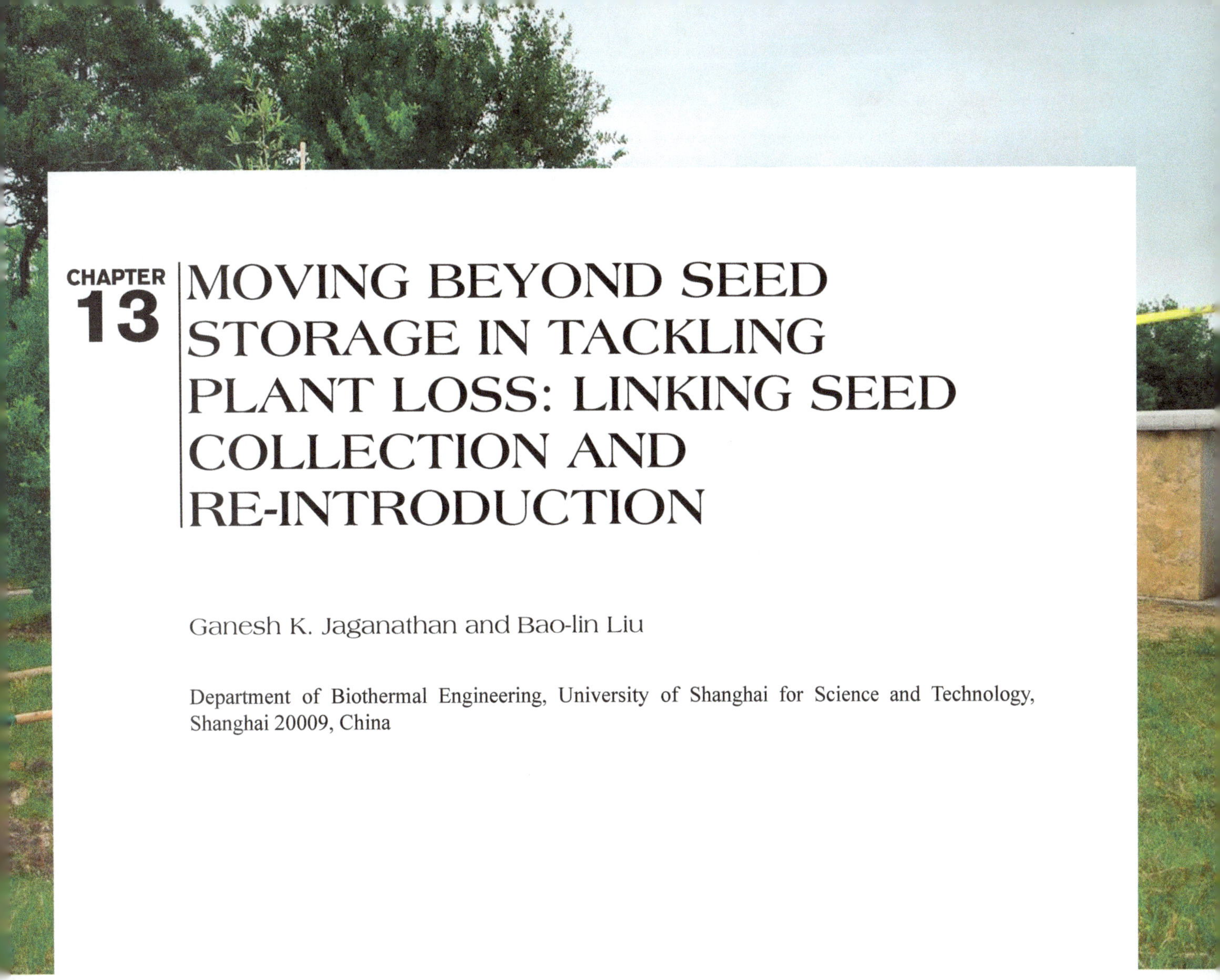

CHAPTER 13

MOVING BEYOND SEED STORAGE IN TACKLING PLANT LOSS: LINKING SEED COLLECTION AND RE-INTRODUCTION

Ganesh K. Jaganathan and Bao-lin Liu

Department of Biothermal Engineering, University of Shanghai for Science and Technology, Shanghai 20009, China

ABSTRACT

Globally, plants are becoming extinct at an unprecedented rate. Seed banking is considered as the most cost-effective technique to preserve many conservation concern species. From a practical standpoint, however, banking seeds is essentially just the first step in tackling plant loss, as the real-life challenge lies in how well these seeds when re-introduced become whole plants and maintain species count. In the current model, merely very few seed banks extend their efforts in re-introduction problems. Compelling evidence suggests that many re-introduction efforts are less than successful. Such axiom convicts seed banks must also participate in the 'post-storage' process and engage working with other ecologists, physiologists and re-introduction practitioners to devise a robust re-introduction practice and exchange information about the seed lot including collection site, maternal environment etc. Bridging these gaps would facilitate enhanced restoration practice and pave way for efficient reconstruction of the ecosystem.

INTRODUCTION

Globally, plants are becoming extinct at an unprecedented rate. Despite the growing awareness on conserving threatened flora, the pace at which plant loss occurs has intensified 10-to 100 fold in the recent years leading to vibrant challenges for conserving at least the remnant species (Corvalan et al., 2005). To preserve plant diversity, several conservation measures are being planned often one approach complementing other- amongst which banking seeds is possibly the most cost-effective, yet promising technique to preserve many species (Hay and Probert, 2013). Seeds collected from wild and dried to lower moisture contents (3-10% moisture on fresh weight basis) can be held at low temperatures (e.g. -196°C) for centuries without any alterations to genetic makeup or viability, thus offering great advantage both in space and time (Walters et al., 2013). The fact that changing climate may impose serious threats to any species has underpinned the need for banking seeds of possibly all species. In order to meet this elaborate goal, seed banks around the world are ambitiously attempting to collect large quantities of seeds and a great deal of money exceeding several million dollars is spent to ensure the global flora is preserved (Walck and Dixon, 2009; Hay and Probert, 2013).

A timely question tied to the seed storage practice is does successful banking of seeds imply the species are effectively conserved? Current state-ofart promises that the stored seeds containing future plants, once stored at low temperatures do not undergo any metabolic activity, thereby providing genetic materials for species even if the species had gone extinct

Citation: Ganesh K. Jaganathan and Bao-lin Liu, Moving beyond seed storage in tackling plant loss: linking seed collection and re-introduction, DOI: http://dx.doi.org/10.14719/pst.2016.3.2.202.

in wild (Hay and Probert, 2013). However, the crux of this problem goes beyond this simplified acceptance. Seed storage is essentially just the first step in embarking the unabridged problem. Although gene banks aim to store seeds for longer time as an insurance policy, these seeds are not meant to be stored interminably (Hamilton, 1994). Further, there are many seeds which are stored for short-time and supplied when required for sowing in wild. In real world scenario, the tangible challenge lies in successfully re-introducing the stored seeds in suitable habitats, in a way that those seeds grow to full plants, reproduce and produce next generation seeds without any assistance to maintain species count (Hamilton, 1994; Guerrant Jr and Pavlik, 1998).

When viewed from this perspective, the extent to which seed banks fulfil the intended tasks of conservation per se depends on how well the reintroductions are planned (Husband and Campbell, 2004). Most seed banks consider collection, cleaning, processing and storage of seeds as their pivotal roles, with viability assessments during storage also form an integral part. Thus, seed banks take less participation in the 'tail-ended' events of converting stored seeds in to whole plants in the wild (Maunder et al., 2004). However, seed banks have distinctive roles in ensuring the stored seeds are utilized in a more valuable way to restore the species that had been or becoming lost, otherwise the efforts are futile. Thus, one of the tenacious issues for seed banks is to efficiently formulate highly resourceful reintroduction protocols in conjugation with other re-introduction practitioners. This is a daunting task especially when re-introductions are planned at community level because every species require certain amount of space and specific demand to complete life-cycles (Margules and Pressey, 2000). Nevertheless, to achieve such as irksome goal, seed banks must tackle various constrains and approach the problem from systematic multidimensional standpoints, if not extinction of reintroduced species will be the consequence.

Some hundreds of hectares are cleared every year for creating living space and agricultural purposes for the ever increasing population (Achard et al., 2002). Due to the effects of global warming, more vegetation are becoming patchy and numerous species are susceptible to substantial biotic and abiotic pressure (Achard et al., 2002; Malhi et al., 2008). It has become a common practice to prioritize endangered species collection in many countries, where funding for plant conservation particularly to seed storage is extremely limited (Farnsworth et al., 2006). However, practicing these approaches requires reassessment, because the area of deforestation is extremely high and many species having extremely specific growing conditions, present only in a smaller region will be lost (Pressey and Taffs, 2001; Husband and Campbell, 2004; Bachmann et al., 2013). Most of the accessions held in seed banks are skewed towards horticultural crops, leaving rare or endangered species more vulnerable (Godefroid et al., 2011). This means seed banks must systematically devise when, where and how to collect the seeds of multiple species. Since the seed maturation time varies with species and habitats, it is important that more than one collection time per year is planned based on the location and seeds must be immediately transformed to storage conditions. More importantly, care must be taken not to disturb the natural re-establishment process, especially if the collected species is endangered or threatened, which has lesser number of plants and limited turn-over of seed counts (Menges, 1998). Knowledge on re-introductions are lagging behind the seed storage practice (Guerrant Jr and Kaye, 2007). One of the important bottleneck is to envisage when and where the species stored in the form of seeds will be re-introduced (Fiedler and Laven, 1996; Guerrant Jr and Kaye, 2007). To this end, economic and political motives often influence site selection for re-introduction. Notably more re-introduction sites are often established in remote or landscapes not suitable for agriculture (Osborne and Seddon, 2012). However, re-introduction site must have the identical or at least similar conditions on which the collected seeds could have otherwise germinated (Maunder, 1992). In addition, reintroduction site needs to be undisturbed by direct or indirect human activities for long-time. One way to do this is by expanding the 'in-situ reserves' already established. This approach would minimize the risk of species threatened by patchy vegetation. Nevertheless, the underlying problem with this practice is the introduction of species to new habitat, presumably as invasive species, which might affect the endemic species of introduced ecosystem. Thus, not all species are suited for growing in or around reserves, as some 70-85% of the reserves are established near the mountains or other drier landscapes (Achard et al., 2002). This raises an important challenge, if a particular forest space is converted for other purpose, whether the seeds collected from that region bears any significance in restoring the vegetation.

These facts underpin seed banks to take more responsibilities by engaging with policy-makers and government authorities to optimize the site selection and re-introduction methodology. First, it may seem tedious, seed banks must start working with ecologists, physiologists and conservationists to explore why a particular species that merits storage actions is becoming conservation concern. This involves complex understanding of species interactions at ecosystem level, reproduction biology, germination ecology, environmental stress etc. Second, seed banks must team up with academic researchers and other practitioners who are enthralled with re-introduction to conduct a 'miniature experiment' in the field, possibly involving complex-network of species, using stored seeds to understand the problems of the reintroduced species might face in wild during establishment before large-scale re-introductions are planned. When planning re-introductions based on the study results, necessary steps should be taken to minimise the risks, failure to do this will inevitably push the re-introduced species becoming extinct again in the near future. One case study offered by Johnson et al. (2004) is that in a Douglas-fir provenance plantation, plants grown from re-introduced seed sources from different sites suffered more damage during an unusual cold spell compared with local populations. Our ongoing studies on tropics also reveal similar conclusion. Briefly, our results demonstrate that plants growing from seeds from local vegetation survive severe drought than seeds from slightly moderate conditions. These pattern invokes a special attention when planning reintroduction of stored seeds.

Examples pertaining to whether or not seeds of conservation concern species effectively reintroduced in wild are scarce. Most of the seed banks aim to store seeds for longer time. According to Walck and Dixon (2009) when the seeds collected in the present environmental conditions are banked for longer time and reintroduced in the wild after several decades or centuries the environment at which these seeds have to germinate and establish would be very different than these seeds were adapted to. Hence, they recommended collecting seeds of each species at regular intervals to ensure more seeds are establishing to full plants in future climatic conditions. As such, this notion questions the storage of seeds for centuries as an insurance policy against plant loss, because these seeds having lesser ability to grow in future climate. The destruction of forest results in forced reintroduction of seeds in to other sites. Theilade and Petri (2003) noted that the seeds of Sophora toromiro, an extinct tree of Easter Island, has been successfully collected and stored from the last tree in 1960. These seeds were stored initially in Naturhistoriska Riksmuseet in Stockholm and later transferred to Gothenburg Botanical Gardens. The success of re-introduction of this tree will largely depend on many factors such as the present environmental conditions, the co-existing species etc. The change in vegetation from the seed maturation time and the time of re-introduction would have serious impact on the re-introduced seeds (Cogoni et al., 2013). Therefore, information on the seed maturation site must be taken in to account when planning the re-introduction protocols. Seed banks must engage in transforming the information of seed origin and play a role in site selection for re-introduction. In case the stored seeds are used for re-introduction, the lesser involvement of seed bank in these process could often lead to less successful results.

While it may seem that the issue of reintroduction is less inclined to the overarching aim of 'seed storage' by seed banks, the steady increase in plant loss due to the climate change calls for an extensive participation of seed bank in converting the stored seeds in to whole plants and further collect seeds from the re-introduced species. Seeds collected from one country and stored in a different country can be re-introduced in a very different place. Demand for the species often results in using stored seeds of any origin when reintroduction projects are initiated. As the present model for re-introduction does not require the direct participation of seed banks in reintroduction, as the 'post-bank' period is handled by different expertise, results of many reintroduction projects are often less successful (Maunder, 1992; Faulk et al., 1996). With lesser information available to the re-introduction practitioners about the seed maturation site, climate etc., there is an increased chance seeds collected from a particular site end up germinating and growing in a less hostile site, presumably posing serious risk for successful plant growth. This situation convinces the joint action of seed banks in re-introduction projects. Needless to mention, coping up with all these goals will be lamentably arduous in the present situation, but if the precious money and effort being spent on banking seeds for decades are to be exploited in reconstructing our ecosystem, seed banks must inevitably take steps to meet these challenges.

REFERENCES

1. Achard, F., H. D. Eva, H.-J. Stibig, P. Mayaux, J. Gallego, T. Richards, and J.-P. Malingreau. 2002. Determination of deforestation rates of the world's humid tropical forests. Science 297: 999-1002. doi: 10.1126/science.1070656

2. Bachmann, P., M. Köhl, and R. Päivinen. 2013. Assessment of biodiversity for improved forest planning. Springer Science & Business Media Vol 51.

3. Cogoni, D., G. Fenu, E. Concas, and G. Bacchetta. 2013. The effectiveness of plant conservation measures: the Dianthus morisianus reintroduction. Oryx 47: 203-206. doi: 10.1017/S003060531200169X

4. Corvalan, C., S. Hales, and A. McMichael. 2005. Assessment, Millennium Ecosystem (2005). Ecosystems and human well-being: Biodiversity Synthesis. World Resources Institute, Washington, DC. USA Vol 5.

5. Farnsworth, E., S. Klionsky, W. Brumback, and K. Havens. 2006. A set of simple decision matrices for prioritizing collection of rare plant species for ex situ conservation. Biol Conserv 128: 1-12. doi: 10.1016/j.biocon.2005.09.010

6. Faulk, D., C. Miller, and M. Olwell. 1996. Restoring diversity: strategies for reintroduction of endangered plants. Island Press.

7. Fiedler, P. L., and R. D. Laven. 1996. Selecting reintroduction sites. Restoring diversity Island Press, Washington, DC: 157-169

8. Godefroid, S., S. Rivière, S. Waldren, N. Boretos, R. Eastwood, and T. Vanderborght. 2011. To what extent are threatened European plant species conserved in seed banks? Biol Conserv 144: 1494-1498. doi: 10.1016/j.biocon.2011.01.018

9. Guerrant Jr, E. O., and T. N. Kaye. 2007. Reintroduction of rare and endangered plants: common factors, questions and approaches. Aust J Bot 55: 362-370. doi: 10.1071/BT06033

10. Guerrant Jr, E. O., and B. M. Pavlik. 1998. Reintroduction of rare plants: genetics, demography, and the role of ex situ conservation methods. In: Conservation biology for the coming Decade. P. L. Fiedler and P. M. Kareiva, Eds. Springer, London. p. 80-108.

11. Hamilton, M. B. 1994. Ex situ conservation of wild plant species: time to reassess the genetic assumptions and implications of seed banks. Conserv Biol 8: 39-49. doi: 10.1046/j.1523-1739.1994.08010039.x

12. Hay, F. R., and R. J. Probert. 2013. Advances in seed conservation of wild plant species: a review of recent research. Conservation Physiology 1: cot030

13. Husband, B. C., and L. G. Campbell. 2004. Population responses to novel environments: implications for ex situ plant conservation. In: Ex situ plant conservation: supporting species survival in the wild. Island Press, Washington. E. O. Guerrant, K. Havens, and M. Maunder, Eds. Island Press, Washington. p. 231-266.

14. Johnson, G., F. C. Sorensen, J. B. St Clair, and R. C. Cronn. 2004. Pacific Northwest Forest Tree Seed Zones A template for native plants? Native Plants Journal 5: 131-140. doi: 10.2979/NPJ.2004.5.2.131

15. Malhi, Y., J. T. Roberts, R. A. Betts, T. J. Killeen, W. Li, and C. A. Nobre. 2008. Climate change, deforestation, and the fate of the Amazon. science 319: 169-172

16. Margules, C. R., and R. L. Pressey. 2000. Systematic conservation planning. Nature 405: 243-253. doi: 10.1038/35012251

17. Maunder, M. 1992. Plant reintroduction: an overview. Biodiversity & Conservation 1: 51-61. doi: 10.1007/

BF00700250

18. Maunder, M., K. Havens, E. O. Guerrant Jr, D. A. Falk, and D. Falk. 2004. Ex situ methods: a vital but underused set of conservation resources. In: Ex situ plant consetvation. Supporting species survival in the wild. Island Press, Washington. E. O. Guerrant, K. Havens, and M. Maunder, Eds. Island Press, Washington. p. 3-20.

19. Menges, E. S. 1998. Evaluating extinction risks in plant populations. Springer, 49-65pp.

20. Osborne, P. E., and P. J. Seddon. 2012. Selecting suitable habitats for reintroductions: variation, change and the role of species distribution modelling. Reintroduction biology: integrating science and management 1

21. Pressey, R., and K. H. Taffs. 2001. Scheduling conservation action in production landscapes: priority areas in western New South Wales defined by irreplaceability and vulnerability to vegetation loss. Biol Conserv 100: 355-376. doi: 10.1016/S0006-3207(01)00039-8

22. Theilade, I., and L. Petri. 2003. Conservation of tropical trees ex situ through storage and use. Guidelines and Technical Notes No. 65. Danida Forest Seed Centre, Humlebaek, Denmark.

23. Walck, J., and K. Dixon. 2009. Time to future-proof plants in storage. Nature 462: 721-721. doi: 10.1038/462721a

24. Walters, C., P. Berjak, N. Pammenter, K. Kennedy, and P. Raven. 2013. Preservation of recalcitrant seeds. Science 339: 915-916. doi: 10.1126/science.1230935

CHAPTER 14 GENETIC DIVERSITY AND POPULATION STRUCTURE OF AN IMPORTANT WILD BERRY CROP

Laura Zoratti[1], Luisa Palmieri[2], Laura Jaakola[3,4] and Hely Ha¨ggman[1]

[1] Department of Genetics and Physiology, University of Oulu, FI-90014 Oulu, Finland

[2] Department of Food Quality and Nutrition, Research and Innovation Center, Fondazione Edmund Mach, Via E. Mach, 1-38010 San Michele all'Adige (TN), Italy

[3] Department of Arctic and Marine Biology, UiT The Arctic University of Norway, Climate Laboratory, 9037 Tromsø, Norway

[4] Norwegian Institute of Bioeconomy Research, NIBIO Holt, 1431 A° s, Norway

ABSTRACT

The success of plant breeding in the coming years will be associated with access to new sources of variation, which will include landraces and wild relatives of crop species. In order to access the reservoir of favourable alleles within wild germplasm, knowledge about the genetic diversity and the population structure of wild species is needed. Bilberry (Vaccinium myrtillus) is one of the most important wild crops growing in the forests of Northern European countries, noted for its nutritional properties and its beneficial effects on human health. Assessment of the genetic diversity of wild bilberry germplasm is needed for efforts such as in situ conservation, on-farm management and development of plant breeding programmes. However, to date, only a few local (small-scale) genetic studies of this species have been performed. We therefore conducted a study of genetic variability within 32 individual samples collected from different locations in Iceland, Norway, Sweden, Finland and Germany, and analysed genetic diversity among geographic groups. Four selected inter-simple sequence repeat primers allowed the amplification of 127 polymorphic loci which, based on analysis of variance, made it possible to identify 85 % of the genetic diversity within studied bilberry populations, being in agreement with the mixed-mating system of bilberry. Significant correlations were obtained between geographic and genetic distances for the entire set of samples. The analyses also highlighted the presence of a north–south genetic gradient, which is in accordance with recent findings on phenotypic traits of bilberry.

INTRODUCTION

The success of plant breeding over the past century has been associated with a narrowing of the available genetic diversity within elite germplasm of species. New sources of variation include landraces and wild relatives of crop species, and although exploiting wild relatives as a source of novel alleles is challenging, it has provided notable successes in crop improvement (Tester and Langridge 2010). Most crop geneticists agree that the enrichment of the cultivated gene pool will be necessary to meet the challenges that lie ahead associated with global environmental changes (Feuillet *et al.* 2008). However, many advances are still needed to access the extensive reservoir of favourable alleles within wild germplasm. These include increasing our understanding of the molecular basis for key traits and expanding existing phenotyping and genotyping of germplasm collections (Feuillet *et al.* 2008). Therefore, knowledge of the genetic diversity and the population structure of wild species is crucial

Citation: Laura Zoratti, Luisa Palmieri, Laura Jaakola and Hely Ha¨ggman, Genetic diversity and population structure of an important wild berry crop, doi: 10.1093/aobpla/plv117.

for their management as well as conservation (Burdon and Wilcox 2007; Zhao *et al.*2014).

Vaccinium is a genus of ~450 plant species in the family Ericaceae that are widely distributed in the Northern Hemisphere and also in the mountains of tropical Asia and Central and South America (Song and Hancock 2011). The species within this genus present different levels of ploidy (2*x*, 4*x* and 6*x*; *x* = 12), which results in evident morphological differences. Regarding domestication and commercial fruit crop production, the most important species are *V. corymbosum* (highbush blueberry), *V. virgatum* (rabbit-eye blueberry), *V. angustifolium* (lowbush blueberry), *V. macrocarpon*(cranberry) and *V. vitis-idaea* (lingonberry). The genus also contains the wild *V. myrtillus* (bilberry) and a number of other currently non-cultivated*Vaccinium* species that show great potential as new berry crops (Song and Hancock 2011). Bilberry belongs to the section *Myrtillus*, and it is a diploid species (2*n* = 2*x* = 24; Song and Hancock 2011). The plant is a deciduous woody dwarf shrub, and it grows typically in pine and spruce heath forests and old peat bogs in Europe, North America, Greenland and northern parts of Asia, including Japan and Greenland (Nestby *et al.* 2011). Bilberry reproduces clonally through rhizomes and also sexually, with an outcrossing rate ranging from 0.66 to 0.75, and it is therefore considered to belong to the group of mixed-mating species (Jacquemart 1993).

Bilberry is an important wild fruit crop, especially in Northern and Eastern European countries, where the berries are picked from the wild and are either sold on the fresh market or frozen for use in food industries to make jams, juices and flavourings. The fruits are also important to the pharmaceutical industry, as they are naturally rich in polyphenols and other antioxidant compounds, which have potential beneficial effects on human health. These berries contain great amounts of flavonoids, in particular anthocyanins, which can reach up to 500 mg/100 g fresh weight (Lätti *et al.* 2008); they also produce carotenoids (Bunea *et al.* 2011) and lower amounts of ascorbic acid (Cocetta *et al.* 2012).

There is an increasing demand for these berries due to their high nutritional value (Martinussen *et al.* 2009), although they are still poorly exploited from a commercial point of view. Despite the fact that the average wild berry yield in Scandinavia has been estimated to be approximately 1 billion kg year^{-1}, only ~5–10 % of the annual crop is utilized for private or commercial consumption (Paassilta *et al.* 2009). Nestby *et al.* (2011) underlined the need for developing an improved production system, in which high yields of good-quality bilberries are produced at manageable costs. So far, cultivation of the species has been very limited and the berries used for commercial purposes are mainly harvested from forests. Therefore, development of forest management systems is considered a good option to achieve this purpose (Nestby *et al.*2011). Forest management systems will initially require efforts to identify areas in which plants produce high yields and high-quality bilberries. Plants with high-quality characteristics can be identified by phytochemical content or phenotypic traits of interest (e.g. plant productivity, fruit antioxidant content and fruit shelf life) and by genotypic-based methods, whereby the detected molecular polymorphisms are correlated to phenotypic traits. The genotypic-based methods are generally effective, they only need a small amount of DNA (Tanya *et al.* 2011) and they are not affected by environmental factors or developmental stages of the plants. Inter-simple sequence repeats (ISSRs) have shown to be good markers for assessing the genetic diversity of wild *Vaccinium* species from wide geographical areas of collection, in particular, lingonberry (*V. vitis-idaea*;Debnath 2007) and lowbush blueberry (*V. angustifolium*; Debnath 2009). Moreover, ISSR markers were able to detect more polymorphisms than random amplified polimorphic DNA in the same species (Debnath 2009). Therefore, ISSR markers were chosen for our study where the aims were to (i) test the applicability of ISSR markers on bilberry; (ii) determine genetic relationships and diversity among bilberry populations derived from biomes in Northern Europe and (iii) find markers to be used in conservation and management of bilberry in forests of Northern Europe.

METHODS

Study sites and sampling

Thirty-two individual bilberry samples derived from different seeds collected at different latitudes in several Nordic countries were included in this study (Table 1, Fig. 1): Iceland (IS1, IS2), Norway (N2, N4, N7), Sweden (R), Finland (S, P, M, L) and Germany (K). The plants were established in 2003 from bilberry seeds harvested from a pool of ripe berries collected in an area of 10 × 10 m and micropropagated *in vitro* (Jaakola *et al.* 2001) at the Botanical Gardens of the University of Oulu (Finland). Plantlets were grown in growth rooms under controlled conditions (+22 °C under 16 h photoperiod, white fluorescent Osram 18 W, 1.8 W m^{-2}).

Table 1: Provenances of bilberry genotypes and genetic diversity parameters based on ISSR markers. Number of samples analysed (*N*), number of different alleles (Na), number of effective alleles (Ne), number of private bands (Np), percentage of polymorphic loci (P %), expected heterozygosity (He) and Shannon's Information index (*I*)

Provenance	ID	Country	Latitude (°N)	Longitude (°E)	Altitude (m above sea level)	Genotype ID	*N*	Na	Ne	Np	P %	He	*I*
Kleifarveugr	IS1	Iceland	66°07′	−18°38′	178	IS1_a, IS1_3, IS1_4	3.000	0.646	1.158	2	24.40	0.092	0.137

Strandav-egur	IS2	Iceland	65°47′	−21°22′	10	IS2_a, IS2_1, IS2_5	3.000	0.803	1.161	0	29.92	0.102	0.156
Storfjord	N2	Norway	69°23′	20°16′	3	N2_2, N2_5, N2_6	3.000	0.661	1.142	0	23.62	0.086	0.129
Trondelag	N4	Norway	63°32′	10°53′	420	N4, N4_3, N4_5	3.000	0.614	1.111	0	22.05	0.072	0.112
Storgata	N7	Norway	60°54′	10°44′	173	N7. N7_5, N7_6	3.000	0.772	1.193	0	31.50	0.115	0.173
Kvikkjokk	R	Sweden	66°57′	17°43′	327	R1, R2, R3	3.000	0.693	1.182	2	25.98	0.103	0.151
Sodankylä	S	Finland	67°25′	26°35′	189	S1, S3	2.000	0.630	1.128	2	18.11	0.075	0.110
Muhos	M	Finland	64°48′	25°59′	39	M, M1, M5	3.000	0.835	1.198	2	34.65	0.122	0.184
Parkano	P	Finland	62°02′	23°02′	117	P, P_1, P_10	3.000	0.961	1.251	6	39.37	0.148	0.220
Lapinjärvi	L	Finland	60°37′	26°11′	21	L2, L3, L6	3.000	0.803	1.199	2	33.86	0.121	0.182
Kiel	K	Ger-many	54°20′	10°08′	14	K2, K6, K10	3.000	0.945	1.276	2	43.31	0.162	0.242

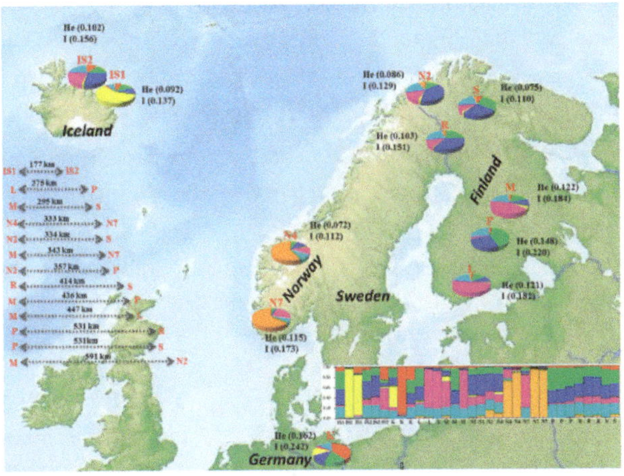

Figure 1: Map of sampling sites in Northern Europe, including ID (according to Table 1). The pie chart represents rather average coefficients of membership resulting from the genetic structure analysis (best fit model, $K = 7$). Each colour represents a different gene pool. The barplot represents each accession as a single vertical bar broken into K colour segments, with lengths proportional to the estimate probability of membership in each inferred cluster. Spatial autocorrelation analysis results, and geographical distances of correlated populations, are reported on the left of the figure (ID, grey arrows and geographical distance in kilometres).

Genotyping

Genomic DNA was isolated from shoot tips of actively growing *in vitro*-cultured bilberry shoot tips, using the EZNA™ SP Plant DNA Mini Kit (Omega Bio-tek Inc., Norcross, GA, USA) following the manufacturer's instructions. The concentration of DNA was estimated with the NanoDrop N-1000 spectrophotometer (NanoDrop Technologies, Thermo Scientific, Wilmington, DE, USA) at 260 nm. Fifteen primers representing di-, tri-, tetra- and pentamer repeats, previously used to characterize other*Vaccinium* species (Debnath 2007), were considered for the study. Of these, UBC-825, UBC-857, UBC-873 and UBC-881, which gave clear banding patterns, were used for the final study (Table 2). Polymerase chain reaction (PCR) was performed in a final volume of 25 μL including 10 ng of DNA template, 1× Optimized DyNAzyme™ buffer (10 mM Tris–HCl pH 8.3, 1.5 mM $MgCl_2$, 50 mM KCl, 0.1 % Triton X-100; Finnzyme, Espoo, Finland), 0.3 μM of each primer, 200 μM dNTPs, 0.8 U of DyNAzyme™ II DNA Polymerase (Finnzyme). The thermal profile consisted of 10 min at 94 °C, followed by 31 cycles of 1 min at 94 °C, 1 min at 46.5 °C and 2 min at 72 °C and a final extension at 72 °C for 10 min. The PCR reaction was purified using sodium acetate–ethanol DNA precipitation. One microlitre of the purified sample was analysed on a capillary electrophoresis system, Agilent 2100 Bioanalyzer with DNA7500 kit (Agilent Technologies, Santa Clara, CA, USA) according to the manufacturer›s instructions. Each primer–clone sample combination was repeated at least two times and only replicated bands were included in the analyses. Fragments of similar size across individuals were assumed to be homologous.

Table 2: Molecular ISSR primers used for bilberry genotypes' discrimination. Y = (C or T) in ISSR primer sequences; repeat motif and the data on DNA profile and polymorphism generated in 32 bilberry samples; total number of bands (NB), number of polymorphic bands (NPB), proportion of polymorphic bands (PPB), rank of molecular weights (RW); resolving power (Rp)

Primer name	Sequence	NB	NPB	PPB (%)	RW (bp)	Rp
UBC-825	$(AC)_8T$	33	31	93.9	310–2100	8.625
UBC-857	$(AC)_8YG$	37	37	100	80–6600	13.87
UBC-873	$(GACA)_4$	25	25	100	80–2600	9.18
UBC-881	$(GGGTG)_3$	32	32	100	60–1400	14.31

Data analysis

Genetic diversity

The amplification product sizes were scored using 2100 Expert Software (Agilent Technologies). The results were transformed into a binomial matrix as present (1) or absent (0) for each marker. Since the ISSR marker is dominant, we assumed that each band represented the single bi-allelic locus (Debnath 2007). Band patterns were analysed in order to determine the level of polymorphism (total number of bands), number of polymorphic bands, proportion of polymorphic bands and the resolving power (Rp) detected for each primer. Resolving power was calculated according to Prevost and Wilkinson (1999); this measure is based on the distribution of alleles among the genotypes and it estimates the discrimination capacity of each primer. Thus, the resolving power of a primer is defined as $Rp = \sum Ib$, where Ib (band informativeness) takes the value of $1 - [2 \times |0.5 - p|]$ and p is the ratio of genotypes sharing the band. Moreover, the binomial matrix was used to produce the input matrix following GenAlex version 6.1 software manual instructions (Peakall and Smouse 2006) and analysed using the same software to estimate genetic diversity parameters, i.e. number of different bands, number of different bands with frequency ≥5 %, number of private bands, number of locally common bands frequency ≥5 % found in ≤25 and ≤50 % of populations, mean of expected heterozygosity (He) and the Shannon›s Information Index (I) calculated as $I = -1 \times (p \times \ln(p) + q \times \ln(q))$, where p and q are the estimated allele frequencies.

Analysis of molecular variance (AMOVA) was used to partition the total genetic variance into 'within-populations› or 'among-populations' levels. The software GenAlex was used to generate a matrix of pairwise genetic distances between individuals and to calculate the following variance components: degrees of freedom, sum of squares, mean sum of squares, estimated variance and conversion of estimated variances to percentage of total variance. The number of permutations for significance testing was set at 9999. Canonical correspondence analysis (CCA) was done using Past software v. 2.17c (Hammer *et al.* 2001), to determine the relative importance of geographical factors in the spatial organization of genetic diversity among genotypes. This analysis, originally designed for relating species composition to different predictive variables (Ter Braak 1986), has been successfully used to describe the relationship between environmental variables and genetic composition (Angers *et al.* 1999; Girard and Angers 2006; Dell›Acqua *et al.* 2014). The analysis was performed using a geographical variables/genetic data matrix where longitude, latitude and altitude were used as geographic factors. Here, we consider individuals as sites, and alleles at outlier loci as objects. The first three input file columns contain geographical variables, following the CCA Past software v. 2.17c instructions.

Spatial genetic analysis

Spatial analysis was conducted using the genetic spatial autocorrelation (SA) (Smouse and Peakall 1999; Peakall *et al.* 2003) option in GenAlex version 6. The pairwise geographical distance matrices were calculated considering them as the crow flies distances in kilometres and were used with the previously obtained genetic distance matrices to generate an autocorrelation coefficient r for each distance class using two different options. The autocorrelation coefficient (r) is similar to Moran's I (Moran 1950) and ranged from −1 to 1. The significance level was tested by constructing a two-tailed 95 % confidence interval around the null hypothesis of no spatial genetic structure, which is $r = 0$. The autocorrelation analysis was performed using a multiple distance class simulation. Since the distance classes have to be small enough to capture the spatial pattern of interest, while large enough to include an adequate number of pairwise comparisons for statistical testing, we characterized the spatial genetic relationship kilometre intervals from 0 up to 2400. This allowed us to determine the strength of autocorrelation and to what extent the autocorrelation decays with increasing distance. To test the null hypothesis of no spatial structure, confidence limits were calculated using permutation and bootstrapping (999 interactions).

Directional autocorrelation analysis was carried out by testing for the direction of maximum genetic correlation using the bearing procedure implemented in PASSAGE version 2 (Rosenberg and Anderson 2011). The bearing method analysed the correlation coefficient (r) between geographical distance and genetic relatedness under fixed bearing angles (degrees north of due east). For each sector, r is calculated from each distance pair weighted by the cosine of its angle with respect to the angle of the centre arc of the respective sector. Geographical locations were imported with associated genetic data. PASSAGE calculates a distance matrix from genetic data and geographical co-ordinates and an angular matrix from geographical co-ordinates. The correlation was calculated for each 10° sector, from 0° to 180°, and the significance was estimated using 999 randomizations.

Grouping of bilberry individuals by STRUCTURE and Cluster analyses

The software STRUCTURE 2.3.3 (Pritchard *et al.* 2000; Falush *et al.* 2003), which, by means of iterative algorithms, identifies clusters of related individuals from multi-locus genotypes, was used to examine the genetic structure of populations. Ten independent runs of STRUCTURE were performed for each K value from 1 to 11. Each run consisted of a burn-in period of 100 000 steps, followed by 1 000 000 Markov Chain Monte Carlo replicates, assuming an admixture model and correlated allele frequencies. No prior information was used to define the clusters. The most likely K was chosen comparing the average estimates of the likelihood of the data, $\ln(\Pr(X|K))$, for each value of K (Pritchard *et al.*2000), as well as calculating the *ad hoc* statistics ΔK, based on the rate of change in the ln probability of data between successive K values (Evanno*et al.* 2005). Furthermore, the Past 2.17c (Hammer *et al.* 2001) software was used to generate a matrix using the Dice similarity index. This matrix was used to construct a Ward's dendrogram tree.

RESULTS

Genetic diversity

Polymerase chain reaction assays using four primers selected in the initial tests allowed 127 ISSR loci to be amplified from the DNA samples derived from 32 bilberry individuals. The detected loci ranged between 60 and 6600 bp (within the limits of the Agilent DNA7500 kit that allowed the detection of fragments between 50 and 7500 bp). The average number of loci per primer was 31.75, with the highest number of loci ($n = 37$) detected by the UBC-857 primer and the smallest number ($n = 25$) by the UBC-873 primer (Table 2). Of the 127 amplified loci, 126 (99.24 %) were polymorphic. The greatest discrimination power among samples was obtained with primer UBC-881 yielding an Rp value of 14.31, while the lowest Rp value of 8.62 was yielded with primer UBC-825. Except for the primer pair of UBC-825, other primers produced 100 % polymorphic bands (Table 2). The number of bands within populations ranged between 50 and 72, with a number of private bands that ranged between 0 and 6 and a mean expected heterozygosity value that ranged between 0.072 and 0.162. The highest percentage of polymorphic loci (P %) was established in Kiel individuals (43.31 %), while the lowest was observed in S individuals (18.11 %). Moreover, the mean number of different alleles over all loci (Na) for each population ranged between 0.614 and 0.961, the mean number of effective alleles (Ne) ranged between 1.111 and 1.276, and the Shannon›s Information Index ranged between 0.110 and 0.242 (Table 1). Analysis of molecular variance indicated that 15 % of the total genetic variance was attributable to among-populations diversity and the rest (85 %) to within-populations diversity. The value of ΦPT (0.190 with a maximum of 0.795) indicated a great level of genetic differentiation among populations. We used a CCA to investigate the possible aggregation or differentiation of analysed genotypes. Results of the CCA (Fig. 2) revealed the strong influence of the geographical position on genotype aggregation (Axis 1 = 63.3 % of variance; Axis 2 = 35.7 %). The correlation biplot, which considers both the direction and the relative length of the vectors, underlines clustering for most of the individuals having the same origins. For instance, the clustering of Finnish individuals (P, L, M and one S sample, Fig. 1) occurred according to longitude, and the clustering of the same Finnish samples with Norwegian individuals derived from the closer latitudes (Fig. 1) was evident. The individuals from Norway (N4 and N7) overlapped and were in close proximity to Swedish individuals probably due to the same influence of all geographical variables on all these samples. IS1, IS2 and K individuals are clustering alone as previously shown, but IS1 and IS2 seem to be strongly influenced by negative longitude values while the *K* individuals seem to be most influenced by lower values of latitude.

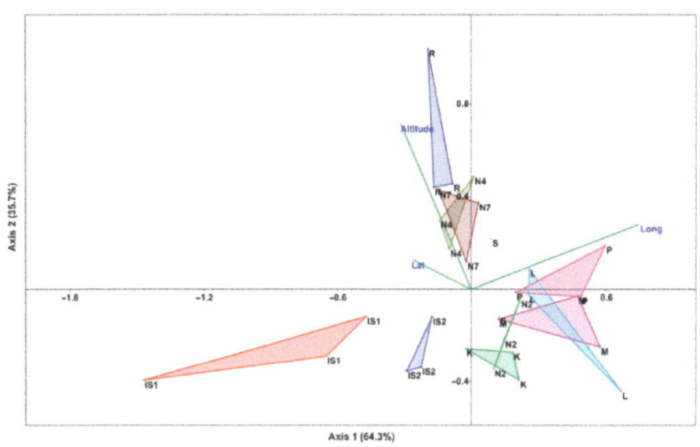

Figure 2: Canonical correspondence analysis ordination biplot representing genotype aggregation and geographical variables (solid arrows). The CCA explained 64.3 and 35.7 % of the variation on the first two axes.

Spatial genetic structure

In the correlogram that resulted from SA analysis, the *y*-axis average SA coefficient, *r*, has a function of distance class on the *x*-axis. The maximum number of distance classes obtained using the even sample classes options was 52. It is apparent in Fig. 3 that there is a highly significant positive SA (the *r* value falls above or at the 95 % confidence interval, close to a correlation of zero) at distance classes 0–100 evidencing non-random spatial genetic structure within populations. Moreover, the positive correlation decreases until 600 km ($r = 0.028$). Figure 1 reports the IDs and the geographical distances of populations showing positive *r* values. The bearing analysis from PASSAGE indicated the strongest correlation occurred along a north–south axis ($r = 0.856$, $P = 0.001$), while the weakest occurred in the east–west direction ($r = -0.188$, $P = 0$).

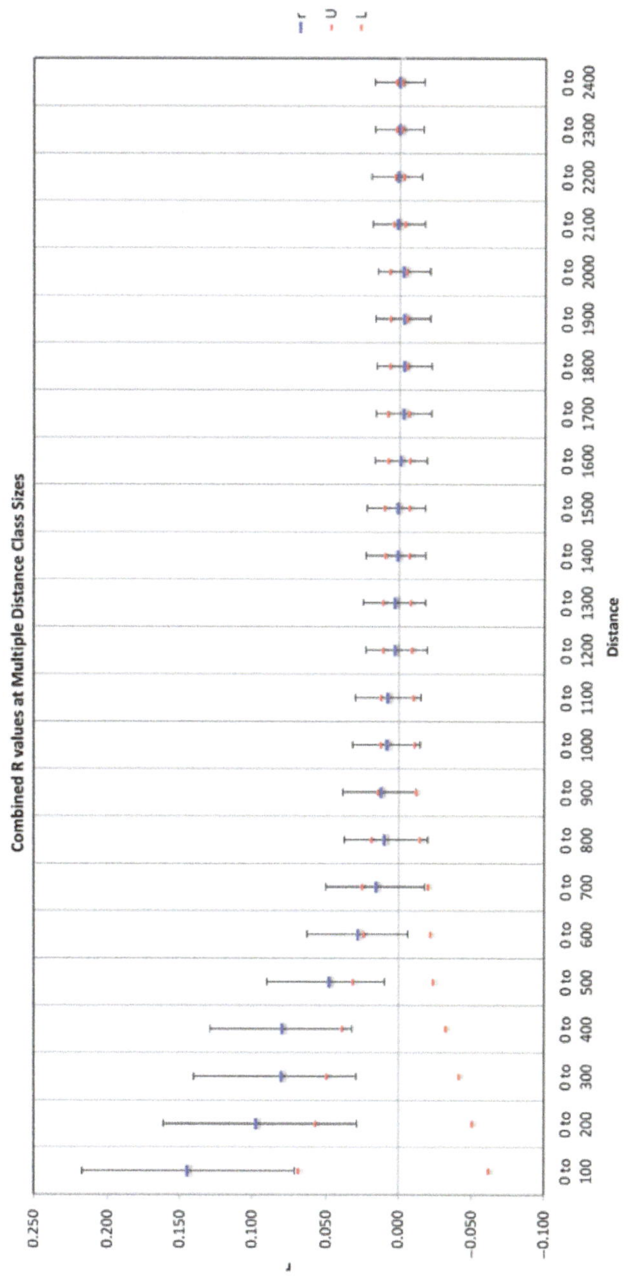

Figure 3: Results of multiple SA analyses for increasing distance class sizes to determine SA. Confidence limits for the *r* values are indicated and were estimated by permutation (999 interactions). Upper (U) and lower (L) confidence limits were generated for the null hypothesis of no SA ($r = 0$) by bootstrap (999 interactions).

Clustering of bilberry individuals by STRUCTURE analysis

STRUCTURE analysis assigned genotypes into respective groups on the basis of their allele frequencies. According to the user-defined settings, the programme assumed that genotypes are admixed and the allele frequencies are correlated as a consequence of shared ancestry and/or migration. Bayesian cluster, based on an admixture model, presumes that each individual has inherited some proportion of its ancestry from each of the *K* genotypes (Pritchard *et al.* 2000). According to the Evanno›s method, the STRUCTURE analysis indicated $K = 7$ as the most likely number of gene pools. These pools represented most geographical groups but with a substantially different proportion of membership (*q*) of each group in each gene pool. Considering the mean value for each geographical group of genotypes (*Q*), six different patterns of genetic makeup were evidenced. The first type of pattern included two Norwegian groups (N4 and N7) out of three with the highest *Q* in gene pool 7 (0.572 and 0.636, respectively). Finnish populations represented two distinct patterns with the highest *Q*value in gene pool 5 for L and M groups (0.508 and 0.420, respectively) and in gene pool 2 for P and S groups (0.411 and 0.295, respectively). The IS1 population had the maximum *Q* value in gene pool 4 (0.59) and the *K*group in gene pool 1. Finally, individuals from Iceland (IS2), Norway (N2) and Sweden (R) had similar *Q* values in gene pools 3, 6 and 7 (Fig. 1).

The Dice distances between pairs of populations were calculated based on the 127 analysed bands. The dendrogram was built using Ward›s method (Fig. 4) and showed clusters comparable with clusters evidenced from the STRUCTURE analysis, except for P and R genotypes, which clustered alone as previously assessed.

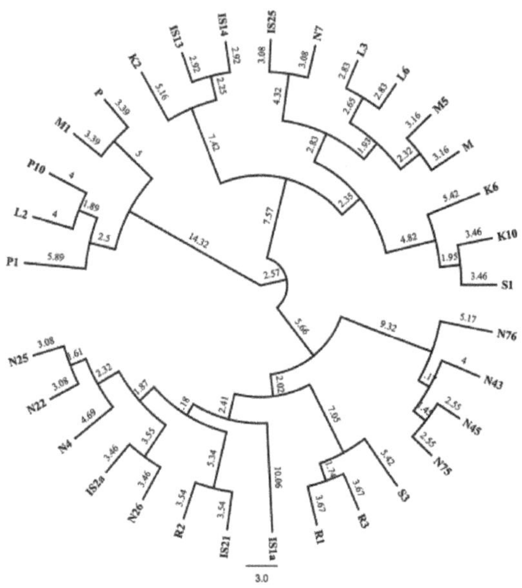

Figure 4: Dendogram of bilberry individuals using Ward's method. Numbers above branches indicate bootstrap values based on 10 000 replicates.

DISCUSSION

Plant genetic resources are essential for sustainable agriculture and food security. One of the best ways to preserve them is *in situ* management, which becomes particularly important in cases of wild crop species such as bilberry. Molecular markers can provide important information regarding genetic polymorphism and essential knowledge for development and improvement of plant populations; however, to date only a few studies have been carried out on the assessment of bilberry germplasm genetic diversity (Raspé *et al.* 2004; Albert *et al.* 2008).

In our study, ISSR markers were used to evaluate genetic variation among 32 bilberry individuals located in Fennoscandia and Germany. Four primers detected significant genetic variation among the genotypes thanks to their high polymorphism level (99 %). A similar high level of diversity in the same genus (80.4 %) was reported in highbush and rabbit-eye blueberry (Garriga *et al.* 2013). The ability to discriminate bilberry samples varied between different ISSR primers, in agreement with the findings in lingonberry (Debnath 2007) and lowbush blueberry (Debnath 2009).

The high percentage value of genetic diversity within-population obtained from AMOVA can be explained as a natural selection mechanism to reduce fitness costs due to geitonogamous self-pollination in bilberry as previously reported by Albert *et al.* (2008). The same results were also found in other wild *Vaccinium* species such as lingonberry (Persson and Gustavsson 2001; Garkava-Gustavsson *et al.* 2005; Debnath 2007), bog bilberry (*V. uliginosum*; Eidesen *et al.* 2007) and wild lowbush blueberry (Debnath 2009). Moreover, the great genetic differentiation among geographically distant populations, determined from the positive value of ΦPT, is in accordance with the theory that the level of genetic heterogeneity among individuals is greater in species with geographically disjunctive populations than in species with more continuous distributions (Hamrick and Godt 1996; Premoli *et al.* 2001). Therefore, in the present study, we focussed on the genetic diversity of bilberry plants separated by long distances, clones of which are discontinuous and isolated from one another by mountains or seas. The effect of these natural barriers together with common results of different clusters and spatial and population structure analyses provided evidence of a great association between the Finnish and the Norwegian N2 populations on one side, the Norwegian populations N4 and N7 on the other side, and split IS and K populations from all the others. Moreover, the presence of a north–south genetic gradient is in accordance with recent findings on bilberry phenotypic traits (Lätti *et al.* 2008; Åkerström *et al.* 2010; Uleberg *et al.* 2012). These results are also supported by other studies carried out on different species that show how intra-specific genetic variation increases or decreases in relation to the physical distance separating the individuals, and is also showing the influence of the geographic structure of natural populations (Brooks *et al.* 2015). To date, most conservation activities have focussed on the species level; however, also genetic variation at an intra-specific level needs to be considered to avoid loss of diversity derived from severe inbreeding, resulting in lowered fitness and increasing risk of extinction. Moreover, the determination and the conservation of the within-population genetic diversity of one species could increase evolutionary resilience when different and geographically separated environments are connected. When populations are interconnected along climatic and geographical gradients, there is the potential for *in situ* adaptive evolution (Sgrò *et al.* 2011). Furthermore, plant species conserved in key biodiversity areas are an essential genetic source to develop new varieties for future breeding work and to avoid the diversity loss derived from severe inbreeding. Finally, genetic diversity conservation might become crucial in a biotic or abiotic crisis where only very rare genotypes may be resistant to a new disease, pathogen strain or environmental condition. This was further supported by the highest values for He, Shannon Index and number of polymorphic bands.

Further studies with a higher number of markers and samples from the distribution area of bilberry are needed to identify the key biodiversity areas of the species.

CONCLUSIONS

Advances in genotyping techniques combined with more sophisticated statistical methods provide the means by which among- and within-population genetic diversity can be estimated in the absence of any prior specific information. This valuation is necessary to conserve the biodiversity of specific areas. In this perspective, we assessed the intra- and inter-population genetic diversity of 32 individuals collected from different North European countries. We found the presence of significant correlations between geographic and genetic distances, which placed Norwegian, Finnish, Icelandic and German genotypes in separate groups. The present results indicate how key biodiversity areas of the wild *V. myrtillus* species could be individualized as a useful source of biodiversity for future ecological studies and breeding purposes.

SOURCES OF FUNDING

The study was funded by the Thule Institute (2013–16) (to H.H.) and by Nordic Innovation Centre—New Nordic Food project 'Bilberry: Towards functional food markets' (2007–10) (to H.H. and L.J.).

CONTRIBUTIONS BY THE AUTHORS

L.Z. performed the DNA analyses and the ISSR data scoring and was involved in writing and editing; L.P. performed all the statistical analyses and was involved in writing and editing; L.J. and H.H. provided contribution to the concept and the design of the work. All authors read and approved the final manuscript.

ACKNOWLEDGEMENTS

The partners of the New Nordic Food project are acknowledged for the bilberry seed samples from Iceland, Norway and Sweden. We thank Aaron Bergdahl for English revision of the manuscript.

REFERENCES

1. Åkerström A, Jaakola L, Bång U, Jäderlund A. 2010. Effects of latitude-related factors and geographical origin on anthocyanidin concentrations in fruits of Vaccinium myrtillus L. (bilberries). Journal of Agricultural and Food Chemistry 58:11939–11945. doi:10.1021/jf102407n
2. Albert T, Raspé O, Jacquemart A-L. 2008. Influence of clonal growth on selfing rate in Vaccinium myrtillus L. Plant Biology 10:643–649. doi:10.1111/j.1438-8677.2008.00067.x
3. Angers B, Magnan P, Plante M, Bernatchez L. 1999. Canonical correspondence analysis for estimating spatial and environmental effects on microsatellite gene diversity in brook charr (Salvelinus fontinalis). Molecular Ecology 8:1043–1053. doi:10.1046/j.1365-294x.1999.00669.x
4. Brooks TM, Cuttelod A, Faith DP, Garcia-Moreno J, Langhammer P, Pérez-Espona S. 2015. Why and how might genetic and phylogenetic diversity be reflected in the identification of key biodiversity areas? Philosophical Transactions of the Royal Society B: Biological Sciences 370:20140019. doi:10.1098/rstb.2014.0019
5. Bunea A, Rugină DO, Pintea AM, Bunea CI, Socaciu C. 2011. Comparative polyphenolic content and antioxidant activities of some wild and cultivated blueberries from Romania. Notulae Botanicae Horti Agrobotanici Cluj-Napoca 39:70–76.
6. Burdon RD, Wilcox PL. 2007. Population management: potential impacts of advances in genomics. New Forests 34:187–206. doi:10.1007/s11056-007-9047-6
7. Cocetta G, Karppinen K, Suokas M, Hohtola A, Häggman H, Spinardi A, Mignani I, Jaakola L. 2012. Ascorbic acid metabolism during bilberry (Vaccinium myrtillus L.) fruit development. Journal of Plant Physiology 169:1059–1065. doi:10.1016/j.jplph.2012.03.010
8. Debnath SC. 2007. Inter simple sequence repeat (ISSR) to assess genetic diversity within a collection of wild lingonberry (Vaccinium vitis-idaea L.) clones. Canadian Journal of Plant Science 87:337–344. doi:10.4141/P06-059
9. Debnath SC. 2009. Development of ISSR markers for genetic diversity studies in Vaccinium angustifolium. Nordic Journal of Botany 27:141–148. doi:10.1111/j.1756-1051.2009.00402.x
10. Dell›acqua M, Fricano A, Gomarasca S, Caccianiga M, Piffanelli P, Bocchi S, Gianfranceschi L. 2014. Genome scan of Kenyan Themeda triandra populations by AFLP markers reveals a complex genetic structure and hints for ongoing environmental selection. South African Journal of Botany 92:28–38. doi:10.1016/j.sajb.2014.01.013
11. Eidesen PB, Alsos IG, Popp M, Stensrud Ø, Suda J, Brochmann C. 2007. Nuclear vs. plastid data: complex pleistocene history of a circumpolar key species. Molecular Ecology 16:3902–3925. doi:10.1111/j.1365-294X.2007.03425.x
12. Evanno G, Regnaut S, Goudet J. 2005. Detecting the number of clusters of individuals using the software STRUCTURE: a simulation study. Molecular Ecology 14:2611–2620. doi:10.1111/j.1365-294X.2005.02553.x
13. Falush D, Stephens M, Pritchard JK. 2003. Inference of population structure using multilocus genotype data: linked loci and correlated allele frequencies. Genetics 164:1567–1587.
14. Feuillet C, Langridge P, Waugh R. 2008. Cereal breeding takes a walk on the wild side. Trends in Genetics 24:24–32. doi:10.1016/j.tig.2007.11.001

15. Garkava-Gustavsson L, Persson HA, Nybom H, Rumpunen K, Gustavsson BA, Bartish I. 2005. RAPD-based analysis of genetic diversity and selection of lingonberry (Vaccinium vitis-idaea L.) material for ex situ conservation. Genetic Resources and Crop Evolution 52:723–735. doi:10.1007/s10722-003-6123-4

16. Garriga M, Parra PA, Caligari PDS, Retamales JB, Carrasco BA, Lobos GA, García-Gonzáles R. 2013. Application of inter-simple sequence repeats relative to simple sequence repeats as a molecular marker system for indexing blueberry cultivars. Canadian Journal of Plant Science 93:913–921. doi:10.4141/cjps2013-057

17. Girard P, Angers B. 2006. The impact of postglacial marine invasions on the genetic diversity of an obligate freshwater fish, the longnose dace (Rhinichthys cataractae), on the Quebec peninsula. Canadian Journal of Fisheries and Aquatic Sciences 63:1429–1438. doi:10.1139/f06-047

18. Hammer Ø, Harper DAT, Ryan PD. 2001. Past: paleontological statistics software package for education and data analysis. Palaeontologia Electronica 4:1–9.

19. Hamrick JL, Godt MJW. 1996. Effects of life history traits on genetic diversity in plant species. Philosophical Transactions of the Royal Society B: Biological Sciences 351:1291–1298. doi:10.1098/rstb.1996.0112

20. Jaakola L, Tolvanen A, Laine K, Hohtola A. 2001. Effect of N6-isopentenyladenine concentration on growth initiation in vitro and rooting of bilberry and lingonberry microshoots. Plant Cell, Tissue and Organ Culture 66:73–77. doi:10.1023/A:1010602726215

21. Jacquemart AL. 1993. Floral visitors of Vaccinium species in the high Ardennes, Belgium. Flora 188:263–273.

22. Lätti AK, Riihinen KR, Kainulainen PS. 2008. Analysis of anthocyanin variation in wild populations of bilberry (Vaccinium myrtillus L.) in Finland. Journal of Agricultural and Food Chemistry 56:190–196. doi:10.1021/jf072857m

23. Martinussen I, Rohloff J, Uleberg E, Junttila O, Hohtola A, Jaakola L, Häggman H. 2009. Climatic effects on the production and quality of bilberries (Vaccinium myrtillus). Latvian Journal of Agronomy 12:71–74.

24. Moran PAP. 1950. Notes on continuous stochastic phenomena. Biometrika 37:17–23. doi:10.1093/biomet/37.1-2.17

25. Nestby R, Percival D, Martinussen I, Opstad N, Rohloff J. 2011. The European blueberry (Vaccinium myrtillus L.) and the potential for cultivation. European Journal of Plant Science and Biotechnology 5:5–16.

26. Paassilta M, Moisio S, Jaakola L, Häggman H. 2009. Voice of the Nordic wild berry industry. Oulu, Finland: University of Oulu Press.

27. Peakall R, Smouse PE. 2006. GENALEX 6: genetic analysis in Excel. Population genetic software for teaching and research. Molecular Ecology Notes 6:288–295. doi:10.1111/j.1471-8286.2005.01155.x

28. Peakall R, Ruibal M, Lindenmayer DB. 2003. Spatial autocorrelation analysis offers new insights into gene flow in the Australian bush rat, Rattus fuscipes. Evolution 57:1182–1195. doi:10.1111/j.0014-3820.2003.tb00327.x

29. Persson HA, Gustavsson BA. 2001. The extent of clonality and genetic diversity in lingonberry (Vaccinium vitis-idaea L.) revealed by RAPDs and leaf-shape analysis. Molecular Ecology 10:1385–1397. doi:10.1046/j.1365-294X.2001.01280.x

30. Premoli AC, Souto CP, Allnutt TR, Newton AC. 2001. Effects of population disjunction on isozyme variation in the widespread Pilgerodendron uviferum. Heredity 87:337–343. doi:10.1046/j.1365-2540.2001.00906.x

31. Prevost A, Wilkinson MJ. 1999. A new system of comparing PCR primers applied to ISSR fingerprinting of potato cultivars. Theoretical and Applied Genetics 98:107–112. doi:10.1007/s001220051046

32. Pritchard JK, Stephens M, Donnelly P. 2000. Inference of population structure using multilocus genotype data. Genetics 155:945–959.

33. Raspé O, Guillaume P, Jacquemart AL. 2004. Inbreeding depression and biased paternity after mixed-pollination in Vaccinium myrtillus L. (Ericaceae). International Journal of Plant Sciences 165:765–771. doi:10.1086/422045

34. Rosenberg MS, Anderson CD. 2011. PASSaGE: pattern analysis, spatial statistics and geographic exegesis. Version 2. Methods in Ecology and Evolution 2:229–232. doi:10.1111/j.2041-210X.2010.00081.x

35. Sgrò CM, Lowe AJ, Hoffmann AA. 2011. Building evolutionary resilience for conserving biodiversity under climate change. Evolutionary Applications 4:326–337. doi:10.1111/j.1752-4571.2010.00157.x

36. Smouse PE, Peakall R. 1999. Spatial autocorrelation analysis of individual multiallele and multilocus genetic structure. Heredity 82:561–573. doi:10.1038/sj.hdy.6885180

37. Song GQ, Hancock F. 2011. Vaccinium. In: Kole C, ed. Wild crop relatives: genomic breeding resource. Temperate fruits. New York: Springer, 210–213.

38. Tanya P, Taeprayoon P, Hadkam Y, Srinives P. 2011. Genetic diversity among Jatropha and Jatropha-related species based on ISSR markers. Plant Molecular Biology Reporter 29:252–264. doi:10.1007/s11105-010-0220-2

39. Ter Braak CJF. 1986. Canonical correspondence analysis: a new eigenvector technique for multivariate direct gradient analysis. Ecology 67:1167–1179. doi:10.2307/1938672

40. Tester M, Langridge P. 2010. Breeding technologies to increase crop production in a changing world. Science 327:818–822. doi:10.1126/science.1183700

41. Uleberg E, Rohloff J, Jaakola L, Trôst K, Junttila O, Häggman H, Martinussen I. 2012. Effects of temperature and photoperiod on yield and chemical composition of Northern and Southern clones of bilberry (Vaccinium myrtillus L.). Journal of Agricultural and Food Chemistry 60:10406–10414. doi:10.1021/jf302924m

42. Zhao D, Yang J, Yang S, Kato K, Luo J. 2014. Genetic diversity and domestication origin of tea plant Camellia taliensis (Theaceae) as revealed by microsatellite markers. BMC Plant Biology 14:14. doi:10.1186/1471-2229-14-14

CHAPTER 15

PLANT SPECIES RECOVERY ON A COMPACTED SKID ROAD

Murat Demir[1], Ender Makineci[2] and Beyza Sat Gungor[3]

[1]Istanbul University, Faculty of Forestry, Forest Construction and Transportation Department 34473 Bahcekoy / Sariyer / Istanbul / Turkey

[2]Istanbul University, Faculty of Forestry, Soil Science and Ecology Department 34473 Bahcekoy / Sariyer / Istanbul / Turkey

[3]Istanbul University, Faculty of Forestry, Landscape Architecture Department 34473 Bahcekoy / Sariyer / Istanbul / Turkey

ABSTRACT

This study was executed to determine the plant species of herbaceous cover in a skid road subjected to soil compaction due to timber skidding in a beech (*Fagus orientalis* Lipsky.) stand. Our previous studies have shown that ground based timber skidding destroys the soils extremely, and degradations on ecosystem because of the timber skidding limit recovery and growth of plant cover on skid roads. However, some plant species show healthy habitat, recovery and they can survive after the extreme degradation in study area. We evaluated composition of these plant species and their cover-abundance scales in 100 m x 3 m transect. 15 plant species were determined belongs to 12 plant families and Liliaceae was the highest representative plant family. *Smilax aspera* L., *Epimedium pubigerum* (DC.) Moren et Decaisne, *Carex distachya* Desf. var. *distachya* Desf., *Pteridium aquilinum* (L.) Kuhn., *Trachystemon orientalis* (L.) G. Don,*Hedera helix* L. have the highest cover-abundance scale overall of determined species on compacted skid road.

INTRODUCTION

Roads are critical components of civilization. Developing and maintaining the economic activity that is vital for the quality of modern life would be difficult without roads. Roads provide access for people to study, enjoy, or contemplate natural ecosystems [1-3]. Building and maintaining roads have become controversial, however, because of public concerns about their short and long term effects on the environment [4]. Potential ecological impacts of roads have received a great deal of attention in recent years. Particularly prominent management issues and research problems include effects of roads on soil erosion, hydrology and aquatic ecosystems. Effects of roads on forest plant communities are less well-documented [5]. Logging operations in forests can cause significant and wide spread soil disturbance, including removal, mixing and compaction of various soil layers. Disturbance can adversely affect both soil physical properties and soil nutrient levels to such an extent that severely diminished growth of subsequent tree rotations as well as significant increase in runoff and sediment loads may result [6]. That the production works being carried out in the forest have many negative impacts on the forest ecosystem is very well known. It has also been determined that the production and skidding negatively affect the amounts and variety of forest floor and herbaceous understory as well as youth development and living conditions of the soil organisms

Citation: Murat Demir, Ender Makineci and Beyza Sat Gungor, Plant Species Recovery on a Compacted Skid Road, doi:10.3390/s8053123.

[7-15]. Skidding or yarding on terrain requires the construction of relatively dense network of forest roads including skid roads, haul roads and landings [16]. Skid roads are defined as tertiary roads that are used by skidders and forwarders that move logs from the point of felling and bucking to log landings. Apart from the felling of occasional trees to provide a clear path, few improvements are made to skid roads [5]. The dimension of the impact created by skidding of the products directly on the ground varies according to many factors such as slope, site characteristics, production methods used, planning of skid roads and production season. Ground-based skidding may result in soil compaction and other soil structural changes, influencing soil water retention, and reducing soil aeration, drainage, and root penetration. Studies focused on exotic plant invasions in roadside plant communities along main forest and grassland roads and in adjacent communities have been completed, but changes in plant composition within interior haul roads and skid trails have received little attention. The extent of haul road and skid road networks in managed forests and linkages between understory vascular plant communities and other flora and fauna, the impacts of these features on ecosystem properties and processes could be substantial [5].

In this study, we evaluated survived herbaceous plant species composition and their cover-abundance scales on the skid road subjected to compaction due to ground based timber skidding activities that have been carried out for many years (since 1956) in a beech (*Fagus orientalis* Lipsky.) stand.

MATERIALS AND METHODS

Research area was in the boundaries of zone 82 of Istanbul Belgrad forest. Belgrad forest covering a surface area of 5441.71 ha located in the Marmara geographical region between latitudes 41° 09′ - 41° 12′ N and longitudes 28° 54′ - 29° 00′ E. According to the data given by Bahcekoy meteorology station, average annual precipitation is 1074.4 mm, average annual temperature is 12.8°C, average maximum temperature is 17.8°C and the average minimum temperature is 9°C. The climate of Istanbul Belgrad Forest is close to sea (ocean) climate with medium water deficit in summers. Vegetation period maintains for 7.5 months (230 days) in average. The period of study is October 2006 and research area is oriental beech (*Fagus orientalis*Lipsky) stand having canopy cover as 0.8, average diameter as 23.12 cm (the height of breast height above ground level is 1.30 m), average height as 24.14 m and stand density as 1400 trees/ha. Average altitude of the research area is 140 m, slope is 10-15% with SW aspect. The skidding road passing through the stand in west-east direction has long been used (since 1956) to skid the logs out of the area. It was estimated that 135 m^3 timbers was skidded annually in harvest activities on the skid road. Last using date of skid road is in October 2004. Skid road width is 3.0 m [17]. The harvested timbers are also being towed by tractors with a rope along with skidding by means of manpower and animals.

We evaluated impacts of disturbance in interior skid road on understory herbaceous vegetation by documenting the areal extent of these features and plant composition along 100 m x 3 m belt transect in October 2006. During sampling 100m tapes were streched along transect to delineate skid road boundaries. Transect was oriented accros the road, perpendicular to its main axis (as described in [5]). The species composition on the transect on skid road was characterized by classical phytosociological plots according to Braun-Blanquet cover-abundance values [18]. which means that total coverage for each species (vertical projection onto the ground) was estimated visually and recorded within seven cover classes: r: 1 or 5 individuals; +: few individuals (<20) with cover <5%; 1: many individuals (20– 100) with cover <5%; 2: 5%–25% cover; 3: 25%–50% cover; 4: 50%–75% cover; 5: 75%–100% cover [14, 18]. Easily identified herbaceous plant species on skid road reported on the field, undetermined ones clipped and brought to the laboratory to determine by comparing the samples in Istanbul University Faculty of Forestry Herbarium and guide books. The general site characteristics of herbaceous plant species were reported as described in [19].

RESULTS AND DISCUSSION

The skidding works that have been carried on for many years in beech (*Fagus orientalis* Lipsky) stand caused decreasing of the forest floor and the herbaceous cover on the skid road to a great extent. Furthermore, there were some other important changes in the soil properties at the two examined soil depths (0-5 cm and 5-10 cm) (Table 1). The major impact that occurred down to 10 cm soil depth was the compaction of soil. Higher volume and fine soil weight values compared to those of the undisturbed area were found at both the depths on the skid road subject to the soil compaction. The total porosity and moisture equivalence on the skid road decreased considerably [20]. In addition there were no significant differences for chemical properties of soil at 0-5 cm soil depth (N, P, K, Na, Ca, Mg, Fe, Zn, Cu and Mn). A significant difference between soil samples at 5-10 cm depth, taken from skid road and undisturbed area was found with respect to Mg (143.85 ppm) concentrations [21]. These previous results clearly show that skidding has particularly negative impacts on herbaceous cover, organic layer and soil. These effects of skidding on skid road have been demonstrated to have detrimental impacts on native flora, and herbaceous plant establishing and maintaining were limited. Similarly, Mariani et al. (2006) [22] reported that organic layer removal reduced abundance of herbs and shrubs. Soil compaction can also severely reduce plant growth by restricting root growth may be due to oxygen stress and lower the percentage of water and air space in the soil [23]. Also, Kozlowski (1999) [24] mentioned a reduced total photosynthesis when soils become increasingly compacted, as a result of smaller leaf areas.

On the other hand, likely more adapted (especially to soil compaction) and less herbaceous plant mass exist on skid road. We evaluated these plant species and their cover-abundance scales in 100m × 3 m transect. According to results obtained from this study, 15 plant species were determined belongs to 12 plant family (Table 2). Liliaceae was the highest representative plant family (Table 2). *Smilax aspera* L., *Epimedium pubigerum* (DC.) Moren et Decaisne, *Carex distachya* Desf. var. *distachya* Desf., *Pteridium aquilinum* (L.) Kuhn., *Trachystemon orientalis* (L.) G. Don, *Hedera helix* L. have the highest cover-abundance scale overall of determined species on compacted skid road (Table 2).

Table 1: Investigated soil properties in soil on skid road [20,21]

Characteristics	Units	0-5 cm	5-10 cm
Sand	(%)	58.05	51.35
Silt	(%)	20.74	30.88
Clay	(%)	21.21	17.77
Soil Acidity	pH	5.49	5.07
Electrical Conductivity	(μhos cm^{-1})	81.32	48.02
Fine soil (<2mm) weight	(g cm^{-3})	0.763	906.55
Coarse soil (>2mm) weight	(g cm^{-3})	0.137	147.15
Root mass	(g cm^{-3})	2.76	2.79
Organic Carbon	(%)	10.20	7.32
Moisture Equivalent	(%)	24.37	24.44
Total Porosity	(%)	52.72	50.22
Moisture	(%)	20.86	18.73
Compaction	(kg cm^{-2})	2.17	2.69
Bulk Density	(g cm^{-3})	0.902	1.09
N	(%)	0.358	0.198
P	(ppm)	3.83	1.08
K	(ppm)	105.41	54.72
Na	(ppm)	19.21	15.90
Ca	(ppm)	1706.44	579.13
Mg	(ppm)	263.52	143.85
Fe	(ppm)	1.11	0.74
Zn	(ppm)	112.71	37.86
Cu	(ppm)	0.86	1.19
Mn	(ppm)	250.83	188.72

4 herbaceous plant species have second cover abundance scale (+, few, with small cover) on the skid road including; *Lamium purpureum* L. var. *purpureum* L., *Ruscus hypoglossum* L., *Geranium asphodeloides* Burm. Fil. subsp. *asphodeloides* Burm. Fil., *Rubus discolor* Weihe et Nees (Table 2). And other plant species including *Stellaria holostea* L., *Primula vulgaris* Huds. subsp. *vulgaris* Huds., *Fragaria vesca* L., *Doronicum orientale* Hoffm., *Ornithogalum wiedemannii* Boiss., have the "r" cover abundance scale, and they were solitary with small cover on the skid road (Table 2). Obtaining results show that determined plant species have the ability to grow on extreme soil and site conditions. Some earlier studies in other regions have also proved the survival capability most of these species on degraded, polluted or disturbed habitats which limit plant growth and survival as summarized on Table 3.

Table 2: Survived plant species on compacted skid road, their general site characteristics and the cover-abundance scales

The Plant Family	The Scientific Name of the Plant Species	General Site Characteristics	The Cover-Abundance Scale
CARYOPHYLLACEAE	*Stellaria holostea* L.	Scrubs, roadsides, damp places.	r
PRIMULACEAE	*Primula vulgaris* Huds. subsp. *vulgaris* Huds.	Often damp places in open or shady turf slopes, evergreen or deciduous woodlands, alpine meadows.	r
LILIACEAE	*Smilax aspera* L.	Macchie, scrub, ravines, and rocky limestone slopes.	1
BERBERIDACEAE	*Epimedium pubigerum* (DC.) Moren et Decaisne	Forest clearings.	1
LABIATAE	*Lamium purpureum* L. var. *purpureum* L.	Oak and fir forests, gravelly banks, fields and waste places.	+
LILIACEAE	*Ruscus hypoglossum* L.	Mixed forests, scrubs, ravines, rocky places.	+
ROSACEAE	*Fragaria vesca* L.	Moist places, especially in forest.	r
CYPERACEAE	*Carex distachya* Desf. var. *distachya* Desf.	Dry stony slopes, open forests, roadsides	1
GERANIACEAE	*Geranium asphodeloides* Burm. Fil. subsp. *asphodeloides* Burm. Fil.	Forests, scrubs, meadows, banks.	+
COMPOSITAE	*Doronicum orientale* Hoffm.	Shady paces in forests and scrubs.	r
LILIACEAE	*Ornithogalum wiedemannii* Boiss.	Woods and forests.	r
DENNSTAEDTIACEAE	*Pteridium aquilinum* (L.) Kuhn.	In forest clearings, cleared woodland, dunes.	1
BORAGINACEAE	*Trachystemon orientalis* (L.) G. Don	Fir forests, shady riverbanks, moist ravines.	1
ARALIACEAE	*Hedera helix* L.	Climbing over trees or creeping on the ground in woods.	1
ROSACEAE	*Rubus discolor* Weihe et Nees.	Deciduous forests and scrubs, shady banks, coastal plains.	+

Table 3: The characteristics and place of disturbed-degraded habitat where the plants can growth

Plant Species	The characteristics and place of disturbed-degraded habitat	References
Stellaria holostea L.	under atmospheric nitrogen deposition, N deposited oak forest, Denmark	[25]
	oak dune forest, Denmark	[26]
	recovered on footpaths that have been closed for access for 6 years, Belgium	[27]
	under disturbed *Q. pyrenaica* ecosystems by human intervention, Spain	[28]
	under wooded landscapes human influented, UK	[29]
Primula vulgaris Huds. subsp. *vulgaris* Huds.	restoration of wet grasslands, Germany	[30]
Smilax aspera L.	undergrowth clearings of the coppice oak forests, Spain	[31]
	disturbed forest areas, potential restoration of natural vegetation, Spain	[32]
	forest disturbances (grazing and thinning), Spain	[33]
Epimedium pubigerum (DC.) Moren et Decaisne	-	-
Lamium purpureum L. var. *purpureum* L.	radioactive-contaminated sites around nuclear power plant, Slovak Republic	[34]
	under the Mediterranean dryland conditions, effects of tillage systems, Turkey	[35]
	man made habitats with irregular disturbances, Czech Republic	[36]
Ruscus hypoglossum L.	in a nature park, Croatia	[37]
Fragaria vesca L.	oak dune forest, Denmark	[26]
	restoration of heathland/moorland, UK	[38]
	response to cattle grazing in mesic semi-natural grassland, Finland	[39]
	restoration success in alluvial grasslands under contrasting flooding regimes, Germany	[40]
	norway spruce forests following clearcutting and shelterwood cutting, Sweden	[41]
	contaminated soil of uranium waste depot, Czech Republic	[42]
Carex distachya Desf. var. *distachya* Desf.	plant reestablishment 15 years after the debris avalanche, USA- described as *Carex* sp.	[43]
	restoration of heathland/moorland, UK- described as *Carex* sp.	[38]
	forests following clearcutting and shelterwood cutting, Sweden. described as *Carex* sp.	[41]
Geranium asphodeloides Burm. Fil. subsp. *asphodeloides* Burm. Fil.	around villages in Sakarya province, Turkey	[44]
Doronicum orientale Hoffm.	on basaltic pyroclastic deposits at different altitudes, Italy	[45]
	on mountainous area (Yunt Mountain), Turkey	[46]
Hedera helix L.	recovered on footpaths that have been closed for access for 6 years, Belgium	[27]
	undergrowth clearings of the coppice oak forests, Spain	[31]
	forest disturbances (grazing and thinning), Spain	[33]
	pine forest clearings along the French Atlantic sand dunes, France	[47]
	in reclaimed area contaminated with dioxin, Italy	[48]
	man made urban habitats, urban-industrial ecosystems, Germany	[49]

Rubus discolor Weihe et Nees.	forest disturbances (grazing and thinning), Spain-described as Rubus sp.	[33]
Ornithogalum wiedemannii Boiss.	-	-
Pteridium aquilinum (L.) Kuhn.	Initial recovery after wildfire and clear cut in Jack pine forest, USA	[50]
	pine forest clearings along the French Atlantic sand dunes, France	[47]
	under disturbed *Q. pyrenaica* ecosystems by human intervention, Spain	[28]
	glades, rides and roads in plantation forests, Ireland	[51]
	Short-term understory plant community responses to timber-harvesting intensity, USA	[52]
Trachystemon orientalis (L.) G. Don	after woody vegetation control treatments (bulldozing and hand-grubbing), Turkey	[53]
	around villages in Sakarya province, Turkey	[44]

Effects of skid roads on soil properties and other site factors have been demonstrated to have detrimental impacts on native flora and fauna. And the capacity of vegetation recover depends on the ability of individual species to recover, habitat factors such as soil properties, microclimate and associate species composition. Establishing and maintaining native plant communities on forest roads is vital because until adequate plant cover is established, ripping alone provides only temporary and marginal improvements in soil structure. Herbaceous plant cover on roads can improve soil structure by increasing soil organic matter, reducing erosion and restoring the habitat lost. Successful revegetation may also help to keep roads closed by reducing their visibility. For these reasons, determining and availability of revegetation, recovery and survival success of native plant species have particular importance. Identification of native species would be successful to rehabilitate skid roads. In conclusion, the determined herbaceous species on skid road in this study show the resistance and survival capability on degraded soil and micro environment conditions.

REFERENCES AND NOTES

1. Crisholm, M. The increasing separation of production and consumption. In *The Earth as Transformed by Human Action*; Cambridge University Press: Cambridge, 1990. [Google Scholar]
2. Grübler, A. Technology. In *Changes in land use and land cover: a global perspective*; Cambridge University Press with Clark University: Cambridge; pp. 287–328. 1994. [Google Scholar]
3. Demir, M. Impacts, management and functional planning criterion of forest road network system in Turkey.*Transportation Research Part A: Policy and Practice* 2007, *41*(1), 56–68. [Google Scholar]
4. Cole, D.N.; Landres, P.B. Threats to wilderness ecosystems: impacts and research needs. *Ecological Applications* 1996, *6*, 168–184. [Google Scholar]
5. Buckley, D.S.; Crow, T.R.; Nauertz, E.A.; Schulz, K.E. Influence of skid trails and haul roads on understory plant richness and composition in managed forest landscapes in Upper Michigan, USA. *Forest Ecology and Management* 2003,*175*, 509–520. [Google Scholar]
6. Laffan, M.; Jordan, G.; Duhig, N. Impacts on soils from cable-logging steep slopes in Northeastern Tasmania, Australia. *Forest Ecology and Management* 2001, *144*, 91–99. [Google Scholar]
7. Messina, M.G.; Schoenholtz, S.H.; Lowe, M.W.; Wang, Z.; Gunter, D.K.; Londo, A.J. Initial responses of woody vegetation, water quality, and soils to harvesting intensity in a texas bottomland hardwood ecosystem. *Forest Ecology and Management* 1997, *90*, 201–215. [Google Scholar]
8. Wang, L. Assessment of animal skidding and ground machine skidding under mountain conditions. *Journal of Forest Engineering* 1997, *8*(2), 57–64. [Google Scholar]
9. Bengtsson, J.; Lundkvist, H.; Saetre, P.; Sohlenius, B.; Solbreck, B. Effects of organic matter removal on the soil food web: forestry practices meet ecological theory. *Applied Soil Ecology* 1998, *9*, 137–143. [Google Scholar]
10. Arocena, J.M. Cations in solution from forest soils subjected to forest floor removal and compaction treatments. *Forest Ecology and Management* 2000, *133*, 71–80. [Google Scholar]
11. Marshall, V.G. Impacts of forest harvesting on biological processes in Northern forest soils. *Forest Ecology and Management* 2000, *133*, 43–60. [Google Scholar]
12. Gilliam, F.S. Effects of harvesting on herbaceous layer diversity of a central appalachian hardwood forest in West Virginia, USA. *Forest Ecology and Management* 2002, *155*, 33–43. [Google Scholar]
13. Williamson, J.R.; Neilsen, W.A. The effect of soil compaction, profile disturbance and fertilizer application on the growth of eucalypt seedlings in two glasshouse studies. *Soil and Tillage Research* 2003, *71*, 95–107. [Google Scholar]
14. Godefroid, S.; Koedam, N. The impact of forest paths upon adjacent vegetation: effects of the paths

surfacing material on the species composition and soil compaction. *Biological Conservation* 2004, *119*, 405–419. [Google Scholar]

15. Johnston, F.M.; Johnston, S.W. Impacts of road disturbance on soil properties and exotic plant occurrence in subalpine areas of Australian Alps. *Arctic Antarctic and Alpine Research* 2004, *36*(2), 201–207. [Google Scholar]
16. Ketcheson, G.L.; Megahan, W.F.; King, J.G. "R1-R4" and "Boised" sediment prediction model tests using forest roads in granitics. *Journal of American Water Resource Assessment* 1999, *35*(1), 83–98. [Google Scholar]
17. Anonymous. *Harvesting reports of Istanbul-Bahcekoy Regional Directorate of Forestry*; Istanbul, (in Turkish). 2005. [Google Scholar]
18. Braun-Blanquet, J. *Pflanzensoziologie-Grundzüge der Vegetationskunde.*; Springer Verlag: Wien and New York, 1964. [Google Scholar]
19. Davis, P.H. *Flora of Turkey and East Aegean Islands*; Vol. 1-10, At University Press: Edinburgh; pp. 1965–1985.
20. Demir, M.; Makineci, E.; Yilmaz, E. Harvesting impacts on herbaceous understory, forest floor and top soil properties on skid road in a beech (*Fagus orientalis* Lipsky.) stand. *Journal of Environmental Biology* 2007, *28*(2), 427–432. [Google Scholar]
21. Demir, M.; Makineci, E.; Comez, A.; Yilmaz, E. Timber harvesting effects on chemical properties of topsoil, herbaceous cover and forest floor in a beech (*Fagus orientalis* Lipsky.) stand. *Journal for Nature Conservation* 2008. (in press). [Google Scholar]
22. Mariani, L.; Chang, S.X.; Kabzems, R. Effects of tree harvesting, forest floor removal, and compaction on soil microbial biomass, microbial respiration, and N availability in boreal aspen forest in British Coloumbia. *Soil Biology and Biochemistry* 2006, *38*, 1734–1744. [Google Scholar]
23. Barzegar, A.R.; Nadian, H.; Heidari, F.; Herbert, S.J.; Hashemi, A.M. Interaction of soil compaction, phosphorus and zinc on clover growth and accumulation of phosphorus. *Soil and Tillage Research* 2006, *87*, 155–162. [Google Scholar]
24. Kozlowski, T.T. Soil compaction and growth of woody plants. *Scandinavian Journal of Forest Research* 1999, *14*, 596–619. [Google Scholar]
25. Kristensen, H.L.; Henriksen, K. Soil nitrogen transformations along a successional gradient from *Calluna* heathland to *Quercus* forest at intermediate atmospheric nitrogen deposition. *Applied Soil Ecology* 1998, *8*, 95–109. [Google Scholar]
26. Lawesson, J.E.; Wind, P. Oak dune forests in Denmark and their ecology. *Forest Ecology and Management* 2002, *164*, 1–14. [Google Scholar]
27. Roovers, P.; Bossuyt, B.; Gulinck, H.; Hermy, M. Vegetation recovery on closed paths in temperate deciduous forests.*Journal of Environmental Management* 2005, *74*, 273–281. [Google Scholar]
28. Tárrega, R.; Calvo, L.; Marcos, E.; Taboada, A. Comparison of understory plant community composition and soil characteristics in *Quercus pyrenaica* stands with different human uses. *Forest Ecology and Management* 2007, *241*, 235–242. [Google Scholar]
29. Rotherham, R.I. The implications of perceptions and cultural knowledge loss for the management of wooded landscapes: A UK case-study. *Forest Ecology and Management* 2007, *249*, 100–115. [Google Scholar]
30. Rosenthal, G. Selecting target species to evaluate the success of wet grassland restoration. *Agriculture, Ecosystems and Environment* 2003, *98*, 227–246. [Google Scholar]
31. Camprodon, J.; Brotons, L. Effects of undergrowth clearing on the bird communities of the northwestern Mediterranean coppice holm oak forests. *Forest Ecology and Management* 2006, *221*, 72–82. [Google Scholar]
32. Arévalo, J.R.; Fernández-Palacios, J.M. Gradient analysis of exotic *Pinus radiata* plantations and potential restoration of natural vegetation in Tenerife, Canary Islands (Spain). *Acta Oecologica* 2005, *27*, 1–8. [Google Scholar]
33. Onaindia, M.; Dominguez, I.; Albizu, I.; Garbisu, C.; Amezaga, I. Vegetation diversity and vertical structure as indicators of forest disturbance. *Forest Ecology and Management* 2004, *195*, 341–354. [Google Scholar]
34. Mičieta, K.; Murín, G. Wild plant species in bio-indication of radioactive-contaminated sites around Jaslovske Bohunice nuclear power plant in the Slovak Republic. *Journal of Environmental Radioactivity* 2007, *93*, 26–37. [Google Scholar]
35. Ozpinar, S. Effects of tillage systems on weed population and economics for winter wheat production under the Mediterranean dryland conditions. *Soil and Tillage Research* 2006, *87*, 1–8. [Google Scholar]
36. Lososová, Z.; Chytrý, M.; Kühn, I.; Hájek, O.; Horáková, V.; Pyšek, P.; Tichý, L. Patterns of plant traits in annual vegetation of man-made habitats in central Europe. *Perspectives in Plant Ecology, Evolution and Systematics* 2006, *8*, 69–81. [Google Scholar]
37. Jelaska, S.D.; Antonić, O.; Nikolić, T.; Hršak, V.; Plazibat, M.; Križan, J. Estimating plant species occurrence in MTB/64 quadrants as a function of DEM-based variables—a case study for Medvednica Nature Park, Croatia. *Ecological Modelling* 2003, *170*, 333–343. [Google Scholar]
38. Tong, C.; Le Duca, M.G.; Ghorbani, J.; Mars, R.H. Linking restoration to the wider landscape: A study of a bracken control experiment within a upland moorland landscape mosaic in the Peak District, UK. *Journal of Landscape Urban Planning* 2006, *78*, 115–134. [Google Scholar]
39. Pykälä, J. Plant species responses to cattle grazing in mesic semi-natural grassland. *Agriculture, Ecosystems and Environment* 2005, *108*, 109–117. [Google Scholar]
40. Bissels, S.; Hölzel, N.; Donath, T.W.; Otte, A. Evaluation of restoration success in alluvial grasslands under contrasting flooding regimes. *Biological Conservation* 2004, *118*, 641–650. [Google Scholar]

41. Hannerz, M.; Hånell, B. Effects on the flora in Norway spruce forests following clearcutting and shelterwood cutting.*Forest Ecology and Management* 1997, *90*, 29–49. [Google Scholar]

42. Soudek, P.; Petřík, P.; Vágner, M.; Tykva, R.; Václav, P.; Petrová, S.; Vaněk, T. Botanical survey and screening of plant species which accumulate ^{226}Ra from contaminated soil of uranium waste depot. *European Journal of Soil Biology* 2007,*43*, 251–261. [Google Scholar]

43. Dale, V.H.; Adams, W.M. Plant reestablishment 15 years after the debris avalanche at Mount St.Helens, Washington.*Science of the Total Environment* 2003, *313*, 101–113. [Google Scholar]

44. Uzun, E.; Sariyar, G.; Adsersen, A.; Karakoc, B.; Otuk, G.; Oktayoglu, E.; Pirildar, S. Traditional medicine in Sakarya province (Turkey) and antimicrobial activities of selected species. *Journal of Ethnopharmacology* 2004, *95*, 287–296. [Google Scholar]

45. Fernández Sanjurjo, M.J.; Corti, G.; Certini, G.; Ugolini, F.C. Pedogenesis induced by Genista aetnensis (Biv.) DC. on basaltic pyroclastic deposits at different altitudes, Mt. Etna, Italy. *Geoderma* 2003, *115*, 223–243. [Google Scholar]

46. Ugurlu, E.; Secmen, O. Medicinal plants popularly used in the villages of Yunt Mountain (Manisa-Turkey). *Fitoterapia*(in press). 2007. [Google Scholar]

47. Lemauviel, S.; Roze, F. Ecological study of pine forest clearings along the French Atlantic sand dunes: Perspectives of restoration. *Acta Oecologica* 2000, *21*(3), 179–192. [Google Scholar]

48. Sartori, F.; Assini, S. Vegetation evolution in reclaimed area contaminated with dioxin. *Chemosphere* 2001, *43*, 525–535. [Google Scholar]

49. Bornkamm, R. Spontaneous development of urban woody vegetation on differing soils. *Flora* 2007, *202*, 695–704. [Google Scholar]

50. Le Duc, S.D.; Rothstein, D.E. Initial recovery of soil carbon and nitrogen pools and dynamics following disturbance in jack pine forests: A comparison of wildfire and clearcut harvesting. *Soil Biology and Biochemistry* 2007, *39*, 2865–2876. [Google Scholar]

51. Smith, G.F.; Iremonger, S.; Kelly, D.L.; O'donoghue, S.; Mitchell, F.J.G. Enhancing vegetation diversity in glades, rides and roads in plantation forests. *Biological Conservation* 2007, *136*, 283–294. [Google Scholar]

52. Fredericksen, T.S.; Ross, B.D.; Hoffman, W.; Morrison, M.L.; Beyea, J.; Johnson, B.N.; Lester, M.B.; Ross, E. Short-term understory plant community responses to timber-harvesting intensity on non-industrial private forestlands in Pennsylvania. *Forest Ecology and Management* 1999, *116*, 129–139. [Google Scholar]

53. Yildiz, O.; Sarginci, M.; Esen, D.; Cromack, K. Effects of vegetation control on nutrient removal and Fagus orientalis, Lipsky regeneration in the western Black Sea Region of Turkey. *Forest Ecology and Management* 2007, *240*, 186–194. [Google Scholar]

CHAPTER 16

THE DYNAMICS OF RECOVERY AND GROWTH: HOW DEFOLIATION AFFECTS STORED RESOURCES

R. R. L. Atkinson, M. M. Burrell, K. E. Rose, C. P. Osborne and M. Rees

Department of Animal and Plant Sciences, University of Sheffield, Western Bank, Sheffield S10 2TN, UK

ABSTRACT

Growth rate varies widely among species and the trade-off between growth rate and storage or maintenance traits is a principal axis of variation between species. Many plant species have substantial root stores, but very little is known about how growth rate modifies responses of these stores to defoliation and other stresses. Species with different growth rates are predicted to respond in distinct ways, because of variation in the pre-defoliation allocation to storage. Here, we quantified the dynamics of stored carbohydrates in seven species with varying growth rate, following defoliation in a pot experiment. For faster growing species, there was significant reduction in carbohydrate concentration following defoliation, followed by relatively fast recovery, whereas for slower growing species, carbohydrate concentration levels remained relatively invariant across treatments. Results for total carbohydrates mirrored those for concentration, but were not as significant. Our findings were consistent with the idea that faster growing species respond more rapidly than slower growers to defoliation, through changes in carbohydrate pool concentrations. Growth rate as an indicator of life-history and ecological strategy may therefore be key to understanding post-defoliation recovery and storage strategies.

INTRODUCTION

Growth rate varies widely among species, and part of this variation is linked to variation in allocation to storage, defence and maintenance [1–4]. Previous work has shown that allocation to storage reduces maximum potential growth rates [5,6], which might seem maladaptive as rapid growth allows plants to quickly increase in size and so effectively exploit resources both above and below ground. However, in highly disturbed environments, storage is important in allowing re-growth following destruction of plant tissue, and theoretical models predict higher allocation to storage in productive, highly disturbed environments [7].

Since the allocation of resources to storage leads to a reduction in growth rate, fast-growing plant species typically have smaller stores, both in terms of absolute size and percentage of root metabolites allocated to carbohydrates [6]. This means that we might expect plant responses to defoliation to be mediated by growth rate. However, there are currently no clear theoretical predictions for understanding how stored resources should be used following defoliation, and whether this response should vary between fast- and slow-growing species. Following defoliation, plants deploy stored resources to rebuild photosynthetic material, but how should the use of stores vary among species? Clearly, the size of the store,

Citation: R. R. L. Atkinson, M. M. Burrell, K. E. Rose, C. P. Osborne and M. Rees, The dynamics of recovery and growth: how defoliation affects stored resources, DOI: 10.1098/rspb.2013.3355.

the construction costs of new photosynthetic material and the flexibility of the growth strategy are likely to influence the re-growth strategy. Fast-growing species typically have: (i) low leaf construction costs and (ii) more flexible growth strategies [8–10], and so we might expect them to rapidly deploy stores. This use of stores will, however, be constrained by the small amount of material stored. By contrast, slow-growing species have relatively large stores, but less flexible growth strategies and higher leaf construction costs [11,12].

These observations lead to two distinct but contrary predictions. Fast-growing species will either: (i) more rapidly use their stores as a result of their more flexible growth strategies and lower leaf construction costs than slow-growing species or (ii) have more conservative re-growth strategies as they are constrained by the small amount of material stored. Slower growing plants, with a large store of carbohydrates, may respond to defoliation by rapidly deploying their root stores to re-grow photosynthetic material or, alternatively, use of root stores may be more gradual, in accordance with the idea that slow-growing species have relatively inflexible growth strategies and higher leaf construction costs.

Here, we report the results of an experiment on seven monocarpic (once-flowering) plant species. This relatively simple life history is ideal for understanding links between growth, survival, storage and defoliation, since energy stores are only allocated to reproduction in the final terminal reproductive event. The individuals in this study were sampled from a larger experiment, where a growth–survival trade-off was established for these species following multiple, full defoliation events [13]. This trade-off may be linked to the depletion of root carbohydrate stores owing to defoliation. In order to investigate the influence of growth rate on the responses of root reserves to defoliation, we used a size-standardized measure of species growth [13,14] to classify the species as slow, medium or fast growing. To quantify how stored carbohydrate pools changed following defoliation in slow versus medium and fast growers, we completed an outdoor pot experiment. We defoliated a subset of plants either once or twice, and measured root storage (concentration and total pool size), using a non-targeted metabolomics approach for comparison with non-defoliated controls.

Based on the two contradictory predictions for how growth rate mediates the response to defoliation, we hypothesized that in the weeks following defoliation, species in each growth category would respond differently in terms of changes in (i) root carbohydrate concentrations and (ii) total root carbohydrate. We focused mainly on carbohydrate concentration, rather than total root pools, since concentration is easier to estimate accurately, although the patterns should be similar in both. We considered a reduction in carbohydrate concentration relative to controls following defoliation to be indicative of storage being used for re-growth. Previously, we found that slower growing species had higher survival following multiple defoliation events [13], and so we expect total root carbohydrates to be significantly lower for fast, compared with slow-growing defoliated plants, if stores mediate the growth–survival trade-off.

MATERIAL AND METHODS

The experiment took place at Tapton Experimental Gardens, University of Sheffield. Seeds of*Cirsium vulgare*, *Digitalis purpurea*, *Verbascum thapsus*, *Verbascum blattaria*, *Carduus nutans*, *Arctium minor* and *Senecio jacobaea* were sown between 15 and 21 March 2007 into degradable pots 70 mm in diameter and put into a greenhouse. The pots were filled with a 9 : 1 : 1 mixture of sand: vermiculite : M3 compost. After a few weeks of growth, the plants were transferred into pots 2.2 l in volume, filled with the same sand, vermiculite and compost mixture as before, and placed outside into a randomized, eight block design, roughly balanced by species. Before the first treatment, plants were allocated to a control or treatment group. The latter consisted of one (T1) or two (T2) full defoliations (removal of all aboveground material). The first defoliation took place on 30/06/07 (dd/mm/yy). T1 plants were then left to grow until harvesting. T2 plants were defoliated a second time on 07/08/07, after full or partial re-growth of most of them had occurred (table 1). Plant size was tracked non-destructively by measuring the length of the longest leaf (table 1). Where possible three to four plants per species per treatment were harvested (n = 266) on the following dates: 12/05/07 (H1) (n = 26), 22/06/07 (H2) (n = 22), 7–8/07/07 (H3) (n = 49), 21–22/07/07 (H4) (n = 43), 31/08/07 (H5) (n = 60) and 05/03/08 (H6) (n = 66) (table 1). Harvests H1 and H2 comprised only control plants, H3 and H4 included control and T1 plants, and H5 and H6 encompassed control, T1 and T2 plants (table 1). Total sample size by treatment was: C = 150, T1 = 76 and T2 = 40.

Table 1: Summary of census dates for non-destructive measures of the longest leaf length, (see Rose *et al.*[**13**] for how these measures were used to calculate size-standardized relative growth rate values), defoliation times (vertical lines) and destructive harvests (H1–H6)

year	2007										2008	
census	1	2	3	4	5	6	7	8	9	10	11	12	13
date dd/mm	10/04	17/04	24/04	01/05	12/06	19/06	10/07	17/07	07/08	21/08	28/08	17/03	22/05
harvest				H1		H2	H3	H4			H5	H6	

At harvest, a small cross section of taproot was removed from all plants, and flash-frozen in liquid nitrogen. These samples were stored at −80°C until extraction. The remaining roots were separated into taproot and fibrous root before taking fresh-weight measurements.

(a) Mass spectrometry

Mass spectrometry is a method of compound detection within biological samples, by separating ions according to mass/charge (m/z) ratios. The output is a list of m/z values, each representing an individual ion, alongside the corresponding count for the ion. The relative ion count corresponds to the concentration of the ion in the biological sample relative to all other ions detected. The plant root samples were extracted using a 5 : 2 : 2 methanol : chloroform : water solution [6]. The metabolite profiles were produced over a mass range of 100–1000 Da, using an electro-spray method with a quadrupole mass spectrometer (API sciex III plus, AB Sciex UK Limited,

Phoenix House Lakeside Drive, Cheshire). Samples were analysed in triplicate.

(b) Storage compound analysis

We previously showed that sucrose and the raffinose series of carbohydrates, consisting of raffinose, stachyose and verbascose are the main carbon storage compounds in the seven species [6], and these were targeted for analysis. Peak height values were extracted from the peak centred, unbinned mass spectra data for *m/z* values corresponding to sucrose, raffinose, verbascose and stachyose. We chose the highest peak within ±0.2 Da of the monoisotopic value corresponding to the hydrogen, potassium and sodium adducts of the metabolites. A single value for each of these compounds› ions in a sample was calculated, which represented the number of ion counts for an ion in 0.1 g of taproot ('carbohydrate concentration' in the analyses), averaged over the three replicates. These were the data used for the individual storage compound concentration analyses. To obtain a value for the total carbohydrate concentration, we added the values for sucrose, raffinose, verbascose and stachyose, after correcting for their relative ionization energy values. Relative energy correction was possible since the basic structure of these compounds is conserved; the raffinose series is extended through addition(s) of galactose units onto a sucrose molecule. The total carbohydrates in the taproot data were calculated by scaling up the carbohydrate concentration values to the full taproot, by multiplying by the total fresh weight of the taproot. The mass spectrometer counts shown in the figures are directly proportional to the concentration of compound in the sample.

(c) Species growth categorization

In order to correct for the size-dependence of growth rate, we fitted a nonlinear mixed effects model to the size census data using non-destructive size measurements of longest leaf length (table 1) [6,13]. In the dataset used in the present investigation, growth curves for each individual could not be estimated accurately, because destructive harvesting was limited to a small number of census dates. However, as the individuals in this experiment were sampled from a larger experiment [13], we used this larger sample of individual growth curves (n = 842), to calculate species-average growth rates and their standard errors [6]. These values, therefore, do not represent maximum intrinsic growth rates for each species, but the species-average growth rates for the conditions experienced by the individuals from this experiment.

(d) Statistical analysis

All statistics were completed using the R statistical package [15]. We used ANOVA with growth category (fast, medium, slow), harvest (H1–H6) and treatment (control, one defoliation, two defoliations) and all interactions between them as fixed explanatory variables. The response variables were carbohydrate concentration, and total carbohydrates in the taproot. The total carbohydrates in the taproot data were calculated by scaling up the carbohydrate concentration values to the full taproot, by multiplying by the total fresh weight of the taproot. All response variables were transformed using the Box-Cox transformation so the data conformed to the assumptions of ANOVA. *A priori* hypotheses were tested using planned contrasts from the fitted analysis of variance model. Block was initially added to the model, but was removed, as it was not significant in all analyses.

RESULTS

(a) Species-average growth rates

For the analyses, *S. jacobaea*, *C. vulgare* and *A. minor* were categorized as slow growers relative to *V. blattaria* and *C. nutans*, which were classified as medium growers, and *V. thapsus*and *D. purpurea* were classified as fast growers, based upon species-average growth rate values (table 2). This categorization of species was used in all hypothesis testing.

Table 2: Categorization of species with differing mean size-standardized relative growth rate, i.e. 'fast-', 'medium-' and 'slow-'growing species. In Rose *et al.* [13], we used nonlinear mixed effects models to fit growth curves for individual plants (n = 842), and individual values over species were averaged to achieve a size-standardized measure of species mean growth rate at census 6, before defoliation

species	species mean growth rate (mm/mm/day) ± the standard error around that estimate	growth category
Verbascum thapsus	0.022820 ± 0.0002	fast
Digitalis purpurea	0.020778 ± 0.0002	fast
Carduus nutans	0.019100 ± 0.0004	medium
Verbascum blattaria	0.018027 ± 0.0003	medium
Arctium minor	0.015869 ± 0.0006	slow
Cirsium vulgare	0.015883 ± 0.0006	slow
Senecio jacobaea	0.015804 ± 0.0003	slow

(b) Carbohydrate concentration

Carbohydrate concentration showed a complex pattern across treatments, and between species and harvests with a significant three-way interaction between harvest, treatment and growth category (figure 1 and table 3).

Table 3: Analysis of variance results for root carbohydrate concentration and total root carbohydrates. The response variables were transformed using Box-Cox transformation. *F*-values are given in the body of the table, with their associated levels of statistical significance indicated by asterisks. RGR = relative growth rate

	d.f.	carbohydrate concentration	total root carbohydrates
RGR	2	4.16*	93.83***
harvest	5	6.21***	33.94***
treatment	2	2.53	10.9***
RGR × treatment	4	1.63	0.72
RGR × harvest	10	1.07	1.71
RGR × harvest × treatment	8	2.97**	1.96
R^2 value for model		0.17	0.62
residuals		230	202

*$p < 0.05$, **$p < 0.01$, ***$p < 0.001$.

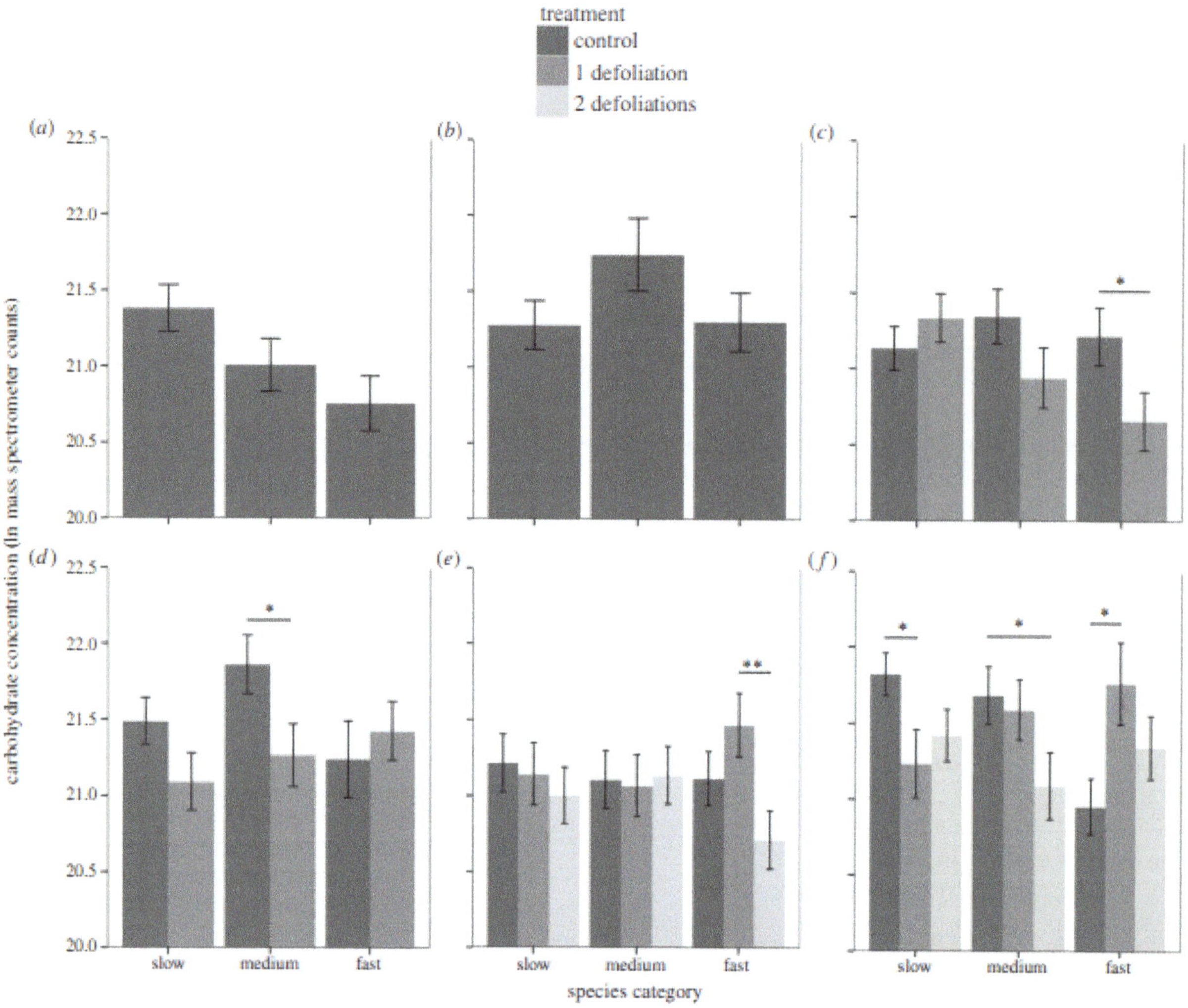

Figure 1: (*a–f*) The relationship between growth rate category and treatment (control, one defoliation and two defoliations) over harvests H1–H6. Significance values $*p < 0.05$, $**p < 0.01$, $***p < 0.001$. Means and standard errors were plotted from the predicted model results. Significance information was extracted from within ANOVA *t*-tests.

Hypothesis: fast- and slow-growing plants would respond differently to defoliation treatments through changes in carbohydrate concentration after defoliation.

To test this hypothesis, in harvests H3 and H4, we compared the carbohydrate concentration between controls versus once-defoliated plants in fast-, medium- and slow-growing species (figure 1 and table 3). In harvest H3 (one week after the first defoliation), defoliated fast growers had reduced carbohydrate concentrations by 43% compared with control plants (d.f. = 230, $t = 2.07$, $p < 0.05$) but both slow- (d.f. = 230, $t = 0.94$, n.s.) and medium-growing (d.f. = 230, $t = 1.51$, n.s.) defoliated plants did not. Therefore, after defoliation faster growers more rapidly mobilized carbohydrate stores from the root than slower growing species. This supports our hypothesis that fast and slow growers would respond differently to defoliation.

In harvest H4 (two weeks after the first defoliation treatment), there were still no differences in carbohydrate concentration for control versus defoliated plants of slow-growing species (d.f. = 230, $t = 1.62$, n.s.). However, in the medium growth rate category defoliation significantly lowered carbohydrate concentrations, by 45% compared with controls, indicating the use of some root stores (d.f. = 230, $t = 2.11$, $p < 0.05$). By contrast, for fast-growing species at harvest H4, defoliation did not significantly affect carbohydrate concentration compared with controls (d.f. = 230, $t = 0.59$, n.s.). This suggests that, within two weeks of defoliation, root stores had been replenished by carbon derived from photosynthesis since defoliation. In summary, fast-growing plants had reduced carbohydrate concentrations immediately after the defoliation at harvest H3, and then medium-growing defoliated plants responded similarly at harvest H4, while slow-growing defoliated plants showed no response at either harvest.

To uncover whether the patterns in harvests H3 and H4 (after the first defoliation) were repeated in harvests H5 and H6 (after the second defoliation), we compared differences in mean carbohydrate concentrations between once-defoliated and twice-defoliated plants for fast-, medium- and slow-growing species. If twice-defoliated plants had lower carbohydrate concentrations than once-defoliated plants left to grow, then this would support the idea that carbohydrate stores had been mobilized from the root for shoot re-growth in the twice-defoliated plants.

In harvest H5 (three weeks after second defoliation), twice-defoliated plants of fast-growing species had significantly reduced mean carbohydrate concentrations by 53% in comparison to once-defoliated plants (d.f. = 230, $t = 2.64$, $p < 0.01$), but this was not the case for slow- (d.f. = 230, $t = 0.51$, n.s.) and medium-growing species (d.f. = 230, $t = 0.26$, n.s.).

These results suggest that, after a second defoliation, fast-growing species mobilized some of the reserves that had accumulated after the first defoliation.

In harvest H6 (approx. six months after second defoliation), there were no differences in carbohydrate concentration between once-defoliated versus twice-defoliated plants in any growth category; for slow (d.f. = 230, t = 0.66, n.s.), medium (d.f. = 230, t = 1.68, n.s.) or fast growers (d.f. = 230, t = 1.24, n.s.). This indicates that, for fast-growing species, the twice-defoliated plants had recovered carbohydrate concentrations to that of the once-defoliated plants (six months after the second defoliation), repeating the pattern seen for the first defoliation.

In harvest H6, the comparison between control and defoliated plants within each growth category showed that for slow-growing species, defoliated plants had a significantly reduced mean carbohydrate concentration, for controls versus once-defoliated plants (d.f. = 230, t = 2.19, p = 0.03) and non-significantly between controls versus twice-defoliated plants (d.f. = 230, t = 1.8, p= 0.07). For medium-growing species, once-defoliated plants had similar carbohydrate concentrations to controls (d.f. = 230, t = 0.35, n.s.), but twice-defoliated plants had significantly reduced mean carbohydrate concentrations compared with controls (d.f. = 230, t = 2.05, p = 0.04). For fast growers, defoliated plants had increased carbohydrate concentrations relative to the controls, and this was significant for once-defoliated plants (d.f. = 230, t = 2.53, p = 0.012), but not for twice-defoliated plants (d.f. = 230, t = 1.41, n.s.).

(c) Total root carbohydrates

Hypothesis: fast- and slow-growing plants respond differently to defoliation treatments through changes in total root carbohydrates.

To test this prediction, the difference in total root carbohydrates between controls and once-defoliated plants was compared in all growth rate categories at harvests H3 and H4 (two and three weeks after the first defoliation, respectively (figure 2 and table 3)). At harvest H3, fast growing once-defoliated plants had reduced total root carbohydrates compared with controls, although this was not significant (d.f. = 202, t = 1.74, p = 0.084). This trend was not seen in the medium growth rate (d.f. = 202, t = 0.76, n.s.) nor for slow growth rate species, where once-defoliated plants had more (but not significantly so), total carbohydrates compared with controls (d.f. = 202, t = 0.55, n.s.). At harvest H4, once-defoliated plants of medium-growing species had reduced total carbohydrates (d.f. = 202, t = 1.87, p = 0.063). This mirrors our findings for carbohydrate concentration, where once-defoliated plants of medium-growing species had a significantly reduced carbohydrate concentration compared with controls (figure 1). In harvest H4, the difference between the total root carbohydrates of controls and once-defoliated plants was not significant for species in the slow- and fast-growth rate categories.

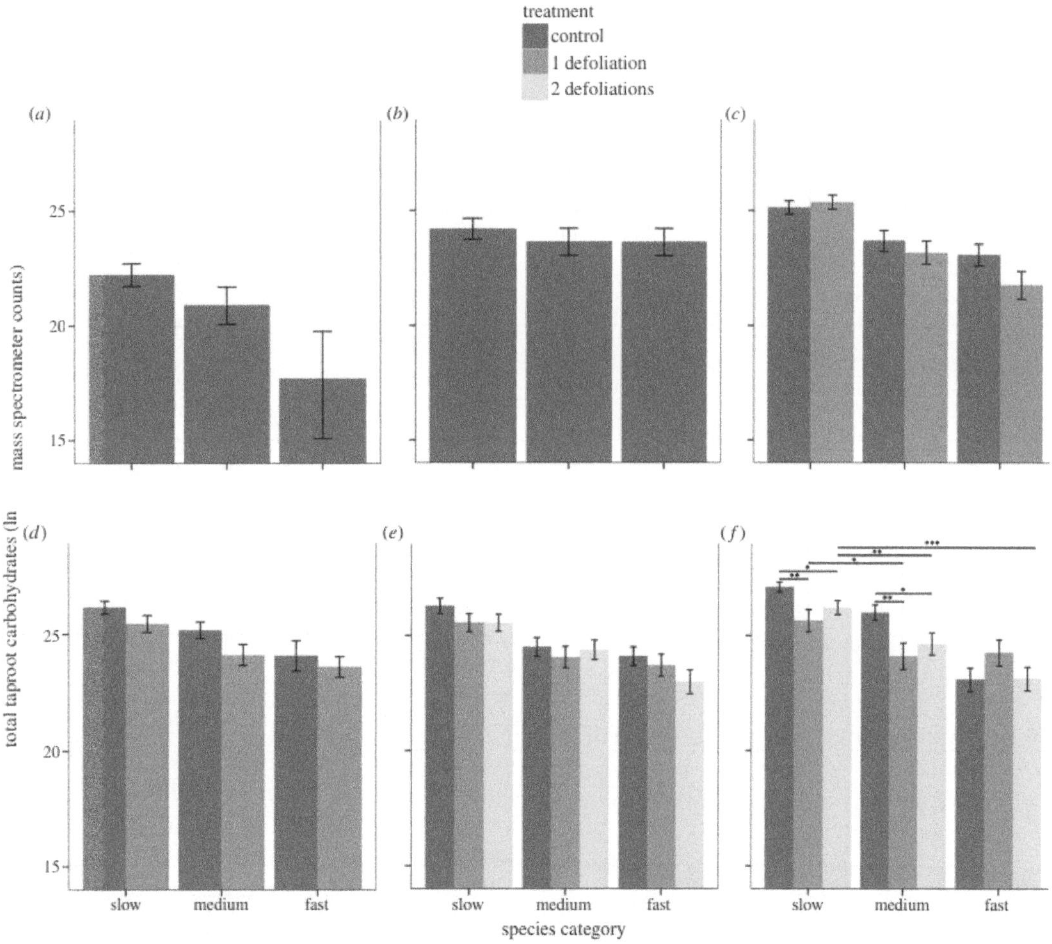

Figure 2: (*a–f*) The relationship between total root carbohydrates, growth rate category and treatment (control, one defoliation and two defoliations) over harvests H1–H6. Significance values $*p < 0.05$, $**p < 0.01$, $***p < 0.001$. Means and standard errors were plotted from the predicted model results. Significance information was extracted from within ANOVA t-tests.

To test whether the second defoliation resulted in an additional reduction in carbohydrate stores, we compared once-defoliated and twice-defoliated plants within each growth category at harvests H5 and H6. There was no difference in total root carbohydrates between once-defoliated and twice-defoliated plants for any growth category at either harvest.

In general, results for total root carbohydrate mirrored results found at the root carbohydrate level, as the effects were in the same direction, but the comparisons were less statistically significant.

Prediction: in the final harvest (harvest H6), we expected that (a) for each defoliation treatment, total root carbohydrates would be lower for fast-growing compared with slow-growing species and (b) in each growth category defoliated plants would have lower total carbohydrates than controls. Higher mortality for fast growers may therefore be associated with lower carbohydrate levels

To test prediction (a), we first compared the total root carbohydrate content of once-defoliated slow-growing species versus once-defoliated medium and fast growers (table 3 and figure 2). The twice-defoliated plants were then compared to test whether slower growing species had a higher total of taproot carbohydrates, as predicted. The medium growing, once-defoliated plants had significantly less, by 79%, total root carbohydrates than once-defoliated slow growers (d.f. = 202, $t = 2.09$, $p < 0.05$), and fast growers had 75% less than slow growers (d.f. = 202, $t = 1.89$, $p = 0.06$). The same comparison for twice-defoliated plants showed that medium-growing species had significantly less total root carbohydrates, by 79% than slow-growing species (d.f. = 202, $t = 2.86$, $p < 0.01$) and fast growers had 95% less than slow-growing species (d.f. = 202, $t = 5.50$, $p < 0.001$).

To test prediction (b), we compared total root carbohydrate content of control plants *versus* once-defoliated and twice-defoliated plants within each growth category. Slow- and medium-growing once-defoliated plants had reduced stores compared with respective controls, for slow growers, once-defoliated plants had 76% less (d.f. = 202, $t = 2.90$, $p < 0.01$), and for medium growers once-defoliated plants had 85% less (d.f. = 202, $t = 3.02$, $p < 0.01$). However, this comparison for fast growers showed no significant change in total stores (d.f. = 202, $t = 1.54$, p = n.s.). Slow- and medium-growing twice-defoliated plants also had reduced stores compared with controls; slow growers had 59% less (d.f. = 202, $t = 2.44$, $p < 0.05$) and medium growers had 74% less (d.f. = 202, $t = 2.34$, $p < 0.05$). Again, this comparison for fast growers was not significant (d.f. = 202, $t = 0.062$, n.s.).

DISCUSSION

The changes in carbohydrate storage over harvests and between treatments were complex, but underpinned by consistent patterns within growth category groupings. We found that the concentration of root carbohydrates in the slow growth category did not alter significantly immediately following defoliation (figure 1 and table 3). By contrast, the carbohydrate concentration of species in the high growth rate category declined significantly shortly after defoliation (one week), but had recovered rapidly by two weeks after defoliation. Species in the medium growth category showed an intermediate response, with a dip in storage carbohydrate concentration two weeks after defoliation. Overall, these patterns are consistent with the idea that faster growing species have a more rapid response to defoliation than slower growing species, and our prediction that species in different growth categories would exhibit distinct post-defoliation recovery strategies.

For slow-growing species, total carbohydrate concentrations did not alter substantially within the first week following a defoliation event (control versus one defoliation in harvest H3 and one defoliation versus two defoliations in harvest H5). By contrast, species in the fast growth category had significantly reduced root carbohydrate concentrations a week following defoliation, and this trend was also apparent for medium growers. Strikingly, in subsequent harvests, fast-growing species had increased carbohydrate concentrations after defoliation relative to controls, which was not observed in other growth rate categories. We have demonstrated that re-growth strategy and the dependence on root carbohydrate pools following shoot loss may differ based on whether species are relatively fast or slow growing, and, in turn, life-history strategy. In general, slow-growing species exhibit more conservative survival strategies, for instance, through the development of organs with high tissue densities and longer turnover times [16]. A high initial investment to root storage, as opposed to shoot storage, maintains reserves in a comparatively safe below-ground compartment that excludes most herbivores. We found that this conservative approach is maintained after defoliation. The results for species in the high growth category were consistent with the idea that faster growing species generally have riskier initial allocation strategies, combined with a more prominent allocation change in response to defoliation. This meant that storage was both depleted faster after defoliation, and recharged more rapidly to higher levels in fast- than slow-growing species.

The final March harvest (harvest H6) was completed just before spring re-growth, several weeks before bolting and subsequent flowering of most individuals. A disparity in carbohydrate storage over treatment groups and growth categories at this time of year may explain why we found previously that faster growing species had higher mortality and lower flowering probabilities than slower growers [13]. We showed that the slower growing species had greater total root carbohydrate pools than the faster growing species in a comparison within each of the defoliation treatments. However, the picture is not simple at the final harvest, as *within* the fast growth category, there were no detectable differences in total carbohydrate pool sizes between defoliated plants and controls. Therefore, it seems the reduced survival in the defoliated, fast-growing species [13], cannot be explained solely by reduced total carbohydrate stores. One plausible explanation is that, in large, fast-growing species, a significant proportion of whole plant nutrient content is likely to be present in above-ground material. Our defoliation treatments removed all of the above-ground material, leading to greater resource loss in fast compared with slower growing species. This greater loss of resources in above-ground material may explain the observation in the companion paper by Rose *et al.* [13] that, following defoliation, faster growing species were subjected to higher mortality and lower flowering rates, despite having comparable total root stores.

In summary, our results suggest that the faster growing species have more flexible allocation responses to defoliation and that growth strategy underpins post-defoliation and recovery strategies. Since our results are based upon measures that allow us only to infer allocation changes, a dynamic whole-plant approach is now necessary to distinguish the effects of nutrient availability and re-allocation from different plant organs. Our study focused on monocarpic perennials, since these species have a life history that considerably simplifies the analysis and interpretation of growth and storage patterns after defoliation. In order to assess the generality of these results, further experiments with a broader range of plant life histories are required.

ACKNOWLEDGEMENTS

We would like to thank Bob Keen for technical assistance.

REFERENCES

1. Kitajima K, Myers JA. 2008 Seedling ecophysiology: strategies towards achievement of positive carbon balance. In Seedling ecology and evolution (eds MA Leck, VT Parker, RL Simpson), pp. 172– 188. Cambridge, UK: Cambridge University Press.
2. Myers JA, Kitajima K. 2007 Carbohydrate storage enhances seedling shade and stress tolerance in a neotropical forest. J. Ecol. 95, 383– 395. (doi:10. 1111/j.1365-2745.2006.01207.x)
3. Sterck FJ, Poorter L, Schieving F. 2006 Leaf traits determine the growth –survival trade-off across rain forest tree species. Am. Nat. 167, 758– 765. (doi:10. 1086/503056)
4. Dı´az S et al. 2004 The plant traits that drive ecosystems: evidence from three continents. J. Veg. Sci. 15, 295– 304. (doi:10.1111/j.1654-1103.2004. tb02266.x)
5. Poorter L, Kitajima K. 2007 Carbohydrate storage and light requirements of tropical moist and dry forest tree species. Ecology 88, 1000– 1011. (doi:10. 1890/06-0984)
6. Atkinson RRL, Burrell MM, Osborne CP, Rose KE, Rees M. 2012 A non-targeted metabolomics approach to quantifying differences in root storage between fast- and slow-growing plants. New Phytol. 196, 200– 211. (doi:10.1111/j.1469-8137. 2012.04274.x)
7. Iwasa Y, Kubo T. 1997 Optimal size of storage for recovery after unpredictable disturbances. Evol. Ecol. 11, 41 – 65. (doi:10.1023/A:1018483429029)
8. Grime JP, Hunt R. 1975 Relative growth-rate—its relative range and adaptive significance in local flora. J. Ecol. 63, 393– 422. (doi:10.2307/2258728)
9. Robinson D, Rorison IH. 1988 Plasticity in grass species in relation to nitrogen supply. Funct. Ecol. 2, 249 – 257. (doi:10.2307/2389701)
10. Christie EK, Moorby J. 1975 Physiological responses of semiarid grasses. 1. Influence of phosphorus supply on growth and phosphorus absorption. Aust. J. Agric. Res. 26, 423– 436. (doi:10.1071/ ar9750423)
11. Atkin OK, Botman B, Lambers H. 1996 The causes of inherently slow growth in alpine plants: an analysis based on the underlying carbon economies of alpine and lowland Poa species. Funct. Ecol. 10, 698– 707. (doi:10.2307/2390504)
12. Grime J. 1979 Plant strategies and vegetation processes. Chicester, UK: Wiley.
13. Rose KE, Atkinson RL, Turnbull LA, Rees M. 2009 The costs and benefits of fast living. Ecol. Lett. 12, 1379– 1384. (doi:10.1111/j.1461-0248.2009. 01394.x)
14. Paul-Victor C, Zust T, Rees M, Kliebenstein DJ, Turnbull LA. 2010 A new method for measuring relative growth rate can uncover the costs of defensive compounds in Arabidopsis thaliana. New Phytol. 187, 1102– 1111. (doi:10.1111/j.1469-8137. 2010.03325.x)
15. R Development Core Team. 2009 R: a language and environment for statistical computing. Vienna, Austria: R Foundation for Statistical Computing.
16. Ryser P. 1996 The importance of tissue density for growth and life span of leaves and roots: a comparison of five ecologically contrasting grasses. Funct. Ecol. 10, 717 – 723. (doi:10.2307/2390506)

INDEX

T

U

W